Sustainability and Digital Engineering Management

Sustainability and Digital Engineering Management

African case studies in sustainable digital transformation

Edited by
Arnesh Telukdarie

Published by Emerald Publishing Limited, Floor 5, Northspring, 21–23 Wellington Street, Leeds LS1 4DL.

ICE Publishing is an imprint of Emerald Publishing Limited

Other ICE Publishing titles:
Digital Transformation in Engineering Management: Case studies of digital transformation in Africa
Edited by Arnesh Telukdarie. ISBN 978-2-83608-559-1
Li Wan, Timea Nochta, Junqing Tang and Jennifer Schooling.
ISBN 978-0-72776-600-7
Digital Twins for Smart Cities: Conceptualisation, challenges and practices
Li Wan, Timea Nochta, Junqing Tang and Jennifer Schooling.
ISBN 978-0-72776-600-7
Energy and Mobility in Smart Cities
William J Nuttall, David V Gibson, Dariusz Trzmielak and Alejandro Ibarra-Yunez. ISBN 978-0-72776-425-6

A catalogue record for this book is available from the British Library

ISBN 978-1-83662-645-9

Cover photo: courtesy of Jay_Zynism/iStock.com

Commissioning Editor: Michael Fenton
Content Development Editor: Cathy Sellars
Production Editor: Benn Linfield

Typeset by KnowledgeWorks Global Ltd.
Index created by Lyn Nesbitt-Smith
Printed and bound by CPI Group (UK) Ltd, Croydon, CR0 4YY

This comprehensive book, *Sustainability and Digital Engineering Management*, serves as a valuable guide for individuals dedicated to incorporating sustainable digital practices into the field of engineering management. Drawing from a wide range of real-life case studies within diverse African industries, the book offers a rich and multifaceted perspective on the seamless integration of sustainable digitalisation into engineering management. By actively tackling sustainable development goals through digital transformation, this book provides actionable solutions and knowledge for scholars, policymakers, and industry professionals.

Hlobisile Natasia Mahure,
University of Cape Town, South Africa.

Dedication

This book pays tribute to the pioneers and visionaries committed to creating a sustainable future for future generations. It honours the efforts of numerous organisations, academics, and industry professionals who are dedicated to leading the way in digital transformation while prioritising the preservation and protection of the planet. Their innovative ideas and unwavering commitment to sustainable digital transformation in Africa have played a significant role in paving the way for a future that embraces sustainability through digital innovation. The case studies showcased in this book are intended to inspire the next generation of leaders.

Contents

Acknowledgements

The Chemical Industries Education & Training Authority (CHIETA): The authors extend their heartfelt gratitude to the team at the CHIETA for their valuable cooperation and insights that helped explore the implementation of business intelligence in the chemical and manufacturing industries.

University of Johannesburg, Johannesburg Business School: We extend our heartfelt appreciation to the University of Johannesburg, Johannesburg Business School for their valuable contributions to this book. The Business School has played a pivotal role in enhancing the content and insights presented in this book.

Johannesburg Water: The authors extend special thanks to Joburg Water for their candid and insightful contributions. This organisation provided valuable insights into the role of digitalisation in this crucial domain.

Department of Science and Innovation (DSI): The authors express their gratitude to the Department of Science and Innovation for their valuable contributions and expertise in the green hydrogen sector.

Preface

Sustainability and Digital Engineering Management: African case studies in sustainable digital transformation delves into the intricate relationship between sustainability and digitalisation within the field of Engineering Management. It extends beyond the traditional focus on digitalisation in engineering, providing practical insights for integrating sustainability strategies into digitalisation efforts. Drawing from real-world industry experience, this book offers solutions to current industry challenges while addressing sustainability concerns, ensuring that development meets present needs without compromising future requirements. It takes an interdisciplinary approach, promoting collaboration between industry experts and academic professionals to provide comprehensive knowledge. This book is a valuable resource for students, offering practical discussions and case studies of implemented solutions for a variety of sustainability challenges. It is also beneficial for academics and professionals at master's or MBA levels, as well as practitioners such as engineers and project managers seeking to integrate sustainability principles into digitalisation efforts within engineering projects.

About the editor

Professor Arnesh Telukdarie is a well-known expert in digital businesses and a professor at the Johannesburg Business School, University of Johannesburg. In addition, he holds the position of Adjunct Professor at Johns Hopkins University and serves as a Visiting Scholar at London South Bank University. Professor Telukdarie is also affiliated with the University of the Witwatersrand and Kaunas University of Technology, Lithuania. At the Johannesburg Business School, he serves as the research chairperson, focusing on areas such as Industry 4.0, Society 5.0, sustainable development, business optimisation, and manufacturing systems. His impressive body of work includes the publication of over 350 peer-reviewed works and extensive industry consulting.

About the contributors

Professor Megashnee Munsamy: Megashnee Munsamy is a distinguished professor specialising in digital business processes, environmental, social, and governance (ESG) reporting, systems for healthcare operations management, engineering, and digital healthcare at the Johannesburg Business School. She has earned a PhD in Engineering Management and holds a master's degree in Chemical Engineering, graduating with cum laude. Currently, Professor Munsamy serves as an associate professor at the esteemed Johannesburg Business School, which is affiliated with the University of Johannesburg.

Dr. Khuliso Murulane: Khuliso Murulane is a highly accomplished individual who holds an MSc and PhD in Mathematical Physics from the University of Johannesburg. He excelled in his academic studies, achieving an outstanding 93% in his MSc and receiving the Chancellor's Medal for his exceptional master's study in the Faculty of Science in 2018. Currently, he is a research coordinator, in the systems research group, at the Johannesburg Business School and a lecturer at the University of Johannesburg, Department of Physics. Dr. Murulane is deeply passionate about data science and utilises his expertise to explore the convergence of physics, technology, and data analysis. His research interests are extensive and encompass areas such as nuclear structure, nuclear reactions, dynamical systems, artificial intelligence, and substantiality.

Dr. Michael Sishi: Michael Sishi is a highly distinguished professional employed as senior manager in the mining sector. He holds a PhD in Engineering Management. His research focuses on the internet of things (IoT), artificial intelligence, and sustainable practices in the mining industry. He has played a significant role in digital transformation, with particular emphasis on manufacturing execution systems for IoT. Dr. Sishi is also a Fellow at the Johannesburg Business School, University of Johannesburg. He has contributed extensively to various technology journal articles and book chapters in academia.

Dr. Joseph Mwanza: Joseph Mwanza graduated from the National University of Science and Technology in Zimbabwe with a Bachelor of Engineering degree in

Industrial Engineering. He later completed his Master of Engineering degree in Manufacturing Engineering and Operations Management at the same institution in 2013. Currently, he holds a PhD in Engineering Management at the University of Johannesburg, focusing his research on digital technologies to develop cost-effective optimisation solutions for business systems.

Dr. Makoni Logistic: Makoni Logistic is a research coordinator at the Johannesburg Business School, University of Johannesburg. His research primarily focuses on tourism, business management, and digitalisation, with a keen interest in developmental issues in Africa. Dr. Logistic's research revolves around planning and policy-related matters, community development, digital transformation, and sustainable economies. He is particularly interested in understanding how political systems, laws, and regulations impact the economic dynamics of African countries. Through his research, Dr. Logistic aims to propose strategies for shaping these laws to enable economies to reach their full potential in the region.

Mahure Hlobisile Natasia: Mahure Hlobisile Natasia is a graduate student at the Faculty of Computer Science, University of Cape Town, currently pursuing a Master of Science in Information Technology. She is a software developer and research coordinator at the Johannesburg Business School, University of Johannesburg. Her areas of expertise include software development, human–computer interaction, software architecture, web programming, machine learning, and research methods.

Tatenda Katsumbe: Tatenda Katsumbe holds a degree in Civil Engineering from the University of the Witwatersrand and a master's in Engineering Management from the University of Johannesburg. Currently, he is pursuing a PhD at Johannesburg Business School under the supervision of Professor Telukdarie and Professor Munsamy. His research focuses on integrated urban systems modelling and he is involved in both qualitative and quantitative research. He has worked with Professor Telukdarie and Professor Munsamy on various industry-related projects, covering business process mapping, systems mapping and analysis, value chain analysis, digital business, circular economy, and the food and beverage, education, and services sectors.

Christian Tshukudu: Christian Tshukudu is a researcher at the University of Johannesburg. His expertise lies in digital maturity, software development lifecycle, and business processes. Christian is dedicated to advancing knowledge and practice in these areas, contributing significantly to academic and industry settings.

Noluthando Gamede: Noluthando Gamede is a highly committed junior researcher at the Johannesburg Business School, University of Johannesburg, boasting a solid academic foundation. She holds a bachelor's degree in Biochemistry and Chemistry, an honours degree in Chemistry, and a master's degree in Chemistry. Moreover, she is currently pursuing a PhD in Engineering Data Science. Noluthando's expertise lies at the intersection of organometallic chemistry, physics, and data science, which fuels her drive for groundbreaking research. She has an impressive track record, having authored influential papers in the areas of chemistry and machine learning in agriculture, showcasing her dedication to advancing scientific knowledge and its practical applications.

Thabile Dube: Thabile Dube is a research coordinator involved in various industry projects. Her research primarily focuses on rural development, innovation, and digital technology. She actively participates in workshops and training for small and medium-sized enterprises (SMEs). Thabile's work delves into how digital transformation can contribute to sustainable development in rural and peri-urban areas. Her goal is to connect traditional and digital approaches to foster economic growth in underserved communities.

Nokwethaba Andile Mtshali: Nokwethaba Andile Mtshali is a graduate of the University of Johannesburg with a Diploma and a BTech in Tourism Management. She has experience in higher education, tutoring, content development, research coordination, and office administration. She also has industry expertise in business management, multimedia, marketing, and rates administration. Nokwethaba served as an administrative and research assistant at the University of Johannesburg, involved in various projects, including those focused on digital development and TVET (Technical Vocational Education and Training) college education.

Nonhlanhla G. Sithole: Nonhlanhla G. Sithole is an accomplished administrative assistant at the Johannesburg Business School, University of Johannesburg, holding a diploma in Management Assistance from Ekurhuleni West College. Her areas of expertise include providing administrative support, creating content for virtual reality, and managing projects across diverse fields at the Business School, University of Johannesburg.

Regionald Mongwe: Regionald Mongwe has achieved a cum laude master's degree in Engineering from the Durban University of Technology. His areas of expertise include Python programming, business processes, machine learning, mobile development, and distance learning. Previously, he served as a research assistant at the Johannesburg Business School, University of Johannesburg, focusing on software development.

Inderasan Munien: Inderasan Munien is a specialist in electrical and electronic engineering science at the University of Johannesburg, with expertise in electrical engineering, engineering business processes, machine learning, and blockchain. He has published research articles on topics such as a framework for a user-focused electronic health record system leveraging Hyperledger Fabric, COVID-19 supply chain resilience modelling for the dairy industry, and the potential future of agriculture for small farms: supervised machine-learning smart irrigation concept for vegetables in *Procedia Computer Science*.

Mangaliso Maduna: Mangaliso Maduna is a dedicated software developer and data scientist who has a strong academic background in Software Engineering, having obtained a BSc (Hons) from Bahçeşehir Üniversitesi. Currently working towards a master's degree in applied data sciences, Mangaliso is known for his use of cutting-edge technology to solve complex problems, especially in the areas of data analysis, application development, and business systems. He has experience as a developer at the University of Johannesburg and has played a crucial role in various projects, including the development of the Skills Supply and Demand Database (SSDD), aimed at enhancing employment outcomes within South Africa's chemical sector. His technical expertise includes back-end and front-end development, and he is proficient in tools such as Python, Django, Flask, SQL, and Power BI. With a passion for using data-driven solutions to tackle real-world challenges, Mangaliso is dedicated to driving innovation in both the technology and education sectors.

Mpho Godfrey Nkadimeng: Mpho Godfrey Nkadimeng holds the position of Data Scientist and Research Coordinator at the University of Johannesburg. In this role, he combines consulting with research to tackle industry challenges. His primary focus is on developing data-driven systems that not only enhance operational efficiency but also address real-world problems impacting people's lives. Mpho's projects range from creating automated data pipelines to utilising machine learning for improved decision-making, all aimed at streamlining processes. However, his commitment goes beyond the technical realm, as he is dedicated to leveraging data for tangible societal benefits, such as enhancing workforce development programmes and optimising service delivery in key industries. With industry experience at Sigmoid Analytica, Mpho has shown proficiency in applying advanced technologies like computer vision and data integration to practical business applications. His driving force is not just technical innovation, but also the belief that data, when used thoughtfully, can significantly improve life by reducing inefficiencies and enabling smarter, more empathetic decisions. Holding a BCom in Information Management and a postgraduate diploma in Data Science, Mpho's research is centred on how artificial intelligence and data science can drive meaningful change, from enhancing workplace strategies to gaining a deeper understanding of the human side of operational issues.

List of figures and tables

Figures

Tables

Prof. Arnesh Telukdarie
ISBN 978-1-83662-645-9
https://doi.org/10.1108/978-1-83662-642-820251001

Chapter 1
An introduction to *Sustainability and Digital Engineering Management: African case studies in sustainable digital transformation*

Arnesh Telukdarie and Mahure Hlobisile Natasia

Introduction

The incorporation of sustainability principles into digital engineering management has become vital in today's rapidly advancing technological landscape. Sustainability entails meeting current needs without compromising the ability of future generations to meet their own needs. It encompasses environmental stewardship, economic profitability, and social responsibility. Digital engineering management, on the other hand, focuses on implementing digital tools and techniques to optimise engineering processes. The intersection of these two fields promotes innovation, competitive advantage, efficiency, and long-term value creation in engineering while minimising environmental impact and advancing sustainable development goals (SDGs). However, despite the consensus on the importance of taking a sustainable approach, there is still a significant gap between awareness and implementation. While digitalisation offers significant benefits, there is a lack of actionable strategies for implementing it in a way that aligns with SDGs. Therefore, *Sustainability and Digital Engineering Management: African case studies in sustainable digital transformation* provides comprehensive case studies from diverse industries, demonstrating how digitalisation can be harnessed to promote sustainable development. The book offers first-hand knowledge on implementing digitalisation to establish sustainable businesses, urban and rural communities, mines, and various other industries, serving as a practical guideline for implementing sustainable digitalisation in these sectors. Furthermore, the book equips professionals with the necessary tools and strategies to effectively integrate digitalisation innovations while incorporating environmental, social, and economic goals. By providing a blueprint for translating theoretical sustainability concepts into real-world applications, the book serves as a manual for implementing SDGs across different industries, making it suitable for professionals in a wide range of sectors.

The book is divided into four main themes. The first theme focuses on industrial digitalisation for sustainable operations, while the second theme delves into sustainable business practices. The third theme includes chapters on sustainable urban and rural development, and the final theme covers digital transformation in education and skills development, as indicated in Figure 1.1. The structure of the book is explained in detail below.

Figure 1.1 Summary of the book's structure

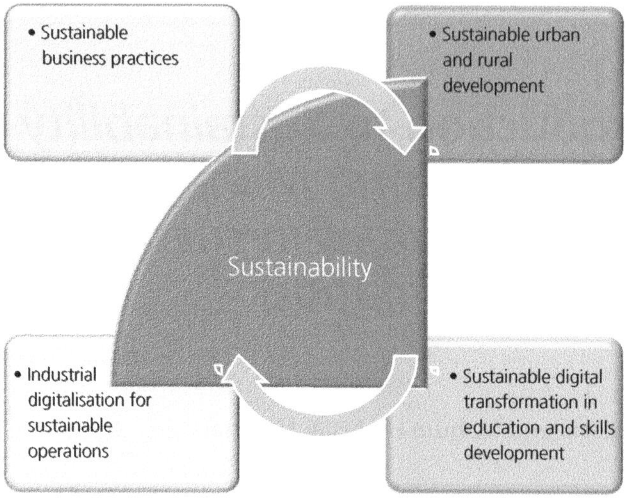

The first theme of the book, 'Industrial digitalisation for sustainable operations', provides practical guidelines for implementing digital strategies to enhance efficiency, sustainability, and productivity in industrial operations (Figure 1.1). The first chapter in this section, titled 'Smart water management for sustainability', delves into the digital management of water resources through smart water management (SWM) systems to promote sustainability. It examines how digital technologies optimise water distribution, reduce losses, and ensure efficient resource use. Chapter 3 of the book explores the integration of green hydrogen production with artificial intelligence (AI). The authors delve into the role of AI in optimising green hydrogen production processes to achieve sustainability. This section explores how AI-driven approaches can enhance efficiency, cut costs, and minimise the environmental impact of green hydrogen production facilities. In addition, the chapter discusses the integration of AI technologies, like machine learning algorithms and predictive analytics, into hydrogen production systems, and their potential to hasten the transition to a low-carbon economy. Chapter 4 of the book, titled 'Sustainability and uranium digitalisation', explores how digitalisation impacts the uranium industry by focusing on improving environmental and operational efficiency. It discusses the integration of digital technologies into uranium mining and processing operations, emphasising the benefits of real-time data analytics, automation, and remote monitoring in enhancing safety, productivity, and sustainability. This chapter also details successful implementations of digital solutions in the uranium sector and their implications in fostering sustainable mining practices.

The fifth chapter, titled 'Satellite imagery for water leak detection', focuses on the application of AI in identifying leaks in various systems to minimise resource wastage and boost sustainability. The chapter explores AI-driven leak detection technologies, such as acoustic sensors, predictive analytics, and anomaly detection algorithms, and their effectiveness in pinpointing and addressing leaks in water pipelines, industrial equipment, and infrastructure. Furthermore, this chapter showcases successful implementations of AI-powered leak detection solutions and their positive impact on resource preservation and minimising environmental risks. The final chapter explores the application of digital tools and strategies to optimise healthcare systems for sustainability. It provides detailed insights into the practical steps involved in integrating digitalisation to enhance operational efficiency and overall sustainability in the healthcare sector.

The second theme of the book, titled 'Sustainable business practices', provides practical strategies for organisations to minimise their environmental and social impact while improving their economic viability. This part consists of three chapters. Chapter 7, 'Business processes for digital sustainability', explores the digital transformation journey of the Chemical Industries Education and Training Authority (CHIETA), one of South Africa's 21 Sector Education and Training Authority (SETA) organisations. The chapter offers insights into the complexities and strategic innovations necessary for businesses to maintain a competitive edge in the digital age. By reading this chapter, readers will gain practical guidelines on how to develop and integrate digitalisation into their organisation's business processes. Chapter 8, 'Strategies for sustainable digitalisation: advancing ESG metrics', offers recommendations for developing an environmental, social, governance (ESG) matrix to address the limitations of traditional ESG reporting techniques, such as lack of transparency, comparability, and consistency. The chapter outlines the development of a digitalised ESG matrix and highlights the importance of incorporating digitalisation to improve sustainability reporting. This case study is particularly relevant for students studying sustainable business practices, environmental management, digital innovation, the fourth industrial revolution, and the evolving landscape of ESG reporting. In Chapter 9 of the book, 'The use of robotic process automation and artificial intelligence in designing a digital system landscape', the focus is on integrating digital strategies to promote sustainable business operations. In today's fast-paced technological landscape and amidst growing environmental concerns, businesses face mounting pressure to adopt digital solutions that not only boost efficiency but also support sustainability. This chapter thoroughly examines best practices for different digital strategies utilised by businesses to reduce environmental impact, improve resource management, and enhance overall sustainability performance, which concludes the second theme of the book.

The third theme of the book, 'Sustainable urban and rural development', offers practical guidelines for sustainable planning and development in urban and rural areas to create healthy, inclusive, and resilient communities. The first chapter of this section is titled 'Enhancing sustainable rural development through collaborative innovation'. This chapter focuses on the barriers and challenges that impede innovation, creativity, and productivity in rural areas. Overall, the chapter provides practical guidelines for implementing digital strategies for sustainable rural development through a system model, paving the way for a more informed, strategic, and impactful approach to addressing challenges and fostering development within rural communities. The last chapter in this theme focuses on urban waste management solutions. It details practical recommendations for digital solutions to urban waste management promoting sustainable city development. The chapter outlines the measures taken by the City of Johannesburg and the University of Johannesburg in response to the waste management challenges faced by the city.

The final theme of the book, 'Sustainable digital transformation in education and skills development', provides practical guidelines based on real-life case studies on how to utilise digital tools to improve the education sector, enhance skills development, and ensure accessibility and effectiveness. The first chapter in this section, 'Green learning management systems', discusses the creation of the skills supply and demand database (SSDD) as a groundbreaking solution to unemployment in the chemical sector. It acts as a centralised database of sector-qualified individuals, enabling seamless interactions among employers, learners, and the CHIETA. This leads to improved employment rates, increased job contracts, and enhanced employability outcomes. This case study not only impacts the chemical sector but also explores the wider implications of the SSDD, representing a change in labour supply and demand systems. Therefore, the chapter offers valuable insights into integrating digitalisation in education, transitioning from paper-based methods to environmentally friendly digital tools, and improving efficiency and sustainability by highlighting unique features and methodologies.

The second chapter of this theme titled, 'Modernising vocational welding education', delves into the use of virtual reality (VR) technology in vocational education and training (VET) programmes to promote sustainable development. This is a response to the growing demand for cost-effective and advanced technical skills training. The chapter showcases successful initiatives that utilise VR simulations to enhance practical training, improve learning outcomes, and fill skills gaps in various industries. It explores the potential of VR to revolutionise VET delivery methods and prepare learners for sustainable careers in an ever-changing digital landscape. The third chapter, 'Sustainable business intelligence in Master of Business Administration programmes', provides insights into how the Johannesburg Business School (JBS) implemented a strategic digital tool to address complex challenges in managing Master of Business Administration (MBA) administration processes. This case study outlines the adoption of modern technology and software engineering practices to streamline operations, ensure compliance with regulations, and enhance stakeholder satisfaction. It highlights the complexities associated with MBA administration, such as cohort management, examination oversight, supervisor appointments, and cohort analytics, emphasising the necessity of a sophisticated digital solution. Overall, the case study details the agile development process, adoption initiatives, testing, and quality assurance measures, as well as the deployment of the digital solution.

The fourth chapter of this theme, titled 'Exploring historical data trend analysis in the Chemical Sector Occupational Skills Forecast Tool', focuses on the use of AI to predict skills needed in the context of sustainable workforce development. This chapter explores AI-driven methods for forecasting future workforce needs, identifying emerging skills trends, and aligning education and training programmes with industry demand. It delves into the implications of AI-driven skills forecasting for promoting sustainable economic growth, reducing skills mismatches, and enhancing workforce resilience in light of technological advancements. The final chapter in this section, titled 'Transformative innovation: harnessing AI for sustainable research practices', provides valuable insights and best practices for integrating AI into research practices to achieve sustainable outcomes.

Each chapter is meticulously structured to include a comprehensive case study, which is paired with detailed teaching notes. These teaching notes are essential resources that provide the relevant information and materials necessary for understanding, teaching, and fully engaging with the case study. They play a crucial role for instructors, enabling them to deliver content in a practical and relatable manner, effectively immersing students in real-world industry experiences. Moreover, industry professionals can leverage these teaching notes to deepen their understanding of the case studies, enhancing their practical knowledge and application in their respective fields. The teaching notes found in each chapter encompass several key components, as indicated in Table 1.1. All of these elements work together to provide invaluable information for both instructors and industry professionals, making each chapter a rich resource for effective teaching and learning.

To sum up, the book is divided into different themes, each with a set of chapters providing practical guidelines on sustainable business practices, sustainable urban and rural development, digital transformation in education and skills development, and industrial digitalisation for sustainable operations. It eloquently explores sustainability principles and their seamless integration into digital engineering processes. The book emphasises the need to reduce environmental impact, promote social equity, and ensure economic viability in various industries. It presents numerous real-world case studies showcasing successful implementations

Table 1.1 Structure of teaching notes in book chapters

Teaching notes	
Synopsis	A brief overview of the case study, highlighting its key points.
Teaching objectives	Clear goals that outline what students should achieve or understand by the end of the lesson.
Research methods	An explanation of the methodologies used in the case study, providing context for the findings.
Keywords and subject code	Important terms and classifications that are relevant to the case, facilitating easier navigation and reference.
Assignment questions	Questions that are designed to stimulate discussion and critical thinking among students and other readers.
Additional materials	Supplementary resources that expand on the topics covered in the case study, allowing for a more thorough exploration.
Suggested core readings	Recommended literature that provides further insight and background, enriching the learning experience.
Teaching plan and timing	A structured outline for instructors on how to deliver the lesson effectively, including time allocations for different activities.
Analysis of assignment questions	Guidance on how to approach the assignment questions, including possible answers and interpretations.
Key learnings	A summary of the essential takeaways from the case study, reinforcing the primary lessons for students.

of sustainability-centric digital engineering practices, demonstrating how digitalisation addresses industry-specific challenges while considering sustainability principles. Table 1.2 provides a summary of the digital tools used in each chapter. In addition, the book highlights the barriers encountered and the opportunities that arose during the implementation process. It offers practical insights into successfully implementing sustainability-centric digital engineering management across various industries, making it an invaluable resource for anyone aiming to achieve SDGs while leveraging digital engineering practices for growth and innovation in each industry.

This book offers a comprehensive exploration of the intersection between sustainability and digitalisation in engineering management, expanding beyond the typical focus solely on digitalisation in engineering. It provides practical guidance for incorporating sustainability strategies into the pursuit of digitalisation in engineering management. The book leverages real-world industry experience and success stories to serve as a practical roadmap. By adopting an interdisciplinary approach and promoting collaboration between industry experts and academic professionals, it serves as a valuable source of comprehensive knowledge. One of its key strengths is the ability to offer solutions to current industry challenges while addressing sustainability issues, ensuring that development caters to present needs without compromising future requirements. The book will be an invaluable resource for students, providing practical discussions and case studies on implemented solutions across a range of sustainability challenges. Additionally, the book is intended for academics, and students pursuing a

Table 1.2 An overview of digital tools utilised in the book

	Sensors	Big data analytics	Internet of things	Artificial intelligence	Cloud computing	Virtual reality
Chapter 2	X	X		X		
Chapter 3		X		X		
Chapter 4		X	X	X		
Chapter 5		X	X			
Chapter 6		X				
Chapter 7				X		
Chapter 8		X		X		
Chapter 9		X		X		
Chapter 10				X		
Chapter 11				X		
Chapter 12		X		X	X	
Chapter 13						X
Chapter 14		X		X	X	
Chapter 15		X		X	X	
Chapter 16				X		

master's or Master of Business Administration (MBA) degree. It also caters to practitioners, such as engineers and project managers, and executives who aim to integrate sustainability principles into engineering management.

REFERENCES

Emina KA (2021) Sustainable development and the future generations. *Social Sciences, Humanities and Education Journal (SHE Journal)* **2(1)**: 57–71.

Farley HM and Smith ZA (2020) *Sustainability: If It's Everything, Is It Nothing?* Routledge, London, UK.

Lim KYH, Zheng P and Chen CH (2020) A state-of-the-art survey of Digital Twin: techniques, engineering product lifecycle management and business innovation perspectives. *Journal of Intelligent Manufacturing* **31(6)**: 1313–1337.

THEME 1

INDUSTRIAL DIGITALISATION FOR SUSTAINABLE OPERATIONS

Prof. Arnesh Telukdarie
ISBN 978-1-83662-645-9
https://doi.org/10.1108/978-1-83662-642-820251002

Chapter 2
Smart water management for sustainability: digital management of water resources through smart water management systems promoting sustainability

Arnesh Telukdarie, Megashnee Munsamy and Noluthando Gamede

A smart metering solution for Johannesburg Water

Sustainable Development Goal (SDG) 6 of clean water and sanitation entrenches water as a basic necessity. Johannesburg, Africa's largest economy, is the economic hub of South Africa. It has a population of approximately 4.5 million, making it the most densely populated city in South Africa. This has led to an increased demand for essential services such as water delivery. In the city of Johannesburg, clean, potable water distribution is managed by Johannesburg Water SOC (JW). JW supplies 1.6 billion litres of potable water per day to both industrial and residential customers by way of 12 406 km of pipes. JW directly contributes to ensuring the availability and sustainable management of water. JW encounters various challenges, including drought, aging infrastructure, technological deployment, water conservation optimisation, and demand management. Thus, in 2022, JW requested proposals for the Blue-Sky Project, which focused on digitalisation through Industry 4.0. Two such initiatives comprising the Blue-Sky Project, as approved by the CEO of JW, were the smart water meter project and the water conservation and water demand management project, hereafter referred to as SWM and WCWDM, respectively. With smart metering technologies, a communication network between the water metering device and the central information system was created. The communication network can be wireless or have fixed connections. Smart metering facilitates remote monitoring, high-resolution consumption metering (hourly or even more frequently), and water flow and quality monitoring. These data provide insights into efficiency improvement and informed decision-making. Application of smart metering facilitates the creation of a cyber-physical system for JW, enabling a smart water network. The WCWDM project focuses on the automation of the existing JW water mass balance and associated reports. These two projects are inherently linked, with the SWM becoming an automated source of real-time information on various measurement indicators, such as water flow rate, pressure, and quality.

The University of Johannesburg (UJ), represented by Professor Telukdarie, was the successful applicant for the project. This was a collaborative research and consulting partnership with the UJ. The project objective was to implement and integrate a proof-of-concept smart water metering system on a subnetwork of the JW water distribution network to quantify water losses, forecast water demand, pinpoint water leakages in the network, and facilitate automation of the water mass balance, ultimately transforming the management of water supply and demand. The implementation of smart water meters facilitates high-resolution consumption metering and water quality monitoring. This helps in maintaining and improving water quality, which is crucial for SDG 6 targets. By automating the water mass balance and integrating smart water meters, JW can monitor water flow rates and pressures and identify leakages, thus improving water use efficiency and reducing losses.

Adopting global best practices together with local laws and policies resulted in the development of a framework guiding the implementation of the selected smart water meters in targeted locations within the Corriemoor to Fairlands area of the JW water distribution network, which would also be the areas for mass balance automation. The JW team is preparing for a new chapter in their innovation quest that aligns with the city's plans for a smart city. In their mission to revolutionise water management, will JW succeed? It is just a matter of time before we find out. But one thing is certain: the journey is just beginning.

Navigating the waters: Johannesburg Water's journey towards a digital water balance protocol and smart water metering

In June 2022, JW stood as a beacon of efficiency, successfully delivering 1.6 billion litres of drinking water daily through its extensive network spanning 12 406 km. Rand Water stores potable water in the network infrastructure of 37 pump stations, 129 reservoirs and water towers, and six wastewater treatment plants strategically distributed across six regions in Johannesburg, forming a vital link in the municipality's water supply chain. In areas characterised by high gravity, the synergy between reservoirs and towers becomes evident, with each equipped with water level meters. However, there is always room for improvement. JW initiated plans to align with the city's mayoral priorities, aiming to create a smart city with intelligent predictive modelling and data for smart decision-making, as demonstrated by the Blue-Sky tender it issued.

Globally, the adoption of smart water metering has become the norm. Countries such as Australia, the Netherlands, and South Korea have made significant progress in implementing SWM. This has been driven by the need for accurate real-time statistics on water usage, precise billing, and efficient resource management. The benefits were clear: reduced water losses, lower operational costs, and improved environmental sustainability. Innovations in metering infrastructure and associated standards are evolving rapidly. Data analytics and leak detection are emerging as crucial components of water distribution networks, aiding in the identification and mitigation of challenges. Advanced communication protocols, such as LoRaWAN and NB-IoT, are revolutionising the way smart meters transmit data, ensuring efficient communication and enhanced security. However, South Africa has laggard adoption of SWM, creating an opportunity for JW to lead digitalisation in the country. As the custodian of Johannesburg's water supply, JW recognised the need to embrace innovation and adapt to the changing landscape of water management to align with the city's vision of a smart city. JW began implementing an SWM system to address current needs and ensure a sustainable future for the city and its residents. SWMs are essential tools for monitoring water levels, ensuring that operators are alerted promptly to prevent potential overflow or depletion issues in the reservoirs, and facilitating the automation of the water mass balance.

Crafting Johannesburg Water's smart metering framework

Professor Telukdarie, an expert on digital transformation in the water sector and a professor of engineering, in collaboration with his team of academics, and JW, embarked on this transformation journey. The first step was to define the global landscape of smart water meters. This was done by way of a comprehensive literature review, which included an analysis of journal publications and white papers. This also included a meticulous examination of international and local regulations and standards to ensure JW adherence to regulations. With a solid foundation in place, the UJ team analysed the water distribution network of JW to gain a comprehensive understanding of the infrastructure at hand.

The team then reviewed the best international practices, identifying those best suited for application for JW's operation. By pinpointing commonalities and differences, they sought to establish a framework that would meet the needs of JW. In parallel, the UJ team identified five specific locations within the Corriemoor-Fairlands area for testing SWMs by way of a proof of concept. The identified areas were selected to test the capabilities of the SWM for broader application across the JW water distribution network. JW selected a diverse range of areas, ultimately identifying the final area. The characteristics of the area ranged from highly populated to high- and low-pressure areas. The team then embarked on the crucial task of selecting optimal smart water meters for JW. To ensure a structured and analytical approach, an evaluation matrix was developed, which enabled the prioritisation of the selected SWM. The matrix defined technical specifications and regulatory compliance requirements. The technical specifications included accuracy, communication protocols, and data management capabilities. Figure 2.1 shows the defining framework.

In the specifications for tenders in metering infrastructure (City of Joburg SOC Ltd, 2021), JW has conscientiously included established standards and regulations, with a particular focus on adherence to the SA Legal Metrology Act 9 of 2014 and its related technical requirements. The South African SANS 1529-1, which outlines requirements for water meters intended for cold potable water, is another critical regulatory requirement. It focuses specifically on the metrological properties of mechanical water meters with a nominal bore of less than 100 mm. The standards for IEC 62055-41 and IEC 62055-51 focus on prepayment tokens. Table 2.1 details the standards reviewed by the team. Standards 1, 2, and 3 were deemed critical for the selection of acceptable SWM, following a comprehensive review.

The matrix also defined the technical requirements of the SWMs. As per JW operations, there are two types of clients: residential and bulk water clients. The selection criteria were similar for both the residential and bulk SWM, but the targets were different, such as the acceptable measurement accuracy for bulk SWMs being more stringent than that of the residential SWM. Some of the criteria considered included expected water volumes influencing the meter size, measurement accuracy, pressure loss, communication protocol, and regulation adherence. Tables 2.3 and 2.4 detail the developed evaluation matrix as shown in the system features, efficient data transfer, and option customisation; these all contribute to comprehensive and successful selection processes. However, selecting the right smart water meter involves balancing several critical criteria.

Area of installation

The subnetwork for testing SWM was done in collaboration with JW. JW identified the geographical area, Corriemoor to Fairlands, as proof of the SWM concept. The UJ team reviewed the area in terms of infrastructure, population density, water usage trends, and water distribution network infrastructure, and identified five possible areas – as detailed in Table 2.2. The final area for the proof of concept for the SMW is to be confirmed by JW.

Smart meters for water balance

The water balance method, grounded in the principle of mass conservation, is a vital tool for quantifying water losses within a water distribution network over a specified period of time. To conduct a comprehensive water balance, data from bulk meters and customer submeters are collected. The water supplied to the system is measured by general meters, while submeters tracking water consumption contribute to determining billed authorised consumption. The estimated unbilled authorised consumption includes water for operational purposes and unmetered households. Water losses are calculated by subtracting the authorised consumption and system input volume. Apparent losses account for inaccuracies owing to unauthorised use, inaccurate meters, and improper data handling. Giurco *et al.* (2008) propose estimating apparent losses at 0.25% of the total input volume, while the International Water Association (IWA) projects apparent losses to be 5% of the billed metered consumption for developing nations and 1.5–2% for developed nations.

Leakage monitoring involves measuring flows into discrete zones to analyse the minimum night flow and detect leaks. Hardware-based methods, particularly acoustic equipment, are employed to pinpoint leak locations, leading to reduced runtime for hidden leaks and minimised water losses post-repair. The creation of a detailed zonal water balance using the standardised IWA water balance as a basis is crucial for effective planning and implementation of water loss management strategies. This internationally recognised water balance provides a systematic framework for water utilities, facilitating comparisons and benchmarking across regions.

The implementation of smart meters in collaboration with the UJ has yielded promising results in enhancing the efficiency and management of JW's water distribution network. The integration of these advanced meters into the Supervisory Control and Data Acquisition (SCADA) system has significantly improved the accuracy and timeliness of data collection. Real-time monitoring of water flow, pressure, and consumption patterns in the water distribution network has provided a comprehensive and dynamic understanding of the water distribution network. This has been crucial in addressing the longstanding challenge of water losses, with smart meters allowing for the prompt identification of anomalies, leakages, and irregularities. The synergy between these two water balance assessments enhances the overall effectiveness of water utility management. It enables a holistic understanding of the entire water supply network while allowing utilities to address localised issues within specific districts. This dual approach not only supports accurate measurements and system optimisation but also contributes to the sustainability and resilience of the water supply infrastructure (Figure 2.2).

The precise measurement of water flow and level is pivotal for JW's effective management of its distribution network. Leveraging data from smart meters enables an accurate assessment of water entering the system, ensuring a comprehensive understanding of supply dynamics from the outset. In conjunction with artificial intelligence (AI), smart meters aid in swiftly identifying and addressing leakages, bolstering the system's efficiency, while bulk meters also contribute by providing precise measurements for billing and network management purposes, as indicated in Tables 2.3 and 2.4.

To reconcile discrepancies, various factors, including meter inaccuracies and unmetered consumption, are considered ensuring a thorough water balance analysis. Smart meters, integrated with SCADA, form the backbone of a digital water balance system. Continuous data transmission facilitates accurate measurement of parameters, leading to proactive leak detection and financial savings. The incorporation of fourth industrial revolution (4IR) technologies enables predictive analytics, fostering proactive management strategies. Collaboration between JW and the UJ has been pivotal in navigating implementation challenges and setting a global precedent for sustainable water management. Through data-driven decision-making and advanced technologies, JW exemplifies efficiency and resilience in water distribution, paving the way for transformative solutions worldwide.

What next?

The team succeeded in creating a protocol for the selection of smart water meters for JW. The development of the smart water metering framework for JW involved a systematic and comprehensive approach. The team initiated the study with a thorough literature review, examining existing studies, publications, and reports on global water utility trends for implementing smart water meters. The team used a thorough selection procedure, standard alignment, and project objectives to help them choose the right smart water meters. International standards, including OIML R 49, ISO 4064/ISO 5814, and EN 13757-1, were considered, and a selection matrix was used to justify the meters selected. To ensure compliance with the Legal Metrology Act and other South African standards, the tender specifications placed a strong emphasis on adhering to legal and technical requirements. The results demonstrate tangible benefits in terms of reduced water losses, financial savings, and environmental sustainability. The precise measurement of authorised consumption, system input volume, and real losses has empowered JW to optimise resource allocation, minimise non-revenue water, and contribute to broader water conservation goals. Furthermore, the successful collaboration between water utilities and academic institutions in navigating the complexities of integrating advanced technologies into existing infrastructure serves as a model for similar initiatives globally. The case study underscores the importance of interdisciplinary partnerships in overcoming implementation challenges and harnessing the potential of 4IR technologies for sustainable water resource management.

As JW continues to refine and expand its digital water balance system, the case study highlights the potential for continued advancements in water management practices through the integration of smart technologies. The journey undertaken by JW serves as an inspiring example for water utilities worldwide, showcasing the pivotal role of innovation, collaboration, and data-driven decision-making in addressing the pressing challenges of water scarcity and distribution inefficiencies.

Exhibits

Figure 2.1 A diagram defining the framework

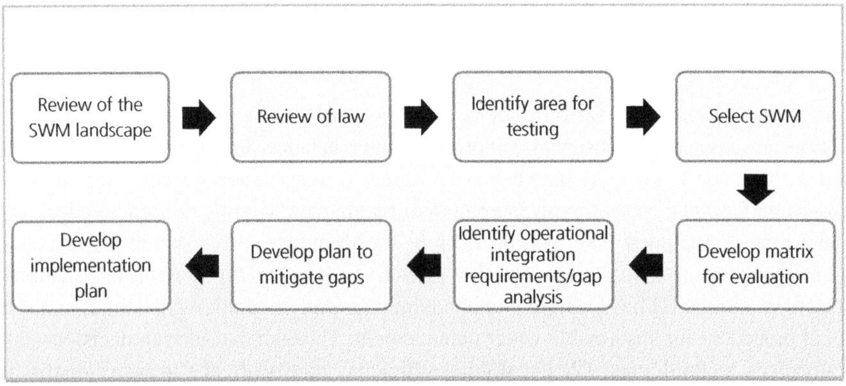

Figure 2.2 Bulk pipeline data input

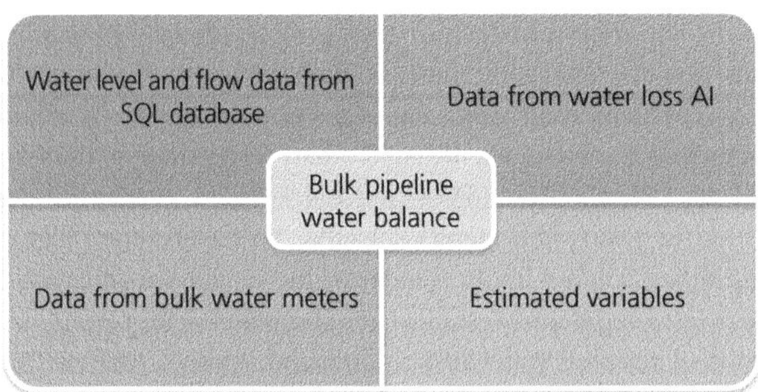

Figure 2.3 Implementation status of smart water metering

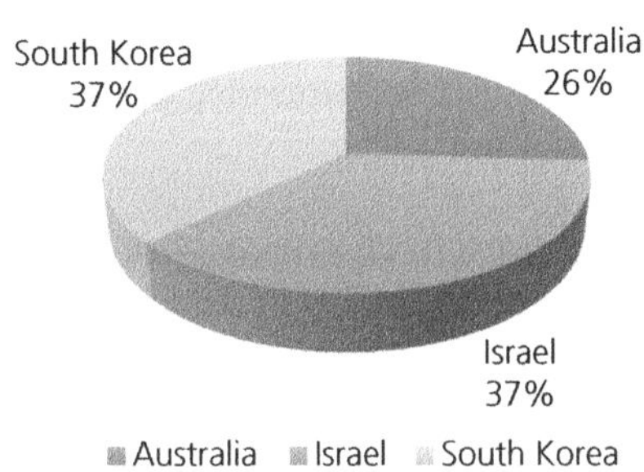

Figure 2.4 Categorisation of leak detection and leak management techniques

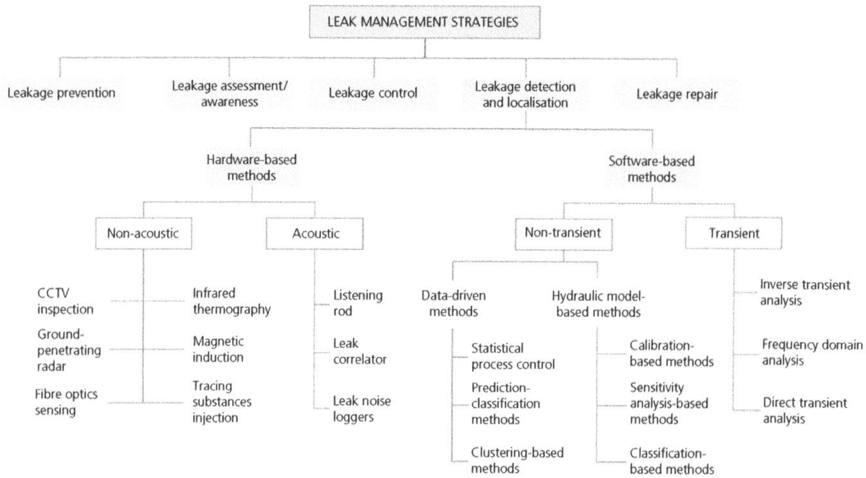

Table 2.1 International standards for smart meters

No.	Standard	Description
1.	ISO 4064/ISO 5814	Outlines the technical and metrological specifications for water meters used to measure the flow of hot and cold potable water through a closed conduit that is completely charged. These water meters are equipped with integrated volume indication devices.
2.	OIML R 49	Pertains to devices used to measure the volume of hot and cold potable water that is based on mechanical principles and electrical or electronic concepts.
3.	EN 13757-1	Explains the data exchange, communication, and remote meter reading for meters.
4.	IEC 62056 (DLMS/COSEM)	The International Electrotechnical Commission has established standards for the interchange of power metering data, known as IEC 62056. The Device Language Message Specification and Companion Specification for Energy Metering is another name for it: DLMS/COSEM.
5.	ANSI/AWWA Standard C750	Transit-time ultrasonic flowmeters for water supply service applications are described in ANSI/AWWA C750, an American Water Works Association (AWWA) standard. There are full-closed pipes that use these flowmeters.

Table 2.2 Suggested area for installation

Area	Rationale
Ebony Dr, Northcliff ext. 2	Its key location within the reservoir's pipeline system guarantees effective water distribution monitoring.
Im Park Retirement Village, 1 Suzanne Cres, Northcliff, Randburg, 2195	It is in the reservoir's pipeline network, and it is also a highly populated area with high static pressure.
Fairland Walk Shopping Centre, Cnr Beyers Naude Drive and Willson St, Fairland, 2030	It is a business block around logging point 21, which shows high pressure.
Ive on Thirteenth Lifestyle Apartments, 5 13th Ave, Fairland, Randburg, 2170	High population and closer to the reservoir.
Laerskool Fairland, 145 Market St, Fairland, Randburg, 2030	It is identified for installation in an area characterised by low pressure, ensuring the inclusion of diverse pressure conditions in the data collection process to enhance result accuracy.

Table 2.3 The residential smart water meters (continued on next page)

Meter name		Sonata ultrasonic residential water meter	Gladiator positive displacement water meter	E-Series ultrasonic meter	RS485 Modbus residential smart ultrasonic water meter
Accepted class		2	1	C	2
Category of meter		Ultrasonic	Positive displacement water meter	Ultrasonic	Ultrasonic
Meter size range: mm		20 and 25	15–25 mm	15–25	15–40
Pressure loss (headloss)		from 0.16 bar at Q3		0.12–0.3 bar	0.04
Minimum working pressure: bar		16	16	12	16
Minimum flow rate, q/min ± 5% m^3/h		5.0–20.0	5.0–16.0	0.01–0.06	0.0125–0.08
Maximum flow rate, q/max ± 2% m^3/h		3.1–12.5	2.0–7.8	5.7–12.5	3.125–20
Nominal flow, qp ± 2% m^3/h		2.5–10.0	1.6–6.3	Not specified	2–16
Transitional flow rate, qt ± 2% m^3/h		8–32	8.0–25.6	Not specified	0.02–0.128
Water temperature range: °C		0–50°C	0–50°C	7…50	50
Body		PPS, highly reinforced composite material	BSP, NPSM		Brass
Physical requirements (pipe)	Piping			Vertical or horizontal	
	Valve		Non return valve	Not specified	
	Strainer			None needed	
System features				In-line connector for easy connection and installation to AMR/AMI endpoints	

Table 2.3 Continued

Meter name		Sonata ultrasonic residential water meter	Gladiator positive displacement water meter	E-Series ultrasonic meter	RS485 Modbus residential smart ultrasonic water meter
Type power supply		Lithium battery		3.6-volt lithium thionyl chloride battery	Lithium
Battery life		15 years	15	Not replaceable; 20-year battery life	8–10 years
Data communication	Uni-/bi-directional communication			Bi	
	AMR/AMI PRODUCT	Dialog3G Register	Dialog3G Register		
	Which communication technology	LoRaWAN, Dialog3G	LoRaWan, Dialog3G, OE and ER		M-Bus, LoRa, or NB-IoT remote transmission interface
	Gateway			AMR/AMI	
	Transmitter				
	Data logger				
Compliance		MID (OIML R49:2013, EN 14154, ISO 4064:2014) NSF61+NSF372	MID 2014/32/EU (based on OIML R49 EN 14154 and ISO 4064:2014), WRAS 1307073/1504700 NMI 1413131 SAI SMK 40532 ACS	Meet and exceed ANSI/AWWA C715 standards. The meters comply with the lead-free provisions of the Safe Drinking Water Act, are certified to NSF/ANSI Standards 61 and 372 and carry the NSF-61 mark on the housing	

Table 2.4 The bulk water meters (continued on next page)

Meter name		Octave – stainless steel threaded ultrasonic water meter	Octave – ultrasonic water meter
Accepted class		2	2
Category of meter		Ultrasonic	
Meter size range: mm		DN50–DN300 with cast iron body DN40 and DN50 with polymer body DN40–DN150 with stainless steel body*	DN50–DN300 with cast iron body DN40 and DN50 with polymer body DN40–DN150 with stainless steel body*
Pressure loss		0.16 bar	0.16 bar
Minimum working pressure: bar		16 bar	16 bar
Minimum flow rate, q/min ± 5% m^3/h		0.16–2	0.16–2
Maximum flow rate, q/max ± 2% m^3/h		50–1250	50–1250
Nominal flow, qp ± 2% m^3/h		40–1000	40–1000
Transitional flow rate, qt ± 2% m^3/h		0.125–3.2	0.125–3.2
Water temperature range: °C		0.1 up to 50°C	0.1 up to 50°C
Accuracy of flow			
Physical requirements (pipe)	**Piping**		
	Valve		
	Strainer		

Table 2.4 Continued

Meter name		Octave – stainless steel threaded ultrasonic water meter	Octave – ultrasonic water meter
System features		Flexible data formats including flow directions, flow rates and volumes • AMR and cellular networks ready with alerts and statistic features • Multiline LCD • Programmable display (units and outputs resolution) • EMI/RFI protection	
Type power supply		2 D size Li-battery	
Battery life		15 years	
Application		Utilities, waterworks, industrial, and agricultural	Utilities, waterworks, industrial, and agricultural
Data communication	Uni-/bi-directional communication	Bi	Bi
	Supported outputs	Pulse (open drain, SSR, dry contact), 4-20+Pulse (SSR), 4-20, Encoder+Pulse (SSR), Modbus+Pulse (SSR), M-bus+Pulse (SSR)	

	Which communication technology	Encoder (Dialog3G, Sensus), M-Bus, Modbus (by request, polling)	Encoder (Dialog3G, Sensus), M-Bus, Modbus (by request, polling)
	Gateway		
	Transmitter		
	Data logger		Volumes and alarms data (48KB, 4130 data points)
Compliance		MID 2014/32/EU (based on OIML R49:2013, EN 14154, and ISO 4064:2014) AWWA WRAS NSF ACS KTW W-270	

*DN40-DN50 with threads, DN50-DN150 only with floating flanges

Smart water management for sustainability: digital management of water resources through smart water management systems promoting sustainability

Teaching notes

Synopsis

The case study of Johannesburg Water presents a strategic initiative aimed at revolutionising water management through the implementation of a smart water metering system. As South Africa grapples with documented challenges in water distribution and management, with losses exceeding 40%, JW recognises the imperative to modernise its infrastructure. Against this backdrop, JW, under the guidance of Professor Telukdarie and a team of experts, embarked on a journey to enhance efficiency and sustainability in water distribution. Recognising the vital role water plays in human life and development, JW meticulously assesses global trends, regulations, and standards to tailor a framework suited to Johannesburg's unique challenges and objectives. This comprehensive approach involves evaluating smart water meters from various vendors, ensuring seamless alignment with JW's vision for optimised water supply, quality, and distribution efficiency. The transition from theory to practice involves small-scale pilot deployments in selected areas, serving as crucial testing grounds for assessing the performance and reliability of the chosen meters in real-world scenarios. Concurrently, efforts are underway to fortify the necessary infrastructure to support smart meters, encompassing communication networks, data storage, and management systems. In recognition of the transformative potential of 4IR technologies, JW formed a research and consulting partnership with the UJ. This collaboration aims to develop a digital water balance system, leveraging advanced technologies to address longstanding challenges in water management. The holistic approach adopted by JW not only aims to optimise water distribution but also underscores its commitment to fostering a more sustainable future for Johannesburg. Through strategic planning, stakeholder coherence, and a relentless pursuit of successful outcomes, JW navigates the complexities and challenges associated with the implementation of SWM systems in public spaces, setting a precedent for innovative water management practices globally.

Teaching objectives

This case study aims to appeal to a wide range of audiences interested in the development of municipal water by covering several issues such as technology adoption, infrastructure modernisation, and the connection between urbanisation and sustainability. This case study is suitable for a diverse audience, including professionals, engineering students (undergraduate, master's), and decision-makers in various fields. This case also focuses on the challenges that water utilities face when performing water balance in the water distribution network. The case will have guidelines for how water utilities can use digital technologies to perform accurate water balances in real time. The following teaching objectives were formulated.

- **Metering trends, standardisation, and compliance**
  To understand global trends in smart metering as well as adherence to established international and local standards, regulations, and protocols in the selection and deployment of smart water meters.
- **Optimal meter selection and evaluation**
  To identify and select smart water meters that best meet JW's requirements, considering technical specifications, accuracy, and compatibility with existing infrastructure.
- **To explore the challenges faced by water utilities when performing water balance**
  The case study highlights the challenges faced by water utilities worldwide in determining water balance and highlights how digital technologies can mitigate some of these challenges.
- **To explore the use of 4IR technologies in performing water balances in the South African water sector**
  Students will learn how JW adopted 4IR technologies to digitalise the water distribution network to perform water balance in real time.

Research methods

The study conducted a comprehensive literature review focused on challenges associated with traditional mechanical water meters, exploring existing studies, publications, and reports highlighting accuracy degradation over time. This grounded the study, linking identified issues to the transformative potential of smart water meters and emphasising advanced technology's role. Standards identification involved scrutinising industry regulations and norms, conducting a comparative analysis, and establishing selection criteria. The smart meter selection process detailed technical requirements, vendor assessments, and pilot testing. The final selection justified the chosen meters based on standards and project objectives. The installation phase included meticulous site assessments using Geographic Information System (GIS) data for optimal placement, while testing and commissioning ensured functional accuracy and reliability.

Keywords and subject code

Water, metering, smart water metering, sustainability, technology, fourth industrial revolution.

Assignment questions

1. **Identify and discuss the key challenges that are experienced in water metering, showing an understanding of global trends and best practices.**
   Before class, students should thoroughly analyse the identified challenges associated with traditional mechanical water meters discussed in the literature review. Explain how these challenges impact the efficiency of water and sanitation services in urban areas and propose potential solutions.

2. **What are the global and South African water metering standards and what are the selection criteria?**
   Examine the standard identification and smart meter selection processes. Critically evaluate the industry regulations, norms, and selection criteria presented in the case study. Justify the prioritisation of certain standards and the rationale behind the project's selection of specific smart water meters.
3. **What framework was developed for selecting and installing smart water meters for JW?**
   Discuss the considerations outlined for selecting installation areas for smart meters, leveraging GIS data. Evaluate the criteria for prioritising certain regions and the strategic approach to capturing diverse consumption patterns. Justify the chosen areas for study based on population density, water consumption, and existing infrastructure.

Additional materials

The case may not necessarily need additional reading but here are some reading materials that will help understand the field better.

Koech R, Cardell-Oliver R and Syme G (2021) Smart water metering: adoption, regulatory and social considerations. *Australasian Journal of Water Resources* **25(2)**: 173–182.

March H, Morote ÁF, Rico AM and Saurí D (2017) Household smart water metering in Spain: insights from the experience of remote meter reading in Alicante. *Sustainability* **9(4)**: 582.

Weranga KSK, Kumarawadu S and Chandima DP (2014) *Smart Metering Design and Applications*. Springer, Singapore.

Zapata-Sierra AJ, Salmerón-Manzano E, Alcayde A, Zapata-Castillo ML and Manzano-Agugliaro F (2023) The scientific landscape of smart water meters: a comprehensive review. *Water* **16(1)**: 113–123.

Suggested core readings

Amaxilatis D, Chatzigiannakis I, Tselios C *et al.* (2020) A smart water metering deployment based on the fog computing paradigm. *Applied Sciences* **10(6)**: 1965.

Gupta AD, Pandey P, Feijóo A, Yaseen ZM and Bokde ND (2020) Smart water technology for efficient water resource management: a review. *Energies* **13(23)**: 6268.

Majid A, Islam MD, Rabia B, Ghosh D and Ahmed A (2023) *Smart Water Metering and Quality Monitoring*. Doctoral dissertation, American International University–Bangladesh, Dhaka, Bangladesh.

Millington N and Scheba S (2021) Day zero and the infrastructures of climate change: water governance, inequality, and infrastructural politics in Cape Town's water crisis. *International Journal of Urban and Regional Research* **45(1)**: 116–132.

Reid RN (2020) *Water Quality Systems: Guide for Facility Managers*. River Publishers, New York, NY, USA.

Teaching plan and timing

Duration: 110 min

Introduction (10 min)

- Begin with a brief overview of the case study's context, emphasising the challenges faced by JW in managing water and sanitation services.
- Highlight the importance of modernising infrastructure in response to the growing urban population.
- Introduce the case study's central theme: the implementation of smart metering systems as a transformative solution.

Interactive warm-up discussion (15 min)

- Engage students with a brief discussion about their understanding of traditional water metering challenges.
- Encourage students to share any experiences or insights related to water and sanitation services in urban environments.
- Link their responses to the broader challenges outlined in the case study.

Literature review and conceptual framework (20 min)

- Summarise the key points from the suggested core readings, emphasising the challenges of traditional water meters and the benefits associated with smart metering.
- Connect the literature to the case study's grounding phase, outlining how the literature review informs the subsequent discussions on smart metering.

Critical analysis of challenges (15 min)

- Pose the first assignment question: 'Identify and discuss key challenges'.
- Allow students to present their prepared answers, providing a platform for diverse perspectives.
- Facilitate a class discussion to critically analyse the challenges and explore potential solutions, drawing connections to the literature reviewed.

Smart water meter selection and standards (40 min)

- Transition to the second and third assignment questions related to smart water meter selection and standards.
- Discuss the methods outlined in the case study, emphasising the importance of adhering to industry standards and the criteria for selecting smart water meters.
- Facilitate a group discussion where students can evaluate the presented standards and selection criteria, defending their reasoning.

Conclusion and reflection (5 min)

- Summarise key takeaways from the case study discussion.
- Facilitate a brief reflection session where students share insights gained and lessons learned.
- Highlight the broader implications of the case study in the context of embracing technological advancements for urban water management.

Closing remarks (5 min)

- Conclude the session by emphasising the relevance of the case study to real-world challenges in water and sanitation services.
- Encourage students to carry forward the lessons learned and apply them in future studies or professional contexts.

Analysis of assignment questions

The assignment questions cover a spectrum of topics, encouraging students to apply theoretical knowledge to practical scenarios. Responses should reflect critical thinking, the ability to analyse complex situations, and a practical understanding of the challenges and solutions in implementing smart water metering systems.

Assignment question 1. Identify and discuss the key challenges that are experienced in water metering, showing an understanding of global trends and best practices.

The global collective has been working on metering infrastructure in terms of categories. Mitigating certain needs is crucial for ensuring high-quality water management and informed decision-making. The trends range from data-centred to physical infrastructure.

The increased adoption of smart metering has advanced the management of water supply and demand. Numerous water utilities have adopted smart water metering. The need for more precise and real-time statistics on water usage, more accurate billing, and better water resource management prompted this adoption. A study titled 'Role of water policies in the adoption of smart water metering and the future market' shows statistical data on different countries that have adopted SWM (Msamadya *et al.*, 2022). Countries like Australia have high metering implementation rates of about 70%. Australia's water metering system is making progress but compared with the initial phases of the SWM implementation before that year, the rate of spread of SWM has decreased since 2015. At least five Israeli cities have completed their SWM deployments to a perfect 100%. Given Israel's current state of water scarcity, additional SWM implementations nationwide are anticipated to conserve more water resources. Since 2015, South Korea has seen the completion of several SWM implementations, and more than 135 pilot and trial cases are expected to move quickly toward SWM implementations soon (Figure 2.3).

Another emerging trend is the use of data analytics for leak detection. Smart water meters in the water distribution network are increasingly integrating with data analytics platforms to analyse consumption patterns and detect leaks. The advancement of analytics in utilities is helping in the identification of challenges in the networks so that it is easier to address issues, reducing water losses and improving overall efficiency.

Wan *et al.* (2022) conducted a study that categorises various leakage management and detection methods, as illustrated in Figure 2.4, and introduces a framework for leak detection. The figure shows that there are different categories of leak detection, from hardware- to software-based methods, but many studies support the newer technology, and software methods (Britton and Stewart, 2011; Farah and Shahrour, 2017; Huang *et al.*, 2020). By minimising water loss, utilities can reduce the costs associated with treating and transporting water. This leads to financial savings for both water utilities and consumers. Leak detection and repair initiatives also align with environmental sustainability goals by reducing the carbon footprint associated with unnecessary water treatment and distribution.

The communication protocol for smart water metering is also an important aspect of the distribution network. In recent years, there has been a notable trend in the communication protocols of smart water meters, with a shift towards more advanced and interoperable standards. Smart meters transmit data remotely for monitoring and management (Pimenta and Chaves, 2021). These meters also come equipped with data logging capabilities, enabling them to store historical consumption data for analysis and troubleshooting (Alvisi *et al.*, 2019). Owing to network communication, data encryption and security features are necessary to safeguard data from unauthorised access and tampering (Slaný *et al.*, 2020). A case study of a utility company's adoption of a low-range wide-area network (LoRaWAN) for its smart water metering infrastructure exemplifies this trend (Slaný *et al.*, 2020). The deployment of LoRaWAN enables the meters to transmit data over long distances with low power consumption, facilitating efficient communication in challenging environments. Some other protocols include NB-IoT, Sigfox, AMR, and Wi-Fi or cellular networks. These are mainly used in residential metering infrastructure; for larger networks Wi-SUN (Wireless Smart Utility Network) and M-Bus (Meter-Bus) (European standard) are used. Numerous best practices, both at international and local levels, abound in the concept of smart metering. They show a diverse range of strategies and approaches for effective implementation. Efficient water management through comprehensive water metering practices is of paramount importance, encompassing both bulk supply and residential consumption. In the context of bulk water metering, precision holds the utmost significance, necessitating the installation of meticulously calibrated flowmeters and the implementation of regular maintenance protocols to uphold accuracy (Brinkley *et al.*, 2020). Furthermore, integrating advanced remote monitoring systems and data logging mechanisms is essential for real-time tracking, anomaly identification, and leak detection in bulk supply lines (Beal and Flynn, 2015). The data garnered from these measures should be subject to rigorous analysis to discern consumption trends, patterns, and potential inefficiencies, contributing to informed decision-making. Transparent billing systems should accompany these practices, ensuring that customers are billed fairly, and educated about their usage patterns and charges (Randall and Koech, 2019). Pressure management strategies are pivotal to curbing water losses stemming from leaks and bursts, bolstering the overall effectiveness of bulk water metering.

Residential water metering, on the other hand, involves a multifaceted approach that extends beyond accurate measurement. Individual water meters for each residential unit offer the granularity required to track consumption, while the integration of smart meters introduces an element of real-time data transmission to utility providers and customers alike, fostering heightened awareness. Australia, known as the driest inhabited continent with a housing and urban population of around 25 million, is set to install nearly ten million smart water meters using narrow-band IoT networks (NB-IoT) by 2027 (Randall and Koech, 2019). This number is a significant increase from the current count of under 50 000. The majority of individuals comprehend the purpose and function of smart water meters. Not as commonly recognised is the impact of international standards on shaping Australia's water management sector. Facilitating the two-way communication essential for smart water meter systems, low-power wide-area networks (LPWAs) offer the necessary connectivity (Islam *et al.*, 2020). Multiple LPWAs exist, each presenting distinctive advantages. Australia stands out as a notable example, as a significant portion of its smart water meters will operate through NB-IoT. Remarkably, NB-IoT is the sole LPWA that relies on a consistent global standard. Research by the Global Mobile Suppliers Association (GSMA) reveals that 140 operators across 69 countries are involved in this realm (Islam *et al.*, 2020).

Assignment question 2. What are the global and South African water metering standards and what are the selection criteria?

The NRS049:2016 specification outlines the National Smart Metering Standards for South Africa, focusing on advanced metering infrastructure (AMI) for residential and commercial customers in South Africa. This specification mandates the storage of active energy consumption data on the meter, encompassing total register values and half-hourly intervals. The AMI master station, part of the data retrieval system, must facilitate the retrieval of both billing register values and half-hourly data. Municipalities are required to decide whether to bill customers based on meter register values or half-hourly values. Beyond billing data, NRS049:2016 specifies that smart meters must record various events, including tampering, supply outages, under- and over-voltage conditions, the application of disconnect commands, load control through the master station, and meter configuration changes.

Compliance with NRS049:2016 results in a smart metering system with key features.

- There are bi-directional communications between the central server and meters or devices.
- Customers can have a portable customer interface unit for reading meter information and receiving utility data.
- The capability to control up to two relays for load control (e.g. a hot water cylinder and a swimming pool pump).
- Remote load disconnect capability for utility revenue protection.

The South African Smart Grid Initiative (SASGI), an initiative of the South African National Energy Development Institute (SANEDI), has compiled the 'Smart Grid 2030

Vision', as outlined in the national vision document. Developed through the Smart Grid programme, this vision document serves as guidance for smart grid planning and implementation by both Eskom and municipal electricity distributors. SANEDI offers support to municipalities in assessing their readiness for investing in a smart grid network through a Smart Grid Readiness Test (Smart Grid Maturity Assessment).

JW has diligently incorporated established standards and references in the specifications for tenders in metering infrastructure (City of Joburg SOC Ltd, 2021), placing a significant emphasis on compliance with the Legal Metrology Act 9 of 2014 and its associated technical requirements. The meticulous integration of these standards has not only been instrumental in addressing technical concerns within the document but has also served as a comprehensive guide throughout its preparation. It is imperative to note that this document does not seek to override the relevant provisions stipulated in the Legal Metrology Act 9 of 2014 and its accompanying technical requirements as delineated by the South African Bureau of Standards (SABS). Adherence to these regulations remains paramount. One such standard, SANS 1529-1, delineates specifications for water meters designed for cold potable water, particularly focusing on the metrological characteristics of mechanical water meters with a nominal bore not exceeding 100 mm. Emphasising the importance of compliance, the standard stipulates manufacturing requirements and mandates that meter accuracy align with Class C, with the explicit exclusion of Class B meters without SANS approvals. Additionally, SANS 1529-9 provides essential requirements for electronic indicators utilised alongside mechanical water meters, electronic water meters, and electronic prepayment water measuring systems. Furthermore, the incorporation of IEC 62055-41 and IEC 62055-51 standards specifically addresses prepayment tokens within the ambit of the Legal Metrology Act 9 of 2014 and its associated technical requirements. Overall, JW's commitment to these standards underscores its dedication to ensuring precision, reliability, and compliance in the realm of water measurement and metering systems.

Several international standards govern smart water meters. The establishment of these standards guarantees the interoperability, accuracy, and reliability of smart water metering systems. It is essential to note that standards may evolve, so it is always best to check for the latest versions and updates. Some of the relevant international standards for smart water meters are the European Commission (2018), Pepermans (2014), and Blom *et al.* (2010). JW needs to determine which of these standards they need to comply with or are currently complying with.

- ISO 4064: This standard, issued by the International Organization for Standardization (ISO), specifies requirements and test methods for water meters of various types, including smart water meters. ISO 4064 ensures accuracy and performance uniformity across different metering devices.
- ISO 5814: This standard, also issued by ISO, deals with determining the hydraulic performance of water meters, including smart water meters.

- OIML R 49: The International Organisation of Legal Metrology (OIML) issues this recommendation, which provides guidelines for the metrological and technical requirements for water meters, including those with remote reading capabilities (smart water meters).
- IEC 62056: The International Electrotechnical Commission (IEC) has established this standard for electricity metering data exchange: the DLMS/COSEM (Device Language Message Specification/Companion Specification for Energy Metering) standard, which is widely used for various utility meters, including smart water meters.
- ANSI/AWWA Standard C750: The American National Standards Institute (ANSI) and the American Water Works Association (AWWA) have developed this standard, which outlines the requirements for wireless communication systems used in water meters, including those with smart capabilities.
- EN 13757-1: This European Standard, part of the EN 13757 series, specifies the data exchange formats and protocols for meter reading, control, and billing of smart metering systems, including water meters.
- IEC 61107: This standard, issued by the IEC, outlines the physical and electrical characteristics of a water meter with a non-metrological auxiliary output.

Assignment question 3. What framework was developed for selecting and installing smart water meters for JW?

- JW conducted an extensive literature analysis, looking at published papers, reports, and other materials, to identify patterns in the use of SWMs by water utilities globally. This involved a careful review of national and international laws and guidelines to make sure they matched JW's goals. After laying a strong foundation, the team focused on the water network in Johannesburg and investigated how smart meters could improve its effectiveness. They gathered as much information as possible about the current network and metering standards, carefully dissecting each detail to provide a complete picture of the infrastructure.
- After that, the team looked into international protocols to determine which ones would work well with JW's operations. They attempted to identify the similarities and contrasts in order to create a framework that would be tailored to the particular requirements of the Johannesburg setting.
- After laying out technological requirements and thoroughly establishing factors like accuracy, communication protocols, and data management capabilities, the team started the vital process of choosing the best smart water meters. Each selection was thoroughly justified, in accordance with the established standards and project objectives, and was assessed using a selection matrix in order to guarantee that only the best meters were chosen (Table 2.3 and Table 2.4).
- Equipped with a wealth of knowledge and expertise, the team carefully considered the deployment of network components. They carefully chose the locations (Table 2.2) to maximise the effects of the smart water meters in order to

create a more sustainable and effective future for Johannesburg, referencing well-known literature and best practices.

Key learnings

- **Technological innovation for operational efficiency:** The case highlights the transformative impact of embracing technological innovation, specifically through the implementation of smart water metering systems. The shift from traditional mechanical meters to smart meters not only addresses accuracy challenges but also enhances operational efficiency through real-time monitoring and remote data collection.
- **Strategic integration of standards and selection criteria:** The case underscores the importance of strategic decision-making in selecting smart water meters. Identification of and adherence to industry standards, coupled with a meticulous selection process based on clear criteria, are essential for the successful integration of advanced technologies. This learning emphasises the need for organisations to align technology choices with established standards and project objectives.

REFERENCES

Alvisi S, Casellato F, Franchini M, *et al.* (2019) Wireless middleware solutions for smart water metering. *Sensors (Switzerland)* **19(8)**: 1–25, 10.3390/s19081853.

Beal CD and Flynn J (2015) Toward the digital water age: survey and case studies of Australian water utility smart-metering programs. *Utilities Policy* **32**: 29–37.

Blom A, Cox P and Raczka K (2010) *Developing a Policy Position on Smart Water Metering.* Interactive qualifying project, Worcester Polytechnic Institute, Melbourne, Australia.

Brinkley M, Ilemobade AA and Ncube M (2020) Water meter replacement and its impact on water consumption, utility revenue and payback periods. *Journal of the South African Institution of Civil Engineering* **62(3)**: 2–9.

Britton T and Stewart RA (2011) Smart metering as a tool for the remote diagnosis of leakage in residential households. https://www.researchgate.net/publication/228748744 (accessed 13/03/2024).

City of Joburg SOC Ltd (2021) Supply chain management unit request for tender. https://www.johannesburgwater.co.za (accessed 17/06/2024).

European Commission (2018) EU Financial Regulation. https://commission.europa.eu/publications/eu-financial-regulation_en (accessed 15/01/2024).

Farah E and Shahrour I (2017) Smart water for leakage detection: feedback about the use of automated meter reading technology. *2017 Sensors Networks Smart and Emerging Technologies, SENSET 2017*, 10.1109/SENSET.2017.8125061.

Giurco, D, Carrard, N, McFallan, S *et al.* (2008) *Residential End-Use Measurement Guidebook: A Guide to Study Design, Sampling and Technology*. UTS and CSIRO, Victoria, Australia.

Huang Y, Zheng F, Kapelan Z, *et al.* (2020) Efficient leak localization in water distribution systems using multistage optimal valve operations and smart demand metering. *Water Resources Research* **56(10)**: 1–21, 10.1029/2020WR028285.

Islam SM, Gaihre YK, Islam MR, *et al.* (2020) Effects of water management on greenhouse gas emissions from farmers' rice fields in Bangladesh. *Science of the Total Environment* **734**: 139382.

Msamadya S, Joo JC, Lee JM, *et al.* (2022) Role of water policies in the adoption of smart water metering and the future market. *Water (Switzerland)* **14(5)**, 10.3390/w14050826.

Pepermans G (2014) Valuing smart meters. *Energy Economics* **45**: 280–294.

Pimenta N and Chaves P (2021) Study and design of a retrofitted smart water meter solution with energy harvesting integration. *Discover Internet of Things* **1(1)**, 10.1007/s43926-021-00010-x.

Randall T and Koech R (2019) Smart water metering technology for water management in urban areas (analysing water consumption patterns to optimize water conservation). *Water e-Journal, online Journal of the Australian Water Association* **4(1)**, 10.21139/wej.2019.001.

Slaný V, Lučanský A, Koudelka P, *et al.* (2020) An integrated IoT architecture for smart metering using next generation sensor for water management based on lorawan technology: a pilot study. *Sensors (Switzerland)* **20(17)**: 1–23, 10.3390/s20174712.

Wan X, Kuhanestani PK, Farmani R and Keedwell E (2022) Literature review of data analytics for leak detection in water distribution networks: a focus on pressure and flow smart sensors. *Journal of Water Resources Planning and Management* **148(10)**, 10.1061/(asce)wr.1943-5452.0001597.

Prof. Arnesh Telukdarie
ISBN 978-1-83662-645-9
https://doi.org/10.1108/978-1-83662-642-820251003

Chapter 3
Unlocking insights into the green hydrogen industry through natural language processing techniques

Arnesh Telukdarie and Khuliso Murulane

Introduction

Green hydrogen (GH_2) stands as a beacon of sustainable fuel, derived solely from renewable sources rather than fossil fuels. This innovative energy carrier holds enormous potential to drive sustainability across industries, leaving oxygen as its only by-product (Albatayneh *et al.*, 2023). Alongside other African nations, South Africa is increasingly recognising the pivotal role green hydrogen can play in its energy and sustainability plans (Grobbelaar and Ngubevana, 2022). Notably, the nation has made significant strides by establishing a US $1 billion fund in 2023 aimed at propelling the rapid advancement of the domestic green hydrogen sector (Sanlam Intelligence, 2023). With substantial renewable energy (RE) assets at its disposal, particularly in solar and wind resources, South Africa is poised to harness these abundant resources to facilitate the production of green hydrogen at scale. Green hydrogen, as a clean and versatile energy carrier, can be used in various sectors, including transportation, industry, and power generation. However, significant barriers hinder its widespread adoption. These include high production costs, limited infrastructure, and the need for a skilled workforce. Moreover, gaps exist in our understanding of market dynamics, technological advancements, and regulatory frameworks. Addressing these challenges is critical to unlocking the full potential of green hydrogen and accelerating its transition to a mainstream energy source. In May 2023, the Department of Science and Innovation (DSI) brought a critical challenge to Johannesburg Business School (JBS) regarding the lack of comprehensive data on technological advancements, skills requirements, and market trends in the green hydrogen sector. Traditional methods of gathering information, such as expert interviews, proved insufficient in offering a holistic view of the industry landscape, leaving decision-makers with incomplete insights, and hindering strategic planning in the green hydrogen sector. This case study follows Dr. Khuliso Murulane, a researcher under Professor Arnesh Telukdarie at JBS. Dr. Murulane plays a key role in the green hydrogen initiative. He brings expertise in energy and data analytics to address the issue that DSI presented to JBS.

Industry background

The energy industry in South Africa is a critical component of the country's economy, characterised by its heavy reliance on coal, which accounts for the majority of its energy production (Department of Energy, 2021, 2022). This reliance has positioned South Africa as one of the largest producers and consumers of coal in the world. According to the Department of Energy, coal alone makes up 67% of the primary energy supply, while crude oil is about 17%, natural gas and nuclear are about 2% each, and renewables are about 11% (Department of Energy, 2021, 2022). The heavy reliance on coal has significant environmental implications, contributing to high levels of greenhouse gas (GHG) emissions. The South African government has introduced various policies and regulations to promote energy efficiency, reduce carbon (CO_2) emissions, and encourage the adoption of renewable energy. The Integrated Resource Plan (IRP) outlines the country's energy strategy and targets for the coming decades (Department of Energy, 2022).

The industrial sector is South Africa's largest energy consumer, accounting for 41% of the country's total energy consumption (Department of Energy, 2022). Industries such as iron and steel, mining, manufacturing, and the chemical industry depend heavily on energy, particularly electricity and coal. The mining industry, in particular, is a major energy consumer, using large amounts of electricity for operations like ore extraction and processing (Department of Energy, 2022). The transportation sector is another significant energy consumer in South Africa, accounting for 23% of the country's total energy consumption (Department of Energy, 2022). This sector relies primarily on liquid fuels (petrol and diesel), which are mostly derived from imported crude oil, and locally produced synthetic fuels from coal and natural gas (Department of Energy, 2022). The extensive road network and high use of private vehicles contribute significantly to energy consumption in this sector.

The commercial sector, which includes businesses, offices, and public buildings, consumes about 12% of the country's energy (Department of Energy, 2022). Electricity is predominantly used for lighting, heating, cooling, and operating various equipment. The residential sector consumes energy at a rate of about 18% (Department of Energy, 2022). In the residential sector, energy is used mainly for cooking, heating, and lighting. Electricity is the primary energy source for urban households, while rural areas often rely on a mix of electricity, paraffin, wood, and other biomass. The agricultural sector consumes 2% of the country's energy, primarily for activities such as irrigation, processing, and transportation of goods (Department of Energy, 2022). Diesel and electricity are the main energy sources in this sector. Energy efficiency and the adoption of RE technologies are becoming increasingly important for enhancing sustainability in agriculture.

According to the Department of Energy (2021, 2022), South Africa possesses a highly developed electricity network with one of the highest rates of electricity access in sub-Saharan Africa. Eskom, the state-owned power utility, dominates the country's energy sector (Department of Energy, 2021, 2022). Eskom dominates electricity production (generating about 92% of the country's electricity) and manages transmission and distribution to industrial, mining, commercial, agricultural, and residential customers, as well as municipalities that further distribute it to local businesses and households (Department of Energy, 2022). Eskom also buys electricity from independent power producers (IPPs) through various agreements and imports electricity from facilities outside the country (Department of Energy, 2022). Historically, Eskom's coal-fired power plants have been the backbone of South Africa's energy supply. However, aging infrastructure and maintenance issues have led to frequent power outages, impacting economic stability and growth (Department of Energy, 2022). These frequent controlled blackouts have been implemented as a measure to prevent the total collapse of the power grid when demand exceeds supply (Kessides, 2020).

This situation is exacerbated by insufficient investment in new capacity, poor operational performance, and logistical challenges in coal supply, further straining the already overburdened power system (Kessides, 2020). To address these challenges and ensure a more reliable and sustainable energy future, the country is transitioning towards more sustainable energy use. This shift is driven by the need to reduce GHG emissions and mitigate climate change. Central to this transition are increased investments in RE sources such as solar, wind, biomass, and hydropower. Initiatives like the Renewable Energy Independent Power Producer Procurement Programme (REIPPPP) aim to diversify the energy mix and promote cleaner energy alternatives (Department of Energy, 2022). Onshore wind generated 53% of the power under the REIPPPP in 2019, followed by photovoltaic power at 33% (Department of Energy, 2022). Concentrated solar power (13%) and small hydropower (1%) produced the remaining electricity (Department of Energy, 2022).

The push for RE is also supported by the potential of hydrogen production, which can help decarbonise sectors that are difficult to electrify, such as heavy industry and transportation. The development of green hydrogen technology aligns with South Africa's goals to enhance energy security, reduce dependence on fossil fuels, and create economic opportunities in the emerging green economy. The country's geographical location provides it with substantial solar and wind resources, which are crucial for producing green hydrogen. The Northern Cape has been identified as a prime area for RE projects owing to its high solar irradiance and wind speeds (Lebrouhi *et al.*, 2022). Additionally, South Africa holds a substantial reserve of platinum group metals (PGMs), accounting for 80% of the world's reserves, which electrolysers utilise (Lebrouhi *et al.*, 2022).

South Africa's involvement in hydrogen dates back to 2008 with the launch of the National Hydrogen and Fuel Cell Technologies Research, Development, and Innovation Strategy (NHFCTRDIS) by the Department of Science and Technology (DST), now known as the DSI. The strategy outlined ambitious goals, including establishing a foundation for hydrogen production and storage technologies as well as processes (Lebrouhi *et al.*, 2022). It also aimed to lay the groundwork for the development of technologies based on PGMs, leveraging global knowledge to surpass existing fuel cell technologies for targeted applications that address regional development challenges (Lebrouhi *et al.*, 2022). A significant milestone within this strategy was the establishment of Hydrogen South Africa (HySA), a driving force behind innovation throughout South Africa's hydrogen and fuel cell technology value chain (Lebrouhi *et al.*, 2022). HySA, founded by the DSI, was established as a 15-year programme with the mission of developing South African intellectual property, enhancing human knowledge and skills, and fostering the creation of products, components, and processes to encourage South Africa's active participation in the rapidly expanding hydrogen and fuel cell markets (Lebrouhi *et al.*, 2022). HySA has already undertaken various research and development (R&D) as well as demonstration projects (Lebrouhi *et al.*, 2022). Examples of completed projects under HySA include the development of a hydrogen forklift, a hydrogen fuel station, a hybrid fuel cell golf car, and initiatives related to renewable hydrogen production and storage.

Recently, the DSI partnered with Implats (Impala Platinum) to advance hydrogen and fuel cell product development, with Implats contributing US $670 000 to the South African project (Lebrouhi *et al.*, 2022). This initiative focuses on exploring new onboard hydrogen storage systems using local raw materials for small vehicles like forklifts, with Implats evaluating the outcomes. Additionally, in August 2020, the DSI collaborated with the Department of Defense to deploy seven hydrogen vehicles to aid a hospital during the COVID-19 crisis (Lebrouhi *et al.*, 2022). Lastly, the DSI launched the 'Platinum Valley Cluster' project through HySA in October 2020, aiming to identify opportunities for hydrogen activities in key centres and highlight advancements in PGM extraction, processing, and recycling for sustainable development.

While South Africa's RE sector shows promising growth, it also faces several challenges and opportunities that are crucial for its long-term success. Below are some of those challenges.

- Infrastructure development: Developing the necessary infrastructure for green hydrogen production, storage, and distribution is a significant challenge. Investments in electrolysis plants, pipelines, and storage facilities are required to scale up production.
- Technology and skills: Advancing green hydrogen technology and building a skilled workforce are essential for the industry's growth. Partnerships with international stakeholders and educational institutions can help bridge this gap.
- Regulatory framework: Establishing a clear and supportive regulatory framework is important for attracting investment and ensuring the industry's sustainable growth. Policies that incentivise RE use and green hydrogen production will be key.
- Cost competitiveness: The cost of RE is decreasing, but green hydrogen production remains relatively expensive. Achieving cost competitiveness with traditional hydrogen production methods (such as natural gas reforming) will be essential for the industry's long-term viability.

The green hydrogen value chain

The green hydrogen value chain (VC) is made up of a series of processes involved in hydrogen production, storage, distribution, and utilisation, as shown in Figure 3.1. Electricity from renewable resources like wind, solar, and hydropower electrolyses water to produce green hydrogen (Albatayneh *et al.*, 2023). This method stands in contrast to traditional hydrogen production, which relies on fossil fuels and results in significant carbon emissions.

Company background

The DSI stands at the forefront of scientific and technological advancements in South Africa. DSI derives its mandate from the 1996 White Paper on Science and Technology, which introduced the concept of the national system of innovation – a set of interacting organisations and policies through which South Africa creates, acquires, diffuses, and puts into practice new knowledge to help achieve individual and collective sustainable development goals (SDGs). With a mandate to promote research, innovation, and economic development, DSI collaborates with stakeholders to harness the potential of science and technology for sustainable development. Aligned with DSI's vision, JBS has established itself as a leading institution committed to providing cutting-edge education tailored to the dynamic demands of today's business world. With a commitment to innovation and excellence, JBS equips students with the knowledge and skills needed to thrive in the modern era. Through partnerships with industry leaders and research institutions, JBS drives impactful research and fosters collaboration to address pressing societal challenges. In green hydrogen, the DSI's involvement is instrumental in driving research and innovation, paving the way for a greener and more sustainable energy future.

Sustainable development goals

Advocates of sustainable development argue that individuals, institutions, and governments can effectively drive sustainability by efficiently combining social, economic, and ecological objectives (Kopnina, 2016). In 2015, the United Nations (UN) introduced 17 SDGs at the UN conference on Sustainable Development, also known as Rio+20 or Earth Summit 2012 (Kopnina, 2016; Nhamo *et al.*, 2020). These goals form a set of 17 interconnected global objectives designed to address the most pressing challenges facing humanity and the planet. They also form a part of the 2030 Agenda for Sustainable Development, which aims to eradicate poverty, protect the environment, and ensure prosperity and peace for all by the year 2030 (Kopnina, 2016; Nhamo *et al.*, 2020). The SDGs cover a broad spectrum of issues; these include economic growth, social inclusion, and environmental sustainability, as summarised in Figure 3.2 (Kopnina, 2016; Nhamo *et al.*, 2020).

Among the SDGs given in Figure 3.2, green hydrogen emerges as an important solution that intersects with several SDGs owing to its potential to transform the energy landscape and promote sustainable development. For example, green hydrogen directly supports SDG7 by providing a clean, reliable, and sustainable energy source. As countries strive to reduce their dependency on fossil fuels, green hydrogen offers an alternative that can be integrated into existing energy systems. By using RE for hydrogen production, it is possible to ensure a steady supply of clean energy, reduce energy costs over time, and increase energy security. The production and utilisation of green hydrogen play a crucial role in mitigating climate change, thus aligning with SDG 13. Green hydrogen helps reduce CO_2 emissions in multiple sectors, particularly in heavy industries such as steel manufacturing and transportation, where electrification is challenging. By replacing fossil fuels with green hydrogen, GHG emissions can be reduced, contributing to global efforts to combat climate change.

Green hydrogen also fosters innovation and supports sustainable industrialisation, directly contributing to SDG 9. The development and scaling of green hydrogen technologies encourage research and innovation, driving advancements in electrolysis, hydrogen storage, and fuel cell technologies. Also, the establishment of green hydrogen infrastructure – such as hydrogen production plants, refuelling stations, and distribution networks – creates new economic opportunities and jobs, promoting sustainable industrial growth. Green hydrogen can enhance the sustainability of urban areas, aligning with SDG 11. Public transportation systems, such as buses and trains, can use green hydrogen in the context of sustainable cities to reduce air pollution and GHG emissions. Hydrogen-powered vehicles produce zero emissions, improving air quality and public health. Additionally, they can support the development of smart grids and decentralised energy systems, contributing to more resilient and sustainable urban infrastructure. Green hydrogen also supports sustainable consumption and production patterns, which are key aspects of SDG 12. By offering a sustainable alternative to fossil fuels, green hydrogen encourages a shift towards cleaner production processes in industries. It promotes the use of RE and efficient resource management, reducing the environmental footprint of industrial activities. Moreover, the adoption of green hydrogen in manufacturing and transportation promotes circular economy principles by minimising waste and optimising resource utilisation. Green hydrogen production does not directly impact access to clean water and sanitation, but sustainable production practices can ensure minimal water usage and pollution, indirectly supporting SDG 6. The electrolysis process used to produce green hydrogen requires water. However, advances in technology are making it possible to use seawater or wastewater for electrolysis, reducing the strain on freshwater resources. Also, wastewater treatment facilities can integrate green hydrogen production to enhance efficiency and sustainability.

DSI and JBS collaboration

As the demand for comprehensive data on technological advancements, skills, and market trends in the green hydrogen sector became increasingly apparent, Dr. Murulane stepped forward with a transformative vision in June 2023. Recognising the limitations of traditional methods for gathering comprehensive data, Dr. Murulane addressed the challenge presented by the DSI to JBS. Traditional approaches, like expert interviews, failed to provide a complete and holistic view of technological advancements, skills requirements, and market trends in the green hydrogen sector. This gap left decision-makers with insufficient insights, hindering effective strategic planning. To overcome these shortcomings, he proposed a novel approach. His plan was to employ natural language processing (NLP) techniques, such as web scraping and text mining, to extract information from vast datasets of academic literature, industry reports, and online sources. Dr. Murulane's approach is grounded in information retrieval (IR) and information extraction (IE) theories, which focus on the extraction of relevant information from large collections of data. IR theories provide the foundation for understanding how to efficiently and effectively retrieve information that meets specific needs (Singh, 2018). They incorporate a variety of models and algorithms that optimise the search and retrieval process, guaranteeing the identification and utilisation of the most relevant and useful information. IR systems use various methodologies, including NLP techniques like web scraping, text mining, named entity recognition (NER), and semantic analysis to enhance the accuracy and relevance of retrieved information (Singh, 2018).

Web scraping

Web scraping is an important technique in modern research, offering an innovative method to collect vast amounts of data from the web (Ferrara *et al.*, 2014). This practice involves the automated extraction of information from web pages, enabling one to access and collect data that might be time-consuming or even impossible to compile manually. Python stands out as the most widely used programming language for web scraping tasks. Within its rich ecosystem, libraries like Beautiful Soup (Richardson, 2007), Scrapy (Kouzis-Loukas, 2016), and Selenium (Chapagain, 2019) have become indispensable tools for simplifying and streamlining the extraction of data from websites. Beyond just collecting web data, these libraries seamlessly integrate with other Python modules dedicated to data analysis and manipulation tasks.

Text mining

Complementing web scraping is the rising domain of text mining, a subset of data mining that focuses on extracting valuable information from unstructured text data (Hotho *et al.*, 2005). Using Python tools like NLTK (Natural Language Toolkit) (Xu, 2010) and spaCy (Neumann *et al.*, 2019), one can perform advanced linguistic analysis, extract topics, recognise entities, and more, transforming raw web data into meaningful insights. As the digital world continues to expand, the combination of web scraping and text mining offers a promising avenue for researchers to harness the power of the web, ensuring they remain at the forefront of innovation and discovery. By leveraging these techniques, researchers can tap into up-to-the-minute data, reflecting the most recent trends, sentiments, and shifts in various fields.

Dr. Murulane aimed to systematically analyse the collected data and provide DSI with actionable intelligence, empowering them to make informed decisions and drive progress in the green hydrogen sector. Starting in August 2023, Dr. Murulane dedicated his efforts to collecting data on the green hydrogen value chain using NLP techniques. Dr. Murulane's approach involved several steps to ensure thorough and accurate data collection. He began by defining data sources and identifying and selecting relevant information related to the green hydrogen value chain; these included academic journals, industry reports, news articles, government publications, and online databases. To automate data extraction, Dr. Murulane employed web scraping tools such as Beautiful Soup and Scrapy, which allowed for the efficient collection of large volumes of data. The extracted data included textual content, tables, and charts relevant to the green hydrogen industry. In addressing the ethical implications of web scraping, Dr. Murulane ensured that all data collection practices adhered to ethical standards and legal requirements. By using diverse data sources, he was able to address potential biases and ensure the reliability of the extracted information. He also implemented algorithms to detect and mitigate bias. This approach produced balanced and accurate insights, avoiding the reinforcement of any existing prejudices present in the data.

Dr. Murulane employed text-mining techniques to process the collected textual data. He also conducted thorough data cleansing and preprocessing to ensure the accuracy and quality of the collected data. This step involved removing duplicates, handling missing values, and standardising formats to prepare the data for analysis. To ensure data integrity and reliability, validation and verification processes were also implemented. This included conducting consistency checks to identify and rectify discrepancies, as well as verifying the authenticity of the data through source verification. The cleaned and verified data were then integrated and organised into a structured format within a centralised repository. Throughout September of the same year, Dr. Murulane focused on analysing the collected data to derive meaningful insights. He employed methods such as Latent Dirichlet Allocation (LDA) (Blei and Lafferty, 2009) alongside other statistical and text analysis techniques to uncover hidden patterns, topics, and themes within the dataset. He then generated detailed reports to communicate his findings to stakeholders. He also developed a graphic user interface for the centralised database, which is depicted in Figures 3.3 and 3.4. This interface provided an experience for users to query and retrieve data, facilitating better decision-making processes.

In October 2023, Dr. Murulane presented his findings to stakeholders, revealing comprehensive data on technological advancements, skills requirements, and market trends in the green hydrogen sector. His approach, using NLP techniques, proved to provide significantly more detailed information compared with traditional methods of data collection. Tables 3.1 and 3.2 present a summary of the results, with Table 3.1 showing the outcomes from traditional data collection methods and Table 3.2 illustrating Dr. Murulane's enhanced results. The results of Dr. Murulane's research were transformative, as he utilised the power of NLP techniques to uncover insights that inform strategic decision-making in the green hydrogen industry. His findings highlighted key technological advancements that are driving innovation in the sector, as well as the evolving skills requirements needed to support these advancements. Additionally, his analysis of market trends provided stakeholders with a clearer understanding of the current landscape and future opportunities within the green hydrogen value chain. The information obtained through NLP approaches allowed for a more granular analysis of the data, detecting patterns and correlations that traditional methods might have missed.

This level of detail enables stakeholders to make informed decisions, such as identifying specific areas where technological investments would yield the highest returns and pinpointing the skills gaps that need to be addressed to support industry growth. Armed with this knowledge, DSI can now develop targeted strategies to foster the green hydrogen sector. This includes allocating resources more effectively to R&D projects that have the potential to drive significant advancements, as well as implementing training programmes to equip the workforce with the necessary skills to support the green hydrogen industry's evolution. Dr. Murulane's work catalyses progress, inspiring collaboration among industry players, researchers, and policymakers. His findings accelerated progress toward a sustainable energy future by showcasing the practical applications of data-driven insights. The impact of his work highlights the critical role of technology and data in shaping the future of RE and underscores the need for continued investment in these areas.

Exhibits

Figure 3.1 The green hydrogen value chain (GH_2VC)

Production	The production phase of green hydrogen involves splitting water molecules into hydrogen and oxygen using electrolysis. This process is powered by renewable energy, ensuring that the hydrogen produced is entirely green. Innovations in electrolyser technology are important for improving the efficiency and reducing the costs of green hydrogen production.
Storage	Once produced, green hydrogen needs to be effectively stored and transported to its points of use. Hydrogen storage solutions include compressed gas, liquid hydrogen, and chemical carriers like ammonia. Each storage method presents unique challenges and opportunities. Storing hydrogen as a compressed gas involves keeping it at high pressures, typically around 350–700 bar. This method is relatively straightforward and widely used but requires high-pressure storage tanks and safety measures to manage the risks associated with high-pressure gases. The energy density is lower compared with liquid hydrogen, necessitating larger storage volumes for the same amount of energy. Liquefying hydrogen involves cooling it to cryogenic temperatures (–253°C or –423°F), resulting in a much higher energy density than compressed gas. However, the liquefaction process is energy-intensive and requires sophisticated insulation techniques to maintain the low temperatures. Safety concerns also arise from the potential for hydrogen to leak and evaporate, posing risks of explosion and fire. Hydrogen can be stored in chemical carriers such as ammonia or metal hydrides. Ammonia offers a higher volumetric energy density and can be transported using existing infrastructure for liquid fuels. However, converting hydrogen to and from these carriers adds complexity and cost to the storage and distribution process. The toxicity and corrosiveness of ammonia also require careful handling and safety protocols.
Distribution/transportation	Efficient distribution networks are important for delivering hydrogen from production sites to end-users. Pipelines offer a direct and efficient means of transporting hydrogen over long distances but require significant investment in infrastructure. In contrast, transport vessels provide flexibility and can leverage existing logistics networks, although they may incur higher operational costs and energy losses during the loading and unloading processes.
Utilisation	Green hydrogen has diverse applications across various industries. In the energy sector, it can be used for power generation and as a fuel for vehicles. Industrial applications include its use as a feedstock for producing chemicals, refining metals, and manufacturing fertilisers. Additionally, green hydrogen plays a significant role in decarbonising sectors that are hard to electrify, such as aviation, shipping, and heavy industry.

Figure 3.2 The 17 sustainable development goals (SDGs) (UN, 2015)

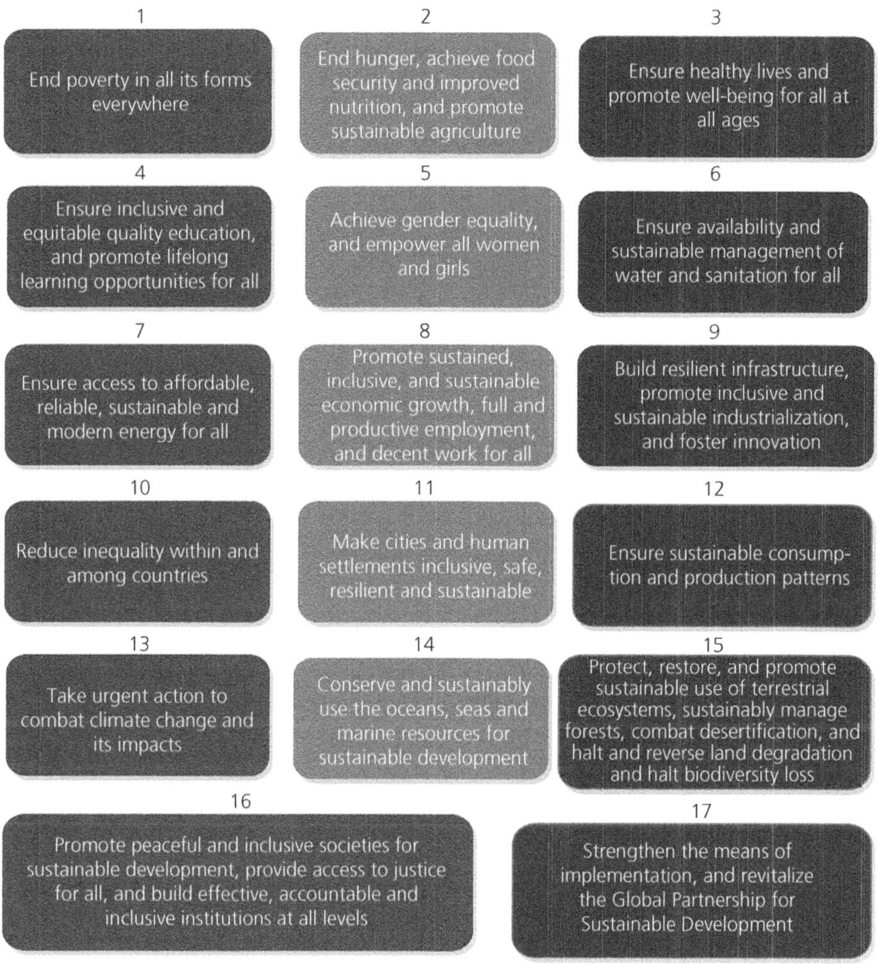

Table 3.1 Expert knowledge results summary (continued on next page)

	Processes	Technologies	Skills
Production	Solar energy electrolysis, wind energy, geothermal energy, biomass energy, hydroelectric energy	Photovoltaic cells, proton exchange membrane (PEM) electrolysers, pyrolysis, photo-fermentation, biogas digesters	Electrical engineering, mechanical engineering, solar panel design and installation, operations and maintenance, energy analysts, construction and installation, project development, financial skills

Table 3.1 Continued

	Processes	Technologies	Skills
Storage	Compressed gas storage, liquid hydrogen storage, underground storage	High-pressure gas tanks, caverns, aquifers	
Distribution	Pipelines, rail, ship	Compression and adsorption	Material sciences, risk and safety practice, inspection and maintenance, operations, project management, logistics, construction engineering
Utilisation	Fuel for vehicles, power generation, industrial processes, energy storage	Fuel cell technology, combustion engines, lithium batteries technology	

Figure 3.3 GH_2 value chain dashboard – top: energy selection under 'Production'; bottom: 'Solar energy' data displayed on selection

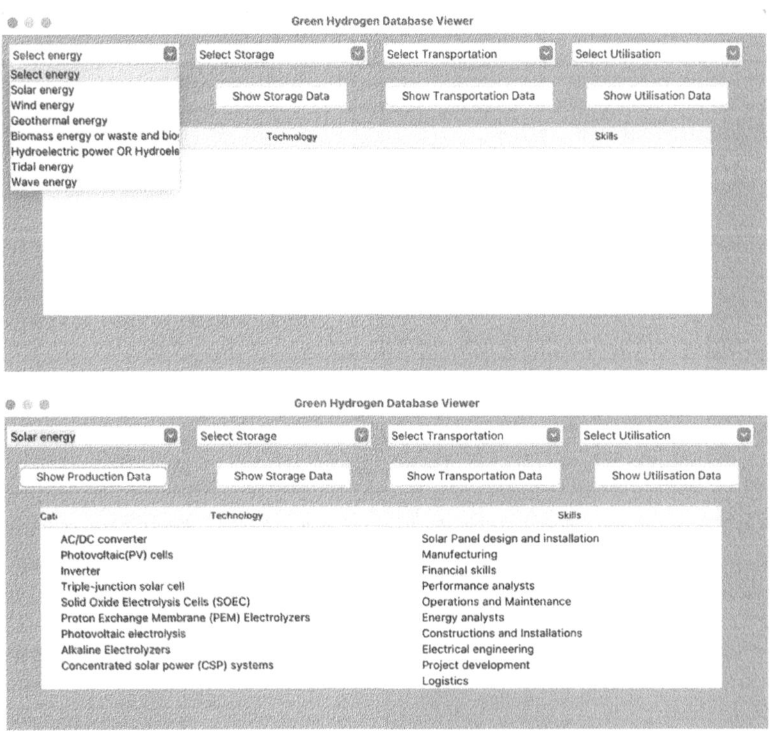

Figure 3.4 GH_2 value chain dashboard – top: storage selection under 'Storage'; bottom: 'Compressed gas storage' data displayed on selection

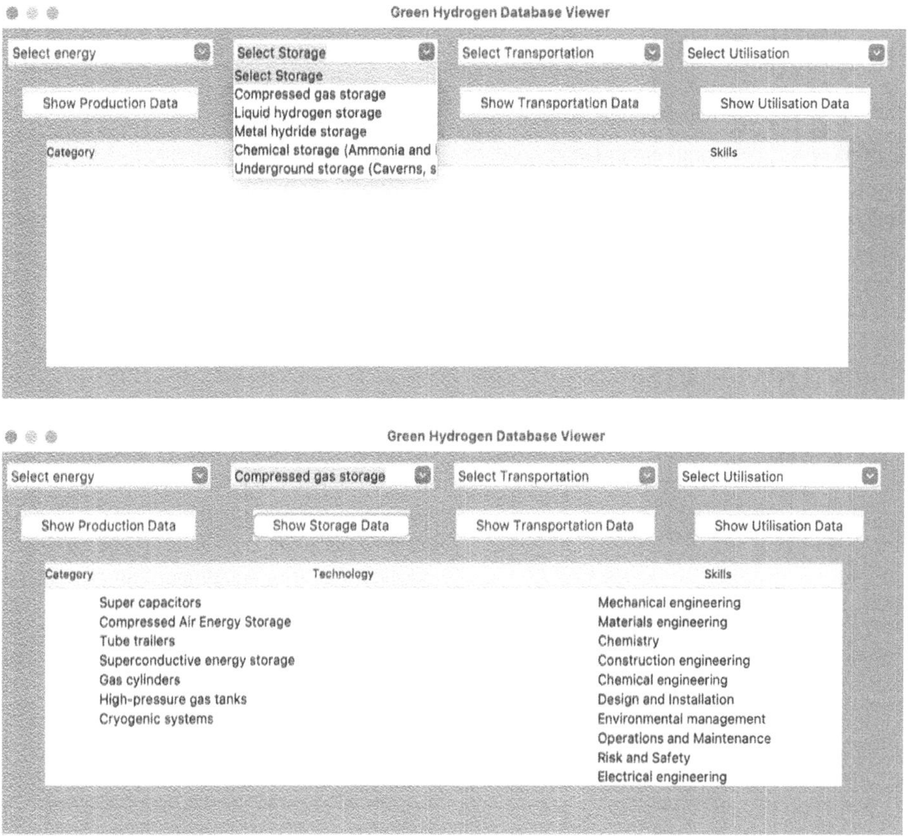

Table 3.2 NLP extraction results summary (continued on next page)

Processes	Extracted technologies	Extracted skills
PV electrolysis	DC/AC converter, proton exchange membrane (PEM) electrolyser, photovoltaic (PV) cells, semiconductors and photoelectrochemical (PEC) devices (Gibson and Kelly, 2008)	Electrical skills and DC/DC converter optimisation
Pyrolysis Thermolysis Solar molten salts	Concentrated solar power (CSP) systems, gasification, and Gema solar tower (Boretti, 2022)	Thermodynamics, geology, geotechnical biotechnology, physics and electrochemistry

Table 3.2 Continued

Processes	Extracted technologies	Extracted skills
Electrolysis Adsorption	Turbine, proton exchange membrane (PEM) electrolyser fuel cells, nanomaterials (Renner *et al.*, 2008)	Design, nanotechnology, operations chemistry, science and technology, engineering, mathematics, logistics
Biophotolysis Steam electrolysis Pipelines	Photofermentation, solid oxide electrolysis cells, liquefaction, compression, inverter heating systems, PEM electrolyser, combustion engines, hydrogenation and dark fermentation (Ahmed *et al.*, 2021)	Project design, transport and logistics, operations, chemistry, electrochemistry, project management, manufacturing, finance and banking, construction
PV electrolysis Wind electrolysis	Solar PV, wind turbines, concentrated solar power (Sooriyaarachchi *et al.*, 2015)	Construction, installation, PV panel production, wholesale, retail, installation and maintenance, software engineers, modellers, manufacturing engineers, building systems specialists, quality assurance experts, logistics operators, equipment transporters, procurement professionals, marketing specialists, project designers (engineers), atmospheric scientists and meteorologists, resource assessment specialists and site evaluators, environmental consultants, developers/facilitators, economic/financial/risk specialists, lawyers (feed-in contracts), electrical, civil and marine engineers, power line technicians, instrumentation and control technicians, recycling specialists, policy-makers and government office workers, educators and trainers, IT professionals, human resources professionals
Water electrolysis Electrochemical water splitting	Solid oxide water electrolysis, PEM water electrolysis, alkaline water electrolysis, AEM water electrolysis, electrocatalyst materials and alkaline water electrolyser nickel (Ni) coated stainless steel electrodes (Kumar and Lim, 2022)	Research and development

Table 3.2 Continued

Processes	Extracted technologies	Extracted skills
Geothermal drilling Power-to-gas Biomass gasification	Alkaline water electrolyser (AWE), high temperature electrolyser (HTE), polymer membrane exchange electrolyser (PEME) (Osman Awaleh *et al.*, 2021)	Thermodynamics, oil refining, metallurgy, chemical engineering, pharmaceutical, meteorological, geothermal resources assessment, economic assessment, drilling and construction, operation and maintenance
Adsorption Thermal processes Photolytic processes Hydrolysis processes Photobiological splitting	Compression and liquefaction, reciprocating pistons, metal hydrides, salt caverns, aquifers, carbon nanotubes, battery/electric vehicles, dark fermentation, hydrogen fuel cell technology (Ratnakar *et al.*, 2021)	Health, safety and environment (HSE), chemical manufacturing, oil refining skills, transport and logistics

Unlocking insights into the green hydrogen industry through natural language processing techniques

Teaching notes

Synopsis

In May 2023, the DSI identified a significant challenge in obtaining comprehensive data on technological advancements, skills requirements, and market trends in South Africa's green hydrogen sector. Conventional data collection methods, such as expert interviews, proved insufficient, resulting in incomplete insights for decision-makers and hindering strategic planning. To address this challenge, Dr. Khuliso Murulane, an esteemed researcher and lecturer at the University of Johannesburg, under the guidance of Professor Arnesh Telukdarie, proposed an innovative solution. They utilised natural language processing (NLP) techniques, including web scraping and text mining, to extract valuable insights from diverse datasets. By integrating NLP into their analysis, the team systematically gathered, processed, and analysed extensive data from academic literature, industry reports, and online sources related to green hydrogen. This collaborative effort enabled them to uncover emerging technologies, essential skills, and market dynamics crucial for green hydrogen integration. The comprehensive data provided actionable intelligence for informed decision-making and strategic planning in the energy sector.

Teaching objectives

This case is intended for postgraduate students in courses related to data science, energy management, sustainability, and innovation. The primary teaching objectives of this case are as follows.

- To illustrate the importance of NLP techniques, specifically web scraping and text mining, in addressing data challenges within the energy sector, particularly in the context of integrating green hydrogen technologies.
- To analyse the collaborative efforts between research institutions and industry stakeholders in driving innovation and sustainability within South Africa's energy landscape.
- To evaluate the benefits and limitations of employing NLP techniques for data collection, analysis, and decision-making processes in complex industry sectors like energy. Various concepts are addressed in this case, of which the most important are data collection and analysis; exploring and mitigating challenges associated with data collection, diversification, and digitalisation.
- Solving the problems related to the case is not related to a right or wrong answer. Students should take each concept addressed and identify the solutions applied in the case.
- Students can generalise the solutions from this case to any business type and any size of organisation, leading to the sustainability of a sound business.

Research methods

This study adopted a mixed-method approach, encompassing several key strategies. Initially, a quantitative analysis was conducted through surveys administered to industry experts and stakeholders. This phase aimed to collect quantitative data on various aspects of green hydrogen technologies, skills requirements, and market trends, employing statistical analysis to identify trends and patterns within the data. Subsequently, a qualitative analysis was undertaken, involving interviews with key stakeholders. This qualitative phase sought to extract insights and perspectives on the challenges, opportunities, and future directions within the green hydrogen sector. Additionally, a systematic literature review was performed to gather existing knowledge, theories, and empirical studies pertinent to green hydrogen, NLP techniques, and data challenges within the energy domain. Lastly, case study analyses were conducted, examining projects that successfully integrated NLP techniques for data analysis in the energy or sustainability sectors. This analysis aimed to identify best practices and delineate implications for future applications within the green hydrogen landscape.

Keywords and subject code

Green hydrogen, natural language processing, digital data collection, energy, innovation, sustainability.

Assignment questions

1. Identify and analyse key challenges faced by researchers and data analysts when collecting and analysing data in complex industry sectors like energy. Propose potential solutions or strategies to overcome these challenges effectively.
2. Discuss the significance of NLP techniques in addressing data challenges within the energy sector. Provide examples of how NLP techniques can be applied to gather insights and trends in the green hydrogen energy sector.
3. Discuss the potential benefits and limitations associated with using NLP techniques for data analysis in the green hydrogen energy domain.
4. Evaluate and discuss the most effective methods for collecting and analysing data within the energy sector, with a focus on green hydrogen technologies and skills. Consider the role of technology, data sources, and data processing techniques in ensuring accurate data analysis.

Suggested core readings

The following articles are recommended for more information.

Department of Energy (2021) *Annual Report 2021*. Department of Energy, Republic of South Africa. https://www.energy. gov.za/files/media/explained/2021-South-African-Energy-Sector-Report.pdf (accessed 22 April 2024).

Department of Energy (2022) *Annual Report 2022*. Department of Energy, Republic of South Africa. https://www.energy. gov.za/files/media/explained/2022-South-African-Energy-Sector-Report.pdf (accessed 22 April 2024).

Lebrouhi BE, Djoupo JJ, Lamrani B, Benabdelazis K and Kousksou T (2022) Global hydrogen development – a technological and geopolitical overview. *International Journal of Hydrogen Energy* **47(11)**: 7016–7048, 10.1016/j.ijhydene.2021.12.076.

Seurin P, Olabanjo O, Wiggins J *et al.* (2022) H2-Golden-Retriever: methodology and tool for an evidence-based hydrogen research grantsmanship. *arXiv preprint arXiv:2211.08614*.

Singh M, Darold DL, Klobasa M, Zielinski A and Frietsch R (2023) Keeping track of cleantech development using innovation clusters and member's website data: evidence from leading energy clusters in Germany. *Energy Reports* **10**: 756–767, 10.1016/j.egyr.2023.07.026.

Yang D, Lee J, Song NC *et al.* (2023) Patent analysis on green hydrogen technology for future promising technologies. *International Journal of Hydrogen Energy* **48(83)**: 32241–32260, 10.1016/j.ijhydene.2023.04.317.

Teaching plan and timing

Introduction (15 min)

- Begin the class with an introduction to the case study. Students should have been provided with the case study as pre-reading material. The instructor should provide an introduction to set the context and importance of the case study.

Challenges in data collection (15 min)

- Assignment Question 1: Facilitate a discussion where students identify and discuss challenges faced by researchers and data analysts while collecting data in complex industry sectors like energy. Encourage critical thinking and analysis of potential solutions.

Discussion on NLP techniques (30 min)

- Assignment Question 2: Students will work in pairs to discuss and explore the importance of NLP techniques, particularly web scraping and text mining, in addressing data challenges within the energy sector.

Presentation of findings (10 min)

- Assignment Question 3: Assign one group to present their findings on the potential benefits and limitations associated with using NLP techniques for data analysis in the green hydrogen sector.

Effective data collection methods (10 min)

- Assignment Question 4: Engage the students in identifying and discussing the most effective ways to collect and analyse data in the green hydrogen energy sector. Encourage creative thinking and collaboration.

Conclusion and reflection (5 min)

- Wrap up the lesson by reflecting on the major challenges identified by students during the discussion and their proposed recommendations. Emphasise the importance of innovative research methodologies, such as mixed methods approaches and NLP techniques, in unlocking insights and driving informed decision-making in the energy sector.

Analysis of assignment questions

Question 1. For this question, the answer should identify and thoroughly analyse key challenges encountered by researchers and data analysts when dealing with data in complex industry sectors like energy.

Possible solutions

Researchers and data analysts face numerous challenges in complex industry sectors like energy, particularly in the emerging field of green hydrogen. The challenges include data heterogeneity, data accessibility, and data quality.

- Data in the energy sector come from various sources, such as government reports, academic literature, industry publications, and real-time sensor data. These data are often stored in different formats and structures, making them difficult to work with. Using AI tools such as NLP can help analyse large and complex datasets efficiently.

- Access to relevant and up-to-date data is often limited owing to proprietary restrictions, high costs, or geopolitical issues. To overcome this, establishing data-sharing agreements and leveraging publicly available datasets can be effective strategies. Collaboration with industry stakeholders and participation in open data initiatives can also enhance data accessibility.
- Ensuring the accuracy, completeness, and reliability of data is important for making informed decisions. However, data collected from diverse sources can vary significantly in quality. Implementing good data validation and cleaning processes can help mitigate this issue. Data preprocessing techniques like automated tools that detect and correct anomalies, outliers, and missing values can improve data quality and reliability.

Question 2. The response to this question should demonstrate an understanding of NLP techniques and their relevance in addressing data challenges specific to the energy sector, particularly in the context of GH_2 value chain. The answer should include examples illustrating how NLP techniques can be used effectively to gather data, identify trends, and analyse large volumes of data related to green hydrogen technologies.

Possible solutions

NLP techniques are instrumental in addressing data challenges within the energy sector. NLP can automate the extraction, analysis, and interpretation of vast amounts of unstructured text data, which are abundant in this sector. Through NLP techniques, one can automatically extract relevant information from large datasets, such as research papers, industry reports, and news articles. This reduces the time and effort required for manual data collection and ensures that analysts have access to the latest information. By analysing text data from various sources, NLP techniques can allow one to identify emerging trends and patterns in the green hydrogen sector. This can help stakeholders stay informed about the latest developments and make data-driven decisions.

Examples of NLP applications

- In the green hydrogen sector, NLP techniques can be used to synthesise information from a large number of academic papers on green hydrogen technologies. Tools like text summarisation and topic modelling can identify key themes and research trends, providing a comprehensive overview of the field.
- Through NLP techniques one can analyse industry reports and news articles to identify market trends, competitive landscapes, and investment opportunities in the green hydrogen sector. Named entity recognition (NER) can be used to extract mentions of key companies, technologies, and projects.

Question 3. For this question, the answer should discuss the potential benefits of using NLP techniques to extract information from textual data sources related to green hydrogen technologies, skills, and market trends. Additionally, the response should address the limitations associated with NLP techniques.

Possible solutions

- NLP techniques significantly enhance the ability to derive insights from unstructured text data. By applying NLP, stakeholders in the green hydrogen sector can stay abreast of technological advancements, market dynamics, skills, and policy changes.
- NLP tools can be customised to focus on specific aspects of the green hydrogen sector, such as production technologies, storage solutions, or skills. This targeted analysis can provide more relevant and actionable insights.
- NLP techniques can process vast amounts of text data quickly and efficiently, making it possible to analyse information at a scale that would be impractical with manual methods.
- Automating data collection and analysis with NLP can reduce costs associated with manual research and analysis.
- NLP models can achieve high levels of accuracy in information extraction and analysis, reducing the likelihood of human error.
- Potential limitations of NLP include data quality, interpretation challenges, and requirements of technical expertise. For example, the accuracy of NLP models depends on the quality of the input data. Poor-quality data can lead to inaccurate or biased results. Also, NLP models can sometimes misinterpret context in text, leading to incorrect conclusions. Lastly, the implementation and maintenance of NLP systems requires specialised technical expertise, which may be a barrier for some organisations.

Question 4. The response should consider the role of technological advancements, such as data analytics platforms and cloud computing, in enabling comprehensive and accurate data collection and analysis. Furthermore, the answer should discuss appropriate data processing techniques to ensure robust data analysis outcomes.

Possible solutions

- Conducting surveys and interviews with industry experts can provide valuable qualitative data and insights that are not available through other sources. These methods can capture expert opinions, emerging trends, and detailed technical information.
- Web scraping tools can automatically collect data from online sources such as industry reports, news articles, and academic publications. This method is particularly useful for gathering large volumes of up-to-date information.
- NLP techniques such as text mining and topic modelling can analyse unstructured text data to extract relevant information and identify trends.

Role of technology

- Cloud-based solutions provide storage and processing capabilities, enabling the analysis of large datasets without the need for significant upfront investment in infrastructure.
- Visualisation tools can help stakeholders understand complex data through an interactive user interface. Effective visualisation can highlight key insights and trends, making it easier to communicate findings. Tools such as Tableau, Power BI, and Python libraries offer powerful capabilities for data analysis and visualisation.

Key learnings

- The case study highlights the role of green hydrogen as a sustainable energy source with applications in transportation, industry, and power generation. While its potential to reduce greenhouse gas emissions is clear, adoption is hindered by high production costs, infrastructure constraints, and regulatory challenges. Workforce development is also necessary, requiring targeted training to meet industry demands. Addressing these challenges through policy support and investment in renewable energy can help integrate green hydrogen into energy systems.
- Employing NLP helped overcome knowledge gaps in the sector. Traditional methods, such as expert interviews, provided insights but were insufficient in capturing industry-wide trends. By analysing academic literature, industry reports, and online data, NLP identified emerging technologies, market developments, and skills requirements. This structured approach enabled better decision-making and strategic planning, supporting the transition to cleaner energy systems.

REFERENCES

Ahmed SF, Rafa N, Mofijur M *et al.* (2021) Biohydrogen production from biomass sources: metabolic pathways and economic analysis. *Frontiers in Energy Research* **9**: 753878, 10.3389/fenrg.2021.753878.

Albatayneh A, Juaidi A, Jaradat M and Manzano-Agugliaro F (2023) Future of electric and hydrogen cars and trucks: an overview. *Energies* **16(7)**: 3230, 10.3390/en16073230.

Blei DM and Lafferty JD (2009) Topic models. In *Text Mining*. Chapman and Hall/CRC, London, UK, pp. 101–124, 10.1201/9781420059458.ch4.

Boretti A (2022) Which thermochemical water-splitting cycle is more suitable for high-temperature concentrated solar energy? *International Journal of Hydrogen Energy* **47(47)**: 20462–20474, 10.1016/j.ijhydene.2022.04.159.

Chapagain A (2019) *Hands-On Web Scraping with Python: Perform Advanced Scraping Operations using Various Python Libraries and Tools such as Selenium, Regex, and Others*. Packt Publishing, Birmingham, UK.

Department of Energy (2021) *Annual Report 2021*. Department of Energy, Republic of South Africa. https://www.energy. gov.za/files/media/explained/2021-South-African-Energy-Sector-Report.pdf (accessed 22 April 2024).

Department of Energy (2022) *Annual Report 2022*. Department of Energy, Republic of South Africa. https://www.energy. gov.za/files/media/explained/2022-South-African-Energy-Sector-Report.pdf (accessed 22 April 2024).

Ferrara E, De Meo P, Fiumara G and Baumgartner R (2014) Web data extraction, applications and techniques: a survey. *Knowledge-Based Systems* **70**: 301–323, 10.1016/j.knosys.2014.07.007.

Gibson TL and Kelly NA (2008) Optimization of solar powered hydrogen production using photovoltaic electrolysis devices. *International Journal of Hydrogen Energy* **33(21)**: 5931–5940, 10.1016/j.ijhydene.2008.05.106.

Grobbelaar N and Ngubevana L (2022) *Ensuring a Just Energy Transition Through Hydrogen: How the G20 Can Support Africa*. South African Institute of International Affairs, Johannesburg, South Africa, pp. 1–19.

Hotho A, Nürnberger A and Paaß G (2005) A brief survey of text mining. *Journal for Language Technology and Computational Linguistics* **20(1)**: 19–62, 10.21248/jlcl.20.2005.68.

Kessides IN (2020) The decline and fall of Eskom: a South African tragedy. *The Global Warming Policy Foundation Report* **45**.

Kopnina H (2016) The victims of unsustainability: a challenge to sustainable development goals. *International Journal of Sustainable Development & World Ecology* **23(2)**: 113–121, 10.1080/13504509.2015.1111269.

Kouzis-Loukas D (2016) *Learning Scrapy*. Packt Publishing, Birmingham, UK.

Kumar SS and Lim H (2022) An overview of water electrolysis technologies for green hydrogen production. *Energy Reports* **8**: 13793–13813, 10.1016/j.egyr.2022.10.127.

Lebrouhi BE, Djoupo JJ, Lamrani B, Benabdelazis K and Kousksou T (2022) Global hydrogen development – a technological and geopolitical overview. *International Journal of Hydrogen Energy* **47(11)**: 7016–7048, 10.1016/j.ijhydene.2021.12.076.

Neumann M, King D, Beltagy I and Ammar W (2019) ScispaCy: fast and robust models for biomedical natural language processing. *arXiv preprint arXiv:1902.07669*, 10.18653/v1/w19-5034.

Nhamo G, Dube K and Chikodzi D (2020) Sustainable development goals: concept and challenges of global development goal setting. In *Handbook of Global Health*, Springer, New York, NY, USA, pp. 1–40, 10.1007/978-3-030-05325-3_79-1.

Osman Awaleh M, Adan AB, Assowe Dabar O *et al.* (2021) Economic feasibility of green hydrogen production by water electrolysis using wind and geothermal energy resources in Asal-Ghoubbet rift (Republic of Djibouti): a comparative evaluation. *Energies* **15(1)**: 138, 10.3390/en15010138.

Ratnakar RR, Gupta N, Zhang K, *et al.* (2021) Hydrogen supply chain and challenges in large-scale LH2 storage and transportation. *International Journal of Hydrogen Energy* **46(47)**: 24149–24168, 10.1016/j.ijhydene.2021.05.025.

Renner M, Sweeney S and Kubit J (2008) *Green Jobs: Towards Decent Work in a Sustainable, Low-Carbon World: Report for United Nations Environment Programme*. UNEP, Nairobi, Kenya.

Richardson L (2007) Beautiful Soup Documentation. https://beautiful-soup.readthedocs.io/_/downloads/en/latest/pdf/ (accessed 05/02/20225).

Sanlam Intelligence (2023) Unveiling the 'SA-H2 Fund': South Africa's Dedicated Green Hydrogen Fund. https://www.sanlamintelligence.co.za/retail/unveiling-the-sa-h2-fund-south-africas-dedicated-green-hydrogen-fund/ (accessed 20 August 2023).

Singh S (2018) Natural language processing for information extraction. *arXiv preprint arXiv:1807.02383*.

Sooriyaarachchi TM, Tsai IT, El Khatib S, Farid AM and Mezher T (2015) Job creation potentials and skill requirements in, PV, CSP, wind, water-to-energy and energy efficiency value chains. *Renewable and Sustainable Energy Reviews* **52**: 653–668, 10.1016/j.rser.2015.07.143.

UN (2015) The 17 goals. https://sdgs.un.org/goals# (accessed 04/02/2025).

Xue N (2010) Book review of Steven Bird, Evan Klein and Edward Loper. *Natural Language Processing with Python. Natural Language Engineering* **17(3)**: 419–424, 10.1017/s1351324910000306.

Prof. Arnesh Telukdarie
ISBN 978-1-83662-645-9
https://doi.org/10.1108/978-1-83662-642-820251004

Chapter 4
Sustainability and uranium digitalisation

Arnesh Telukdarie and Michael Sishi

Introduction

It was at the beginning of the year 2015 when an open-pit uranium mining company began its operations in Johannesburg, South Africa as one of the world's largest uranium mines. A large part of the processing was performed manually, resulting in the absence of a real-time flow of information between systems and stakeholders. Therefore, the Chief Executive (CE) had a responsibility to enable company-wide real-time flow of information and collaboration of stakeholders across the globe. Furthermore, the CE had to ensure that the mine contributed significantly to the sustainable development goals (SDGs). To fulfill these responsibilities the CE had to put together a team of engineers that would research and implement fourth industrial revolution (4IR) techniques to create an integrated enterprise that also forms part of the connected business world. The team had to determine if it was possible to implement 4IR technologies in the mining industry, the protocols for the introduction of 4IR in the mining industry, the benefits of 4IR for the mine, and the contribution of 4IR to the SDGs.

The background of the mining company

The uranium mine was established in 2008 but started operations in 2015. It is one of the world's largest uranium mining companies. This mine had an expected production of 7 000 tons of uranium annually. The mine was designed as an open-pit mine and used drills for creating mining pits and electric shovels for excavation. The shovels and trucks were used for loading and hauling. The rest of the operated equipment included excavators, rock breakers, and dozers. The processing plant produced high-grade uranium from ore extracted from the ground. This was achieved through the utilisation of sulphuric acid that leaches uranium. This process aimed to recover about 88% of uranium from ore. The precipitation was washed, thickened, dried, packed, and thereafter shipped to customers. Figure 4.1 depicts the mine processes from drilling, blasting, excavation, and loading in the open pit. Thereafter, hauling, dumping to stockpiles, crushing, and manufacturing processes are illustrated. The mining corporation used various stand-alone systems and manual processes to manage its operations.

The mining firm used systems, applications, and products (SAP) as the enterprise resource planning (ERP) system, which was not automatically integrated with other systems. Consequently, SAP ERP was not sending and receiving data in real time. Pivotal data such as production planning, master data, and scheduled maintenance were exported into Excel spreadsheets and forwarded to the mine shop floor and plant systems. Data from various sources were captured manually into SAP ERP. The mine made use of the following SAP ERP modules: materials management (MM), production planning (PP), quality management (QM), plant maintenance (PM), asset management (AM), and finance (FI). These modules acquired manual data that were susceptible to errors. As a result, the business was prompted to integrate SAP ERP with shop floor and plant systems at the mine.

A significant amount of money was spent on fuel for the mine equipment. The monitoring of fuel supply and consumption was performed through a software system called fuel management system (FMS). Fuel levels were measured by dipping a rod in the tanks and recording these fuel measurements on pieces of paper. Industry experts and researchers criticise manual data entry because it is susceptible to errors and events are recorded long after they have occurred. Mobile dispensers were filled up from these tanks and thereafter all other equipment that consumed fuel was filled up from mobile dispensers. Once more, fuel consumption for all equipment was recorded on pieces of paper. These paper sheets, containing tank levels and fuel consumption data, were later sent to the fuel management office for capturing in FMS. FMS also required equipment and measuring points master data to be acquired from SAP ERP so that fuel consumption is recorded against corresponding equipment and measuring points. A measuring point is a description of the physical or logical position where a condition is illustrated. Examples of measuring points are fuel sump, speedometer, and operating time of equipment. This integration was done by downloading equipment and measuring points master data into Excel spreadsheets and sending this data for capturing in FMS by email. The integration was usually not done on time, hence the master data inconsistency between SAP ERP and FMS. FMS produced fuel consumption data that had to be sent to SAP ERP for financial reporting purposes. Fuel consumption had to be captured with conforming equipment and measuring points in SAP ERP. The extended challenge was the absence of real-time reporting of fuel tank levels and fuel consumption by the fleet.

The mining corporation used a fleet manager (FM) software system to manage a fleet of mining equipment in the open pit. FM needed master data from SAP ERP to record data such as the running hours of an engine against corresponding equipment. There was no real-time master data integration between SAP ERP and FM at that time. Data were downloaded manually from SAP ERP and sent to FM by email. Thereafter, FM updated local master data from Excel spreadsheets. There was an absence of real-time integration from FM to SAP ERP for production (metres drilled and tonnes hauled), unscheduled maintenance, engine running hours, vehicle operating time, and vehicle locations and movements. Furthermore, in the absence of real-time key performance indicators (KPIs) reports, the mine was only using basic FM reports that were deemed inadequate by management. The mine had just implemented a system for monitoring and managing tyres and rims. Tyre management system (TMS) was intended to monitor the lifespan of every tyre and rim at the mine. Each tyre or rim would have characteristics such as serial number, size, make, type, new, re-tread, and position. Data from TMS were to be imported and stored in a central repository for sharing with other mine systems. Master data synchronisation between TMS and SAP ERP was required for TMS to send data against correct tyres and rims back to the mine (central data repository).

The mine proposed the implementation of a manufacturing execution system (MES) to manage production in the processing plant where uranium was extracted from ore. There existed a business requirement to integrate MES with SAP ERP to provide relevant data to the ERP as required by the business and share ERP data as required by the processing plant systems. The initial steps in the life of any mine pivot around the feasibility of profitably extracting minerals from an identified resource. This is accomplished by exploration and geology and mine planning (GMP) processes. In simplified terms, these functions can be grouped under exploration, exploratory drilling and sampling, resource and reserve definition, geotechnical modelling, shell and pit design, shell optimisation, mine planning (long-, medium-, and short-term), and survey grade control.

Throughout the life of any mining enterprise, these functions remain vital in terms of reassessing the remaining resources and the most effective sequence to extract them. Traditionally, these functions are performed by a wide variety of different systems that create and use complex three-dimensional (3D) and vector files to simulate, model, and record geospatial properties such as grade levels and density. Seeing that these software packages are predominantly desktop applications, and these packages can usually exchange (import and export) files seamlessly with one another, there was not much scope for automated integration of these software packages. However, there did exist value in extracting specific information from sections of this information flow and sharing this information electronically with other systems in the mining landscape such as the ERP. Therefore, there was a business requirement to integrate mine planning with other mine systems to share data such as short- and medium-term planning and survey results.

To fulfil the mandate, the CE appointed a team of engineers to digitally transform the mine and enable company-wide real-time flow of information and collaboration of stakeholders across the globe. The team had to research the prospects of implementing 4IR in the mining industry, protocols for the implementation of 4IR, the benefits of 4IR, the benefits of 4IR for the mine, and the contribution of 4IR to the SDGs.

The prospects of implementing 4IR in the mining sector

The team had to perform thorough literature reviews to determine from the previous case studies whether 4IR technologies apply to the mining sector. Therefore, it was pivotal for the team to have a deeper understanding of the 4IR core definition, constituents, related technologies, and applicability.

Through a study by Sishi and Telukdarie (2018), the team observed 4IR as a strategic inventiveness adopted by the German government as a 'High-Tech Strategy 2020 Action Plan' in 2011. 4IR has gained popularity in many other countries globally and it is viewed as the extension of the previous third industrial revolution that was based on mechanisation. Internet of things (IoT), industrial internet of things (IIoT), big data, and cyber-physical systems (CPS) were thrown into the manufacturing sphere to move manufacturing towards a fourth industrial revolution (Sishi and Telukdarie, 2018). Figure 4.2 depicts industrial revolution in accordance with Industry 4.0 working group elucidations. The first industrial revolution was based on water-and steam-powered mechanical manufacturing. The second industrial revolution was as a result of electrically powered mass production advocating for the division of labour (Almada-Lobo, 2016). The use of electronics and ICT to achieve further manufacturing automation yielded the third industrial revolution. Subsequently, IoT, big data, and CPS ushered in the fourth industrial revolution (Almada-Lobo, 2016).

The team observed that CPS comprises of field devices, machines, production modules, and products that independently make decisions, exchange information, trigger actions, and control each other (Rafael *et al.*, 2020). CPS is realised by the employment of IoT technologies to transform industries into smart environments. IoT is the connection of physical or smart devices that consequently collect huge amounts of data. On the other hand, IIoT is the implementation of IoT in the industrial sector. These devices have embedded mechatronic components, which are sensors for collecting data and actuator systems for manipulating physical processes. Human–machine interfaces are used by CPS to communicate with humans (Monostori *et al.*, 2016). The team of engineers paid special attention to the three facets of 4IR: vertical integration, end-to-end digital integration, and horizontal integration through value networks. Vertical integration and networked industrial systems connect different internal sections of the industrial company such as ICT systems and processes. They define a connection of the value creation factors such as product, human, and process along the various levels of the value creation modules, from production stations through production cells and production lines up to the smart factory. This networking throughout the different aggregation levels also includes data flow between product development to fabrication, logistics, marketing and sales, service, and procurement, thus creating a smart production environment that enables CPS to create flexible and reconfigurable manufacturing systems (Sishi and Telukdarie, 2020).

End-to-end digital integration is the engineering across the entire value chain that describes the cross-linking and digitalisation of the entire product lifecycle. The product is tracked from the raw material status up to the end of life. This integration enables connected stakeholders, products, and equipment to share data in real time during various phases of the product lifecycle. The product lifecycle encompasses phases such as raw material acquisition, product development, production engineering, manufacturing, product use and service, and end of life. This integration answers questions such as how end-to-end business processes like engineering workflow can be delivered by CPS (Sishi and Telukdarie, 2020; Wang *et al.*, 2016).

Horizontal integration creates a value network that enables local and globally dispersed value chain partners to collaborate through the integration of ICT systems, processes, and data flows. The integration of ICT systems, processes, and data flows within the company and between various businesses such as customers, suppliers, and other external partners allows closer cooperation with value chain partners across business boundaries. This integration sheds light on how CPS should be used to sustainably implement and support the company's business strategies, value networks, and business models. Figure 4.3 depicts a connected enterprise that encompasses all the 4IR integration layers (Sishi and Telukdarie, 2020; Wang *et al.*, 2016).

The extensive knowledge acquired through an exhaustive literature review shed light on the prospects of implementing 4IR technologies in the mining industry. It became evident that 4IR technologies could be applicable in several types of industrial sectors such as manufacturing, mining, agriculture, logistics, among others (Sishi and Telukdarie, 2020). This gave the team sufficient confidence to move forward to discover protocols for introducing 4IR in the mining industry (Sishi and Telukdarie, 2020; Wang *et al.*, 2016).

The techniques for introducing 4IR in the mining industry

The team employed systems development lifecycle (SDLC) techniques to implement 4IR technologies at the mining firm. The SDLC process comprised several steps: requirements engineering through interview workshops and observations, data analysis and validation, designs and implementation, testing, deployment, and support (Balaji and Murugaiyan, 2012).

Requirements engineering

The requirements engineering was conducted by gathering and documenting the current state of the mine to determine its future state. The current processes were documented and verified with the mine. Subsequently, these processes were examined to close existing gaps and find better ways to enhance processes. It was important for all relevant stakeholders to be involved in the requirements engineering exercise. The following sections discuss how business requirements engineering was conducted at the mining corporation.

Interview workshops

To learn more about the mine in terms of the current and desired state, different sections and levels of the business were represented during interview workshops. These representatives were senior management (executives), middle management (department managers), external business partners (suppliers), ERP super users, shop floor supervisors, plant maintenance manager, and ICT representatives.

Observations

Observations included observing if there were any existing vertical, end-to-end, and horizontal data flow mechanisms. During these observation exercises, a large number of business processes were captured and documented using a business process modelling tool, Microsoft Visio.

Data analysis and validation

Data analysis and validation revealed the absence of integrated systems and real-time flow of data. Despite having implemented various computer-based systems, these were stand-alone systems that did not communicate with each other. This resulted in several operations being executed manually, especially when sharing data and information between these stand-alone systems. This gap necessitated the implementation of vertical, end-to-end, and horizontal integration layers of 4IR. The requirements for implementing the 4IR integration layers were stipulated in the business requirements specification (BRS) document. Consequently, the team began with the design and implementation of the 4IR solution for the mine.

Designs and implementation

The designs and implementation of the 4IR solution encompassed all the integration layers. The design phase produced functional design specification (FDS) and technical design specification (TDS) documents. The implementation phase produced the actual 4IR solution based on the FDS and TDS artefacts.

Vertical integration

This was the core integration layer for the 4IR solution at the mine. It encompassed SAP ERP, FMS, FM, TMS, and MES as discussed below.

- SAP ERP: SAP ERP was integrated with the mine systems through the central data repository. Data from SAP ERP was published to the central repository to grant all participating mine systems access to the data. This connection also enabled SAP ERP to have access to all data published by the mine systems. The connection to SAP ERP was implemented by way of SAP business application programming interfaces (BAPIs). A BAPI encapsulates internal functioning by exposing only the name of a BAPI, input, and output parameters. Each integration interface called a BAPI by name and passed required input data. A BAPI consumed input data and returned required output data through output parameters, respectively. The outputs from BAPIs were recorded with related inputs in the central data repository. These interfaces ensured that SAP ERP shared data with the low-level systems in the vertical integration.
- FMS: Sensors were installed on fuel (diesel) tanks to automatically measure fuel levels. Sensors were identified by unique IDs and paired with respective fuel tank IDs. Data revealing how much fuel was left in each tank were collected from the sensors into the central repository. A threshold was set in the system to create an alert if fuel reached a certain level in each tank. Consequently, an alert was sent to SAP ERP to create an order for more fuel for the tank in question. Mobile fuel dispensers fed from fuel tanks and all other vehicles refilled from mobile dispensers. Hence, sensors were also installed on these mobile dispensers and vehicles to measure how much fuel was loaded or consumed. Fuel consumption data, for each dispenser and vehicle, were collected and stored from sensors to the central repository. Data collection included the total amount of time a vehicle was used and kilometres a vehicle travelled each day. Fuel tank levels and fuel consumption transactions were extracted and uploaded from the central repository to SAP ERP in real time. FMS updated local equipment and measuring points master data from the central data repository once a day.
- FM: Fleet manager collected and stored data in the central repository such as vehicle engine hours, mining production volumes, unscheduled vehicle maintenance, vehicle operating time, corrective maintenance, ore movement, and ore stockpiles. Some of this data was subsequently sent to the SAP ERP system by way of integration interfaces.
- TMS: Tyre management required the development of integration interfaces to send rims and tyres master data from the central data repository to TMS once a day. The integration interface to send data, regarding which rims and tyres were installed in which positions and on which vehicles, to SAP ERP was also implemented. Lastly, the interface to collect data used for TMS cross-functional KPIs was developed.
- MES: Control systems monitored, managed, and passed shop floor data collected by sensors and actuators to MES. The integration interfaces for getting data such as production actuals, equipment status, material consumption, corrective maintenance, equipment working hours, quality, and KPIs from MES into the central repository were implemented. In return, the integration interfaces that fed MES with data from the central repository such as preventative maintenance, corrective maintenance status, calibration, ore movement, ore stockpile, and PP were developed. Finally, the

development of integration interfaces for sending production actuals, quality, materials consumption, and corrective maintenance data from the central repository to SAP ERP were completed.

End-to-end integration

This was realised through the integration of GMP, FM, MES, and SAP ERP systems. The design and demarcation of blocks for drilling and blasting occurred in the GMP system while the actual drilling, blasting, loading, and hauling were managed by FM. In addition to getting data broadcast by GMP and FM, MES obtained product design information published by SAP ERP for processing ore into uranium and consequently manufactured uranium accordingly. This enabled MES to combine design and planning data from GMP; drilling, blasting, loading, and hauling data from FM; and product design data from SAP ERP during the manufacturing process. SAP ERP modules took care of marketing, sales and distribution, and product support activities after the successful production of uranium.

Horizontal integration

Horizontal integration was accomplished by integrating ERP, FMS, FM, MES, TMS, and GMP systems. Each of the six systems published data to the central repository so that data broadcast were accessible to all mine systems. The communication between SAP ERP modules created a horizontal integration at the enterprise level. The overall horizontal integration made it possible for the mine to collaborate with the customers as well as the suppliers of fuel, tyres, rims, equipment, and raw materials. The overall system architecture, including all integration layers, is provided in Figure 4.4.

Testing

Various studies emphasise the importance of the testing phase of the SDLC process to ensure that the system meets the design requirements and is free of errors before it is deployed in the live environment. The testing exercises employed the bottom-up approach and entailed all three integration layers (vertical, end-to-end, and horizontal) and six systems (SAP ERP, FMS, FM, TMS, MES, and GMP) mentioned in prior sections. Firstly, the integration layers and interfaces for each system were tested individually through unit testing techniques. All errors found were corrected and the functionality was re-tested until the design requirements were met. Thereafter, the integration testing technique was applied–that is, all three 4IR integration layers were tested together as a complete system. The purpose of integration testing was to reveal problems with the interfaces among system components before problems occur in the production environment. The testing phase yielded user acceptance testing (UAT) documents that were approved and signed by business users who participated in the testing exercises.

Deployment

The system was deployed in the live environment as a result of thorough and successful testing operations. Deployment yielded a fully integrated solution that encompassed all six systems at the mine, as shown in Figure 4.5. The mine systems were connected through an industrial network to the central data repository. The industrial network transmitted data from and to these systems and central repository. Thereafter, data published were used for global monitoring, control, and reporting.

Support

According to the various studies, the support phase is important to ensure that the system deployed is stable in the production environment. Following a successful deployment, the 4IR solution at the mine was supported according to the guidelines illustrated in the service level agreement (SLA).

The benefits of 4IR for the mining firm

Mass customisation

The mining team could change any drilling pattern, blast, and pile or feed component for the mine at any time. The movements of the haul trucks were scheduled in near real time. Fuel dispensing alerts were sent in real time if a haul truck was scheduled for production and insufficient fuel was detected in it. The maintenance of each tyre and rim was scheduled and customised based on the nature and magnitude of the damage.

Production monitoring

The mining corporation production was reported through cross-functional KPIs that enabled management to have a clear view of the mining productivity. The production status from the processing plant was reported by displaying real-time work in progress (WIP) and quality data against daily production targets. Displaying WIP allowed shop floor supervisors to intervene if daily production targets were not met. This also relieved operators from recording daily achievements on the pieces of paper and prevented incorrect recording of daily targets and scores at the same time. Having real-time production status allowed decision-makers, for example, to shut down a machine if it kept producing bad quality products.

Fuel consumption monitoring

The reports for fuel monitoring shed light on the amount of fuel available for consumption. It was also possible to compare which vehicle or group of vehicles consumed more fuel. Fuel consumption could also be viewed by several physical areas such as the plant and equipment groups. The implementation of cross-functional KPIs made it possible to measure and display total fuel consumption, fuel consumption per hours worked, and fuel consumption per tonnes hauled. Therefore, management could take a closer look if a sudden or erratic increase of fuel consumption by the plant or equipment group was noticed.

Equipment management

Pivotal KPIs were monitored in order to optimise equipment management. For example, engineers would quicky know if equipment was underperforming using overall equipment effectiveness (OEE) and utilisation KPIs. The mean time before failure (MTBF) and mean time to repair (MTTR) KPIs enabled the mine to avoid, through proactive maintenance, unexpected and expensive downtime. Presenting the amount of corrective and preventive costs of maintenance helped reveal how much time and money the mine spent on maintenance activities. The equipment availability KPI helped divulge how long equipment was available to operate during production time. This helped management determine which equipment is usually unavailable and investigate the underlying reasons.

Operational fleet efficiency

The mine was now privy to truck information such as cycle times, average payloads, tonnes hauled per hour, average fuel consumption per hour, and average tonnes hauled per litre of fuel consumed.

Rims and tyre visualisation

The mine had real-time visibility on rims and tyre costs, average failure rate by rim and tyre type, cost of rims and tyres per tonne hauled, cost of rims and tyres per operating hour, tread utilisation of fitted tyres, comparison between failed and worn tyres, and rims and tyres wear cost per tonne hauled.

Agility

Important information such as time, quality, and price generated by machines, products, and systems at the mine was instantaneously collected and stored centrally. This information made it possible for MES to automatically stop a machine or production line if it yields inferior quality products or makes high-energy consuming machines and production lines unavailable during peak time.

Improved data analysis

All systems (SAP ERP, FMS, FM, MES, TMS, and GMP) at the mine were now integrated and consequently enabled to publish generated data to the central repository. In addition to enabling the mine systems to communicate with each other in real time, this also yielded huge volumes of data being generated by these systems and machines. The resulting data were used for improved analysis and KPI definition, creation, and reporting. Through big data, the mine was simultaneously realising inexpensive operations, high performance, and good quality production.

Remote monitoring

The reports and dashboards showed, among other things, data collected from the shop floor systems and data shared from the enterprise level back to shop floor systems in real time. The reports were also intelligent enough to display the overall status of the mine through real-time dashboards. Real-time stock and material movement from exposed ore, drilling, blasting, loading and hauling, stockpiling, up to crushing were reported in real time. These reports were compatible with any web-enabled fixed and mobile device, making them accessible at any time and from any place. Management used the reports to view the mine operations from their offices instead of physically taking a walk to the shop floor. The weekly and monthly reports were mostly accessed and used for optimisation by the CE while in China. Therefore, management and executives had a real-time view of the mine performance from remote offices, comfort of their homes, or anywhere in the world.

Automation

Automation eradicated manually executed processes that were generally susceptible to errors and took longer to complete. Some of these processes were production orders from SAP ERP; WIP from production systems; fuel tank levels to SAP ERP; fuel consumption to SAP ERP; kilometres, engine hours, and operating to SAP ERP; and master data distribution to mine systems.

Proactive maintenance

The vehicles and other mine equipment leveraged IIoT technologies to publish performance information such as operating time, engine hours, and mileage. Through this information, the vehicles and machines had become smart enough to know when they were nearing the maintenance period. Hence, proactive maintenance (PM) orders were automatically created in the PM module of SAP ERP. Consequently, the parts were ordered and the work was planned, scheduled, and executed during an outage instead of a downtime situation. This resulted in a significant reduction of downtime and costly reactive maintenance.

Integrated supply chain

It was now possible for respective suppliers to have visibility on the lifespan of rims and tyres they provide to the mine. The fuel supplier had real-time visibility on the fuel capacity and transactions at the mine. The suppliers of all types of equipment could collect performance data for continuous improvement purposes. Customers could have real-time visibility on the progress and status of their orders in production.

Energy management

The MES used smart readers to produce energy consumption data from production lines and machines in real time. Knowing which production lines and machines consumed more energy made it possible for the mine to avoid running those production lines and machines during peak time. The mine could also integrate energy data to the production scheduling system.

Summary of additional benefits

The summary of additional benefits is: automated, real-time order management for all inventories; automated fleet movement and the potential for fully autonomous vehicles; real-time product movement dashboard; real-time maintenance; predictive analytics; online QM; upskilling of workers; more occupational safety through augmented automation; better working conditions through ergonomically adapted workstations; improved teamwork in the production environment through consistent data availability; improved environment through optimised used of resources; augmented innovative capabilities through technological possibilities in manufacturing; fully automated real-time balanced scorecard; inventory reporting; and emissions reporting (Table 4.1).

The contribution of 4IR to the SDGs

The team had to investigate the contribution of the implementation of 4IR to the SDGs. SDGs are a collection of 17 global goals set by the United Nations General Assembly in 2015. These goals are a universal call to action to end poverty, protect the planet, and ensure that all people enjoy peace and prosperity by 2030. The SDGs are interconnected and cover a wide range of issues, including poverty, hunger, health, education, gender equality, clean water and sanitation, affordable and clean energy, decent work and economic growth, industry, innovation and infrastructure, reduced inequalities, sustainable cities and communities, responsible consumption and production, climate action, life below water, life on land, peace, justice and strong institutions, and partnerships for the goals. The implementation of 4IR at the mining company contributed to the achievement of several SDGs in the following ways.

- SDG 8: Decent work and economic growth: 4IR increased productivity, efficiency, and competitiveness, leading to economic growth and the creation of high-quality jobs.
- SDG 9: Industry, innovation, and infrastructure: 4IR drove innovation and the development of new technologies, services, and improved infrastructure through smart systems and connected devices.
- SDG 12: Responsible consumption and production: 4IR enabled more sustainable production processes through the use of real-time data, predictive maintenance, and resource optimisation, reducing waste and emissions.

Lessons for the CE and engineering team

The lessons learnt by the CE and engineering team suggested that the mining sector is one of the biggest contributors to the economic and social development of a country. The mining sector has economic benefits such as foreign direct investments, increased exports, government taxes, augmented national income and gross domestic product, improved employment and salaries, and speedy economic development. However, mining is facing, among other problems, significant economic and declining commodity price challenges. The absence of contemporary technological mining and production systems adds more challenges to the mining industry. There are many manually operated activities and disparate systems in the mining sector. Thus, there is a great need to implement such technologies to integrate disparate systems and automate manual processes. The implementation of the 4IR technologies through SDLC at the mine bridged the gaps identified. The vertical integration ensured the automation of manual processes from the shop floor, through MES, up to the SAP ERP and back. This integration also created a real-time view of production status that helped management make necessary interventions on time. The end-to-end integration enabled a smooth product lifecycle management (PLM) process at the mine, where products were traced from the raw material to the finished and consumption state. PLM appeared to be one of the most important achievements of Industry 4.0.

The horizontal integration also enabled suppliers of fuel, rims, tyres, mining and processing equipment, and raw material to be integrated with the mine. This value chain provided suppliers with performance information of their supplies to implement necessary continuous improvements. Customers were able to be integrated with the mine to view the real-time progress of the orders in production. The combination of these three aforementioned integration layers formed CPS, which is about making smart machines, products, and systems communicate like in a social network. The big data philosophy was realised by getting massive amounts of data from different machines, products, and systems into the central repository for sharing with all the mine systems. Big data made it possible for the mine to use available information for improved data analytics, analysis, visualisation, and querying. The establishment of master data management and governance resulted in master data integrity and reliability. The master data management made use of the available big data set-up to distribute the copies of master data to all the mine systems, thus enforcing master data consistency and trustworthiness by proving a single version of the truth. 4IR can contribute to the achievement of several SDGs by enhancing productivity, innovation, sustainability, and inclusiveness.

Some examples of SDGs that can benefit from Industry 4.0 are SDG 8 (Decent work and economic growth), SDG 9 (Industry, innovation, and infrastructure), and SDG 12 (Responsible consumption and production). 4IR also poses some challenges and opportunities for the transition to a more sustainable and equitable society, such as the need for responsible management, skills development, digital inclusion, and social protection. In conclusion, the CE and the engineering team proved that it was possible to implement 4IR technologies in a mining company using the SDLC approach, as is the case with the manufacturing sector. IoT, IIoT, and big data are the technological tools to realise CPS, which is the core of the 4IR philosophy. By implementing 4IR technologies to provide data from the shop floor systems to the decision-makers and back, share information with business partners, and track products from inception up to the end of life, excellence in management, transparency, and sustainability of the business was achieved. 4IR and related technologies form significant contributors to the achievement of the SDGs.

Future work

Virtual twin

Anthes *et al.* (2016) define virtual twinning as a simulation of the physical environment. Changes in the virtual twin are reflected in the real physical environment and vice versa. Virtual twinning models the physical environment at a digital level and yield dynamic operations and the performance of system modules during production in real time. Therefore, future work should encompass the realisation of virtual twinning of the processing plant at the mine. IoT, IIoT, big data, and CPS have already been set up as enablement technologies for the virtual twinning of the processing plant.

Augmented reality and virtual reality

Augmented reality and virtual reality are used for simulating physical processes before they are performed in the real world. These technologies can be used in spheres such as production chain, maintenance, and research and development (Dunleavy and Dede, 2014). Therefore, this work can be extended to explore the methods, tools, and benefits of implementing augmented reality and virtual reality for a mine.

Exhibits

Figure 4.1 Open pit mine

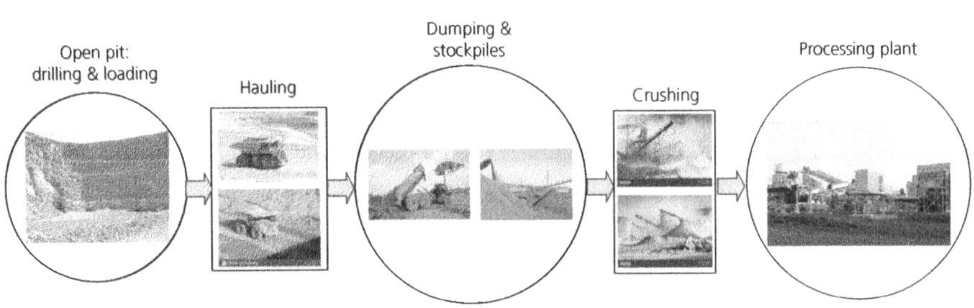

Figure 4.2 The four stages of industrial revolution

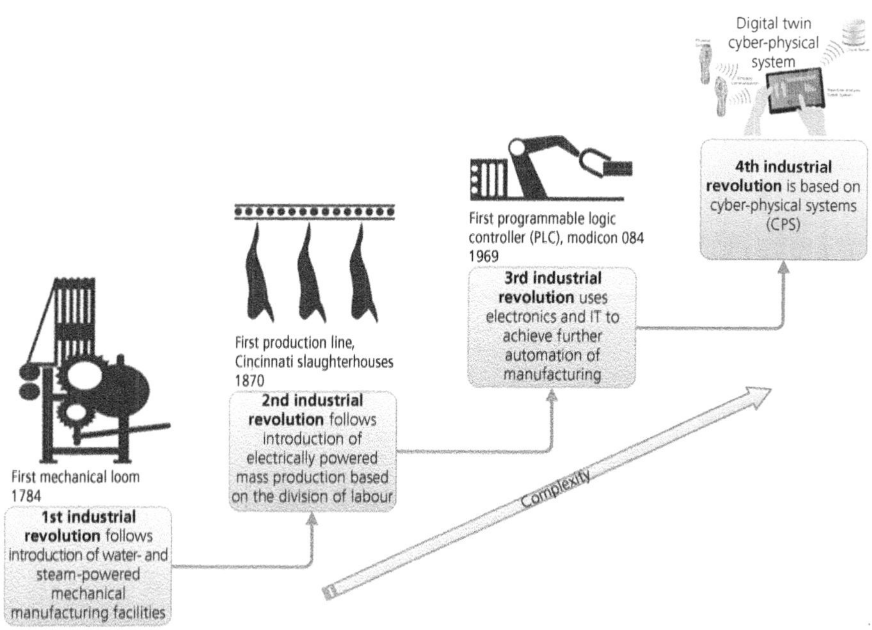

Figure 4.3 4IR integration layers

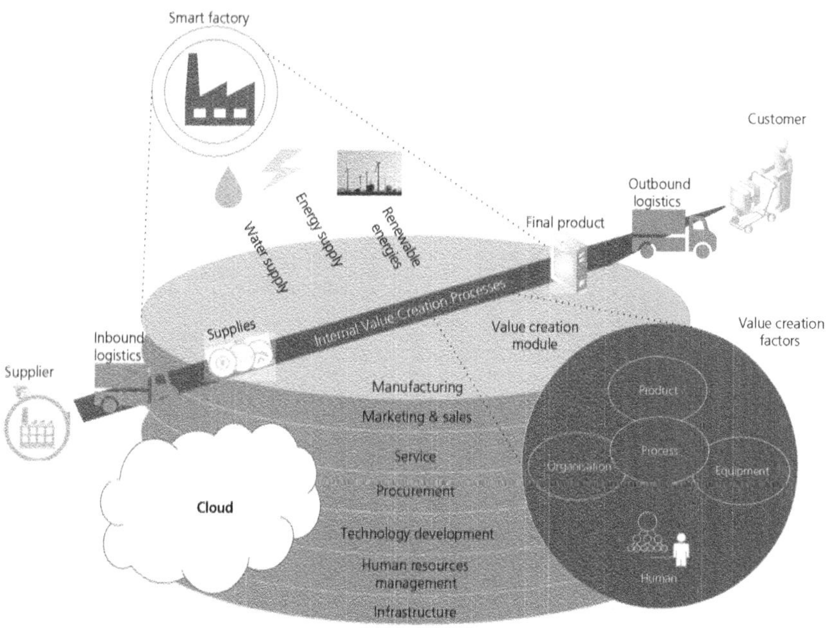

Figure 4.4 Overall systems architecture

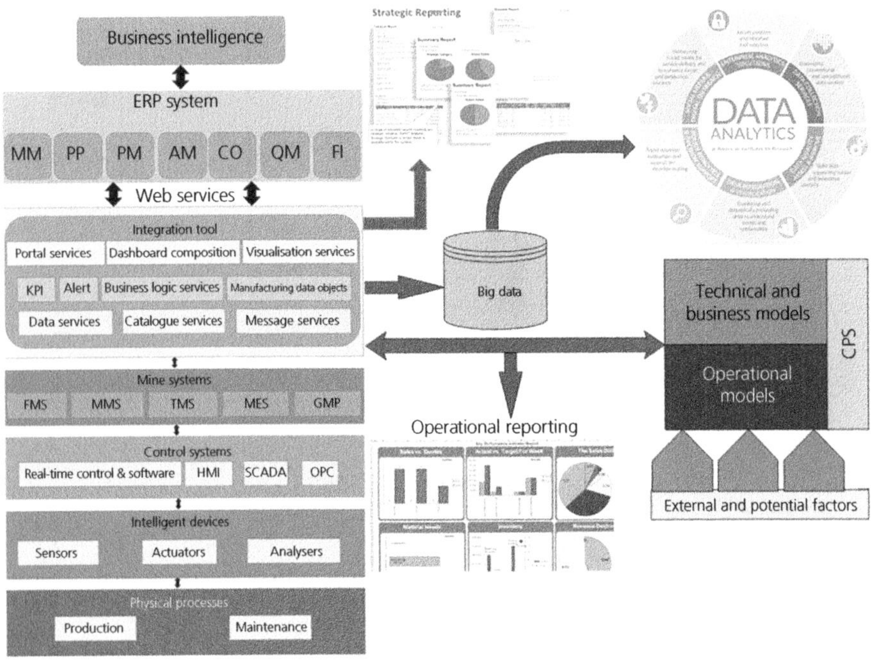

Figure 4.5 4IR smart mine

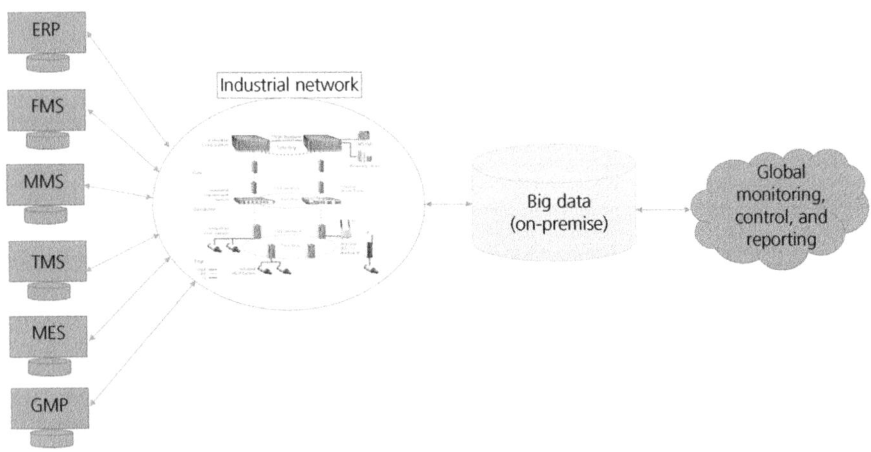

Table 4.1 Service level agreement

Priority	Description	Response service level		Restoration service level: hours ZAR		Resolution time
		Expected	Minimum	Expected	Minimum	
Priority 1 – Critical	Total loss of critical business process and application accessibility creates a serious business and financial exposure. All users cannot access, and all locations are down.	95% response (within 30 min)	90% response (within 1 h)	95% response (within 1 hr	75% response (within 4 h)	Remote – 4 h. Alternatively, next flight out to plant location
Priority 2 – High	Significant loss of critical business process and application accessibility creates a high business and financial exposure. The failure affects many clients, and some locations are down.	95% response (within 1 h)	90% response (within 2 h)	95% response (within 4 h)	75% response (within 8 h)	Remote – 8 h. Alternatively, next flight out to plant location
Priority 3 – Medium	Partial loss of critical business process and application accessibility creates a low business and financial exposure. The failure affects a small number of clients, and some parts of a location are down.	95% response (within 4 h)	90% response (within 6 h)	95% response (within 8 h)	75% response (within 16 h)	Remote – 16 working hours
Priority 4 – Low	Client requires additional services or functionality that are non-critical in nature.	95% response (within 3 working days)	90% response (within 5 working days)	95% response (within 2 weeks)	75% response (within 4 weeks)	Remote – 24 working hours

Sustainability and uranium digitalisation

Teaching notes

Synopsis

In 2015, a leading open-pit uranium mining company, among the world's largest, began operations with largely manual processes, hindering real-time information exchange between systems and stakeholders. The Chief Executive (CE) recognised the need for company-wide real-time information flow and global stakeholder collaboration, while ensuring significant contributions to the sustainable development goals (SDGs). To address these challenges, the CE assembled a team of engineers to research and implement fourth industrial revolution (4IR) techniques, creating an integrated enterprise within the connected business world. The team's objectives included assessing the feasibility of 4IR technologies in mining, developing introduction protocols, evaluating benefits, and determining the contribution of 4IR to achieving the SDGs, ultimately driving sustainability and innovation in the industry.

Teaching objectives

This case study is intended for use in undergraduate, MBA, and executive education courses in Systems Engineering and Digital Transformation. The nature of the teaching objectives suggests that this case is more appropriate for undergraduate programmes or introductory Systems Engineering courses in postgraduate programmes. The case has four pedagogical objectives.

- Critical thinking: Comprehending the strategic choices that companies need to make to achieve the adoption of 4IR.
- 4IR implementation prospect for mines: Determine the prospects of successfully implementing 4IR technologies in the mining industry.
- 4IR implementation approach: Understand and leverage possible approaches and protocols to introducing 4IR technologies in the mining industry.
- Key success factors: Uncover attributes that are exhibited by organisations that have adopted 4IR technologies.
- Contribution of 4IR to the SDGs: Discover the contribution made by the 4IR implementation projects to the SDGs.

Research methods

This case draws on data, information, and insights collected through an exhaustive literature review to determine the prospect of successfully implementating 4IR technologies in the mining industry. This was followed by the adoption of SDLC techniques to implement 4IR technologies at the mine. The SDLC process comprised several steps: requirements engineering through onsite and online interview workshops and observations, data analysis and validation, designs and implementation, testing, deployment, and support.

Keywords and subject code

Industry 4.0, internet of things, cyber-physical systems, enterprise resource planning.

Assignment questions

1. What are the prospects of successfully implementing 4IR technologies in the mining industry?
2. What are the possible approaches and protocols to introducing 4IR technologies in the mining industry?
3. What attributes are exhibited by organisations that have adopted 4IR technologies? Should these attributes be used as drivers of 4IR adoption by organisations of various types?
4. What is the contribution of 4IR to the SDGs?

Additional materials

The authors would recommend the following reading.

Oesterreich TD and Teuteberg F (2016) Understanding the implications of digitization and automation in the context of Industry 4.0: A triangulation approach and elements of a research agenda for the construction industry. *Computers in Industry* **83**: 121–139.

Rafael LD, Jaione GE, Cristina L and Ibon SL (2020) An Industry 4.0 maturity model for machine tool companies. *Technological Forecasting & Social Change* **159**: 120203.

Sishi MN and Telukdarie A (2018) Enterprise definition for Industry 4.0. *2018 IEEE International Conference on Industrial Engineering and Engineering Management (IEEM)*, Bangkok, Thailand, pp. 849–853.

Sishi M and Telukdarie A (2020) Implementation of industry 4.0 technologies in the mining industry–a case study. *International Journal of Mining and Mineral Engineering* **11(1)**: 1–22.

Wang S, Wan J, Li D and Zhang C (2016) Implementing smart factory of Industry 4.0: an outlook. *International Journal of Distributed Sensor Networks* **12(1)**: 3159805.

Teaching plan and timing

For a class of 90 min, the researchers would recommend the following teaching plan:

Introduction (5 min)

- Instructor provides a brief summary of the case.

Assignment question 1 (20 min)

- What are the prospects of successfully implementing 4IR technologies in the mining industry?

Assignment question 2 (20 min)

- What are the possible approaches and protocols to introducing 4IR technologies in the mining industry?

Assignment question 3 (20 min)

- What attributes are exhibited by organisations that have adopted 4IR technologies? Should these attributes be used as drivers of 4IR adoption by organisations of various types?

Assignment question 4 (20 min)

- What is the contribution of 4IR to the SDGs?

Conclusion (5 min)

- Wrap up the case and summarise key learnings.

Analysis of assignment questions

Question 1. What are the prospects of successfully implementing 4IR technologies in the mining industry?

[Teaching suggestion] The instructor can ask students to identify and provide a brief definition of the 4IR facets, and the applicability of these facets in the mining industry that were observed by the engineering team. The 4IR facets and their applicability in the mining industry are as follows.

The team of engineers observed three facets of 4IR: vertical integration, end-to-end digital integration, and horizontal integration through value networks. Vertical integration and networked industrial systems connect different internal sections of the industrial company such as information and communications technology (ICT) systems and processes. They define a connection of the value creation factors such as product, human, and process along the various levels of the value creation modules from production stations through production cells and production lines up to the smart factory. This networking throughout the different aggregation levels also includes data flow between product development to fabrication, logistics, marketing and sales, service, and procurement, thus creating a smart production environment that enables CPS to create flexible and reconfigurable manufacturing systems.

End-to-end digital integration is the engineering across the entire value chain that describes the cross-linking and digitalisation of the entire product lifecycle. The product is tracked from the raw material status up to the end of life. This integration enables connected stakeholders, products, and equipment to share data in real time during various phases of the product lifecycle. The product lifecycle encompasses phases such as raw material acquisition, product development, production engineering, manufacturing, product use and service, and end of life.

Horizontal integration creates a value network that enables local and globally dispersed value chain partners to collaborate through the integration of ICT systems, processes, and data flows. The integration of ICT systems, processes, and data flows within the company and between various businesses such as customers, suppliers, and other external partners allows closer cooperation with value chain partners across business boundaries. This integration sheds light on how CPS should be used to sustainably implement and support the company's business strategies, value networks, and business models. Figure 4.3 depicts a connected enterprise that encompasses all the 4IR integration layers.

The extensive knowledge acquired through an exhaustive literature review shed light on the prospects of implementing 4IR technologies in the mining industry. It became evident that 4IR technologies could be applicable in several types of industrial sectors such as manufacturing, mining, agriculture, logistics and so on.

Question 2. What are the possible approaches and protocols to introducing 4IR technologies in the mining industry?

The engineering team proposed a systematic approach to introduce 4IR technologies for the mine. This approach and method were underpinned by extensive knowledge gathered on the prospect of successfully implementing 4IR technologies in the mining industry.

[Teaching suggestion] The instructor can ask learners to identify and discuss an approach and the protocols mining companies can use to adopt digital transformation through 4IR technologies. The approach and protocols adopted by the engineering team to introduce 4IR technologies at the mine are as follows.

The team employed systems development lifecycle (SDLC) techniques to implement 4IR technologies at the mine. The SDLC process comprised several steps: requirements engineering through interview workshops and observations, data analysis and validation, designs and implementation, testing, deployment, and support.

Requirements engineering: The requirements engineering was conducted by gathering and documenting the existing state of the mine to determine its future state. The existing processes were documented, verified, and examined to close existing gaps and find better ways to enhance processes. The following sections discuss how business requirements engineering was conducted at the mine.

Interview workshops: To learn more about the mine's current and desired state, different sections and levels of the business were represented during interview workshops.

These representatives were senior management (executives), middle management (department managers), external business partners (suppliers), enterprise resource planning (ERP) super users, shop floor supervisors, plant maintenance manager, and ICT representative.

Observations: Observations included observing if there were any existing vertical, end-to-end, and horizontal data flow mechanisms. During these observation exercises, a large number of business processes were captured and documented using a business process modelling tool, Microsoft Visio.

Data analysis and validation: Data analysis and validation revealed the absence of integrated systems and real-time flow of data. Despite having implemented various computer-based systems, these were stand-alone systems that did not communicate with each other. This resulted in a number of operations being executed manually, especially when sharing data and information between these stand-alone systems. This gap necessitated the implementation of vertical, end-to-end, and horizontal integration layers of 4IR. The requirements for implementing the 4IR integration layers were stipulated in the business requirements specification (BRS) document.

Designs and implementation: The designs and implementation of the 4IR solution encompassed all the integration layers. The design phase produced functional design specification (FDS) and technical design specification (TDS) documents. The implementation phase produced the actual 4IR solution based on the FDS and TDS artefacts.

Vertical integration: This was the core integration layer for the 4IR solution at the mine. It encompassed SAP ERP, FMS, FM, TMS, and MES, as discussed below.

- SAP ERP: SAP ERP was integrated with the mine systems through the central data repository. Data from SAP ERP was published to the central repository to grant all participating mine systems access to data. This connection also enabled SAP ERP to have access to all data published by the mine's systems.
- FMS: Sensors were installed on fuel tanks to automatically measure fuel levels. Data revealing fuel levels in each tank were collected from the sensors into the central repository. A threshold was set in the system to create an alert if fuel reached a certain level in each tank. Consequently, an alert was sent to SAP ERP to create an order for more fuel for a tank in question. Mobile fuel dispensers fed from fuel tanks and all other vehicles refilled from mobile dispensers. Hence, sensors were also installed on these mobile dispensers and vehicles to measure how much fuel was loaded or consumed. Fuel consumption data, for each dispenser and vehicle, were collected from sensors and stored in the central repository. Fuel tank levels and fuel consumption transactions were extracted and uploaded from the central repository to SAP ERP in real time. FMS updated local equipment and measuring points master data from the central data repository once a day.

- FM: Fleet manager collected and stored data in the central repository such as vehicle engine hours, mining production volumes, unscheduled vehicle maintenance, vehicle operating time, corrective maintenance, ore movement, and ore stockpiles. Some of these data were subsequently sent to SAP ERP system by integration interfaces.
- TMS: Tyre management required the development of integration interfaces to send rims and tyres master data from the central data repository to TMS once a day. The integration interface to send data, regarding which rims and tyres were installed in which positions and on which vehicles, to SAP ERP were also implemented. Lastly, the interface to collect data used for TMS cross-functional KPIs was developed.
- MES: Control systems monitored, managed, and passed shop floor data collected by sensors and actuators to MES. The integration interfaces for getting data such as production actuals, equipment status, material consumption, corrective maintenance, equipment working hours, quality, and KPIs from MES into the central repository were implemented. In return, the integration interfaces that fed MES with data from the central repository such as preventative maintenance, corrective maintenance status, calibration, ore movement, ore stockpile, and PP were developed. Finally, the development of integration interfaces for sending production actuals, quality, materials consumption, and corrective maintenance data from the central repository to SAP ERP was completed.

End-to-end integration: This was realised through the integration of geology and mining (GMP), FM, MES, and SAP ERP systems. The design and demarcation of blocks for drilling and blasting occurred in the GMP system while the actual drilling, blasting, loading and hauling was managed by FM. In addition to getting data broadcasted by GMP and FM, MES obtained product design information published by SAP ERP for processing ore into uranium and consequently manufactured uranium accordingly. This enabled MES to combine design and planning data from GMP; drilling, blasting, loading, and hauling data from FM; and product design data from SAP ERP during the manufacturing process. SAP ERP modules took care of marketing, sales and distribution, and product support activities after successful production of uranium.

Horizontal integration: Horizontal integration was accomplished by integrating ERP, FMS, FM, MES, TMS, and GMP systems. Each of the six systems publishes data to the central repository so that data broadcasted is accessible to all mine systems. The communication between SAP ERP modules created horizontal integration at the enterprise level. The overall horizontal integration made it possible for the mine to collaborate with the customers as well as the suppliers of fuel, tyres, rims, equipment, and raw material. The overall system architecture including all integration layers is provided in Figure 4.4.

Testing: Various studies emphasise the importance of the testing phase of the SDLC process to ensure that the system meets the design requirements and is free of errors before it is deployed in the live environment. The testing exercises employed the bottom-up approach and entailed all three integration layers (vertical, end-to-end, and horizontal) and six systems (SAP ERP, FMS, FM, TMS, MES, and GMP) mentioned in prior sections. Firstly, the integration layers and interfaces for each system were tested individually through unit testing techniques. All errors found were corrected and the functionality was re-tested until the design requirements were met. Thereafter, the integration testing technique was applied–that is, all three 4IR integration layers were tested together as a complete system. The purpose of integration testing was to reveal problems with the interfaces among system components before problems occur in the production environment. The testing phase yielded user acceptance testing (UAT) documents that were approved and signed by business users who participated in the testing exercises.

Deployment: The system was deployed in the live environment as a result of thorough and successful testing operations. Deployment yielded a fully integrated solution that encompassed all six systems at the mine, as shown in Figure 4.5. The mine systems were connected through an industrial network to the central data repository. The industrial network transmitted data from and to these systems and the central repository. Thereafter, data published were used for global monitoring, control, and reporting.

Support: Following a successful deployment, the 4IR solution at the mine was supported according to the guidelines illustrated in the service level agreement (SLA).

Question 3. What attributes are exhibited by organisations that have adopted 4IR technologies? Should these attributes be used as drivers of 4IR adoption by organisations of various types?

[Teaching suggestion] The instructor can ask learners to identify attributes exhibited by organisations that have adopted 4IR technologies. Students can further suggest if these attributes can be deemed as a benchmark for the digital transformation of organisations through 4IR technologies.

A high-level benefit analysis of adopting 4IR technologies included, but was not limited to:

Mass customisation: The mining team could change any drilling pattern, blast, and pile or feed component for the mine at any time. The movements of the haul trucks were scheduled in near real time. Fuel dispensing alerts were sent in real time if a haul truck was scheduled for production and insufficient fuel was detected in it. The maintenance of each tyre and rim was scheduled and customised based on the nature and magnitude of the damage.

Visibility and optimised decision-making

Production monitoring: The mining production was reported through cross-functional key KPIs that enabled management to have a clear view of the mining productivity. The production status from the processing plant was reported by displaying real-time work in progress (WIP) and quality data against daily production targets. Displaying WIP allowed shop floor supervisors to intervene if daily production targets were not met. This also relieved operators from recording daily achievements on pieces of paper and prevented incorrect recording of daily targets and scores at the same time. Having real-time production status allowed decision-makers for example, to shut down a machine if it kept producing bad quality products.

Fuel consumption monitoring: The reports for fuel monitoring shed light on the amount of fuel available for consumption. It was also possible to compare which vehicle or group of vehicles consumed more fuel. Fuel consumption could also be viewed by several physical areas such as the plant and equipment groups. The implementation of cross-functional KPIs made it possible to measure and display total fuel consumption, fuel consumption per hours worked, and fuel consumption per tonnes hauled. Therefore, management could take a closer look if a sudden or erratic increase of fuel consumption by the plant or equipment group was noticed.

Equipment management: Pivotal KPIs were monitored in order to optimise equipment management. For example, engineers would quicky know if equipment was underperforming using overall equipment effectiveness (OEE) and utilisation KPIs. The mean time before failure (MTBF) and mean time to repair (MTTR) KPIs enabled the mine to avoid, through proactive maintenance, unexpected and expensive downtime. Presenting the amount of corrective and preventative costs of maintenance helped reveal how much time and money the mine spent on maintenance activities. The equipment availability KPI helped divulge how long equipment was available to operate during production time. This helped management determine which equipment is usually unavailable and investigate the underlying reasons.

Operational fleet efficiency: The mine was now privy to truck information such as cycle times, average payloads, tonnes hauled per hour, average fuel consumption per hour, average tonnes hauled per litre of fuel consumed.

Rims and tyre visualisation: The mine had real-time visibility on rims and tyre costs, average failure rate by rim and tyre types, costs of rims and tyres per tonne hauled, cost of rims and tyres per operating hour, tread utilisation of fitted tyres, comparison between failed and worn tyres, and the rims and tyres wear cost per tonne hauled.

Agility: Important information such as time, quality, and price generated by machines, products and systems at the mine was instantaneously collected and stored centrally. This information made it possible for MES to automatically stop a machine or production line if it yields inferior quality products or makes high-energy consuming machines and production lines unavailable during peak time.

Improved data analysis: All systems (SAP ERP, FMS, FM, MES, TMS, and GMP) at the mine were now integrated and consequently enabled to publish generated data to the central repository. In addition to enabling the mine systems to communicate with each other in real time, this also yielded huge volumes of data being generated by these systems and machines. The resulting data were used for improved analysis and

KPI definition, creation, and reporting. Through big data, the mine was simultaneously realising inexpensive operations, high performance, and good quality production.

Remote monitoring: The reports and dashboards showed, among other things, data collected from the shop floor systems and data shared from the enterprise level back to shop floor systems in real time. The reports were also intelligent enough to display the overall status of the mine through real-time dashboards. Real-time stock and material movement from exposed ore, drilling, blasting, loading and hauling, stockpiling, up to crushing were reported in real time. These reports were compatible with any web-enabled fixed and mobile device, making them accessible at any time and from any place. Management used the reports to view the mine operations from their offices instead of physically taking a walk to the shop floor. The weekly and monthly reports were mostly accessed and used for optimisation by the CE while in China. Therefore, management and executives had a real-time view of the mine performance from remote offices, comfort of their homes, or anywhere in the world.

Automation: Automation eradicated manually executed processes that were generally susceptible to errors and took longer to complete. Some of these processes were production orders from SAP ERP; WIP from production systems; fuel tank levels to SAP ERP; fuel consumption to SAP ERP; kilometres, engine hours, and operating to SAP ERP; and master data distribution to mine systems.

Proactive maintenance: The vehicles and other mine equipment leveraged IIoT technologies to publish performance information such as operating time, engine hours, and mileage. Through this information, the vehicles and machines had become smart enough to know when they were nearing maintenance periods. Hence, proactive maintenance orders were automatically created in the PM module of SAP ERP. Consequently, the parts were ordered and the work was planned, scheduled, and executed during an outage instead of a downtime situation. This resulted in a significant reduction of downtime and costly reactive maintenance.

Integrated supply chain: It was now possible for respective suppliers to have visibility on the lifespan of rims and tyres they provide to the mine. The fuel supplier had real-time visibility of the fuel capacity and transactions at the mine. The suppliers of all types of equipment could collect performance data for continuous improvement purposes. Customers could have real-time visibility on the progress and status of their orders in production.

Energy management: The MES used smart readers to produce energy consumption data from production lines and machines in real time. Knowing which production lines and machines consumed more energy made it possible for the mine to avoid running those production lines and machines during peak time. The mine could also integrate energy data to the production scheduling system.

Summary of additional benefits: The summary of additional benefits is: automated, real-time order management for all inventory; automated fleet movement and the potential for fully autonomous vehicles; real-time product movement dashboard; real-time maintenance; predictive analytics; online QM; upskilling of workers; more occupational safety through augmented automation; better working conditions through ergonomically adapted workstations; improved teamwork in the production environment through consistent data availability; improved environment through optimised use of resources; augmented innovative capabilities through technological possibilities in manufacturing; fully automated real-time balanced scorecard; inventory reporting; and emissions reporting.

Question 4. What is the contribution of 4IR to the SDGs?

[Teaching suggestion] The instructor can ask learners to identify the SDGs that the 4IR implementation project at the mine contributed to.

The implementation of 4IR at the mine contributed to the achievement of several SDGs in the following ways.

- SDG 8: Decent work and economic growth: 4IR increased productivity, efficiency, and competitiveness, leading to economic growth and the creation of high-quality jobs.
- SDG 9: Industry, innovation, and infrastructure: 4IR drove innovation and the development of new technologies, services, and improved infrastructure through smart systems and connected devices.
- SDG 12: Responsible consumption and production: 4IR enabled more sustainable production processes through the use of real-time data, predictive maintenance, and resource optimisation, reducing waste and emissions.

Key learnings

The instructor can conclude the case by highlighting the elements that are pivotal to the adoption of 4IR technologies in the mining industry. The absence of contemporary technological mining and production systems causes significant challenges in the mining industry. There are many manually operated activities and disparate systems in the mining industry. Thus, there is a great need to implement such technologies to integrate disparate systems and automate manual processes. The CE and the engineering team proved that it was possible to implement 4IR technologies in a mining company using the SDLC approach, as is the case with the manufacturing sector. IoT, IIoT, and big data are the technological tools to realise CPS, which is the core of 4IR philosophy. By implementing 4IR technologies to provide data from the shop floor systems to the decision-makers and back, share information with business partners, and track products from inception up to the end of life, excellence in management, transparency, and sustainability of the business was achieved. 4IR and related technologies form significant contributors to the achievement of the SDGs.

REFERENCES

Almada-Lobo F (2016) The Industry 4.0 revolution and the future of manufacturing execution systems (MES). *Journal of Innovation Management* **3(4)**: 16–21.

Anthes C, García-Hernández RJ, Wiedemann M and Kranzlmüller D (2016) State of the art of virtual reality technology. *2016 IEEE Aerospace Conference*. IEEE, Big Sky, MT, USA, pp. 1–19.

Balaji S and Murugaiyan MS (2012) Waterfall vs. V-Model vs. Agile: a comparative study on SDLC. *International Journal of Information Technology and Business Management,* **2(1)**: 26–30.

Dunleavy M and Dede C (2014) Augmented reality teaching and learning. *Handbook of Research on Educational Communications and Technology*. Springer, New York, NY, USA, pp. 735–745.

Monostori L, Kdr B, Bauernhansl T *et al.* (2016) Cyber-physical systems in manufacturing. *CIRP Annals-Manufacturing Technology* **65(2)**: 621–641.

Oesterreich TD and Teuteberg F (2016) Understanding the implications of digitisation and automation in the context of industry 4.0: a triangulation approach and elements of a research agenda for the construction industry. *Computers in Industry* **83**: 121–139.

Rafael LD, Jaione GE, Cristina L and Ibon SL (2020) An Industry 4.0 maturity model for machine tool companies. *Technological Forecasting & Social Change* **159**: 120203.

Sishi MN and Telukdarie A (2018) Enterprise definition for Industry 4.0. *2018 IEEE International Conference on Industrial Engineering and Engineering Management (IEEM)*, Bangkok, Thailand, pp. 849–853.

Sishi M and Telukdarie A (2020) Implementation of Industry 4.0 technologies in the mining industry – a case study. *International Journal of Mining and Mineral Engineering* **11(1)**: 1–22.

Wang S, Wan J, Li D and Zhang C (2016) Implementing smart factory of Industry 4.0: an outlook. *International Journal of Distributed Sensor Networks* **12(1)**: 3159805.

Prof. Arnesh Telukdarie
ISBN 978-1-83662-645-9
https://doi.org/10.1108/978-1-83662-642-820251005

Chapter 5
Satellite imagery for water leak detection: a joint research initiative by the city's Water Department and the University of Johannesburg

Arnesh Telukdarie and Mpho Godfrey Nkadimeng

Strategic collaboration and data-driven decision-making

Addressing the complexities of urban infrastructure requires a collaborative effort. Professor Telukdarie's initiative exemplifies this by bringing together the expertise of academic researchers, municipal water authorities, and technology partners. By fostering these strategic collaborations, the project leverages diverse resources and knowledge bases, ensuring comprehensive and sustainable solutions to Johannesburg's water infrastructure challenges. The vast amounts of data generated by these digital technologies require sophisticated data optimisation techniques to transform raw data into actionable insights. Professor Telukdarie and his team are employing advanced data analytics tools to analyse water usage patterns and detect anomalies indicative of leaks. This data-driven approach supports informed decision-making, enabling the development of effective strategies to combat water losses and enhance service delivery. Furthermore, the integration of digital technologies necessitates robust risk assessment frameworks to safeguard the infrastructure's integrity. By implementing proactive risk management strategies, Telukdarie's team aims to secure the digital infrastructure against potential threats and ensure its resilience. This innovative approach not only addresses current inefficiencies but also sets the foundation for future advancements in urban infrastructure management.

Digital integration in urban water infrastructure

To address these challenges, Professor Telukdarie proposed the integration of advanced digital technologies into Johannesburg's water management system. The initiative includes the use of satellite-based monitoring and synthetic aperture radar (SAR) technologies for real-time data collection and precise leak detection. These technologies enable the city to monitor its water infrastructure continuously, identify leaks with high precision, and carry out targeted maintenance efforts. The digital integration aims to reduce water losses, enhance operational efficiency, and ensure a sustainable water supply for Johannesburg's residents. While traditional methods of leak detection, like acoustic methods, have gained widespread use, they have several limitations. Acoustic methods rely on the sound produced by water escaping from pipes to locate leaks. These methods can be less effective in urban environments where background noise interferes with leak detection. Additionally, the accuracy of acoustic methods is influenced by the material and condition of the pipes. For example, plastic pipes dampen the sound of leaks, making it more difficult to detect issues accurately. Another traditional method, the minimum night flow (MNF) analysis, involves measuring the lowest water flow during the night when usage is minimal.

While MNF can indicate the presence of leaks, it does not provide precise locations, requiring further investigation and often resulting in labour-intensive and time-consuming efforts. The challenges presented by different pipe materials also complicate traditional leak detection methods. Older pipes made from materials like cast iron or steel are prone to corrosion, leading to false positives in acoustic leak detection. Newer materials, such as PVC, present different acoustic properties that further complicate detection efforts. The variability in pipe materials and conditions across Johannesburg's extensive water network necessitates a more advanced approach. Professor Telukdarie's initiative addresses these limitations by integrating satellite-based monitoring and SAR technologies, which are not affected by pipe material or background noise. These technologies provide a comprehensive and non-invasive method for detecting leaks, allowing for more precise and efficient maintenance planning. The continuous monitoring capability of these technologies ensures that leaks are identified and addressed promptly, reducing water losses and enhancing the overall reliability of the water supply system.

Data optimisation for decision-making

The integration of digital technologies generates vast amounts of data, which require sophisticated data optimisation techniques to transform into actionable insights. Professor Telukdarie and his team employ advanced data analytics tools to analyse water usage patterns and detect anomalies indicative of leaks. They can identify high-risk areas, prioritise maintenance efforts, and develop targeted strategies to reduce water losses by optimising this data. The data-driven approach supports informed decision-making, enabling the city to allocate resources more effectively and improve service delivery. A significant area of interest for this research is the Corriemoor water network, a critical section of Johannesburg's water infrastructure. For this study, the team utilised geographic information system (GIS) data provided by the Johannesburg Water team. These shapefiles are essential for mapping and visualising the water network, allowing for precise overlaying of satellite images and the identification of key points of interest. To enhance the accuracy of leak detection, the team employed a sensor-synthetic aperture radar (SAR) with an L-band frequency for its depth of penetration capabilities. This choice is particularly effective for penetrating urban asphalt pavements prevalent in the region of interest. The team can detect subsurface anomalies that are indicative of leaks, even in areas with heavy urban infrastructure, using an L-band SAR.

Application of shapefiles and SAR technology

The shapefiles serve as a geographic framework that helps overlay satellite images onto the existing water network infrastructure. This overlay process is crucial for accurately pinpointing potential problem areas within the Corriemoor water network. By integrating SAR data with the shapefiles, the team can identify specific points of interest, such as low-pressure events and hidden leaks (Figure 5.1). The integration of shapefiles and SAR technology enables the detection of typical low-pressure events, which are often early indicators of leaks.

Additionally, this approach helps in detecting hidden leaks, such as those near booster pumps, which can be particularly challenging to identify using conventional methods. By addressing these specific issues, the team can develop targeted maintenance strategies that prevent water loss and enhance the overall efficiency of the water network. The optimised data and advanced analytics tools enable the team to create detailed maps of the water network, highlighting areas with potential leaks and inefficiencies. This comprehensive mapping supports more informed decision-making processes, allowing the city to allocate resources more effectively. For instance, data analysis enables the dispatch of maintenance teams to high-risk areas, thereby concentrating efforts where they are most critical. The project not only improves leak detection and repair but also contributes to long-term sustainability goals by leveraging data-driven approaches. The ability to predict and prevent infrastructure failures reduces the need for costly emergency repairs and extends the lifespan of the water network. Furthermore, improved efficiency in water management enhances the reliability of water supply to residents and businesses, fostering greater public trust in municipal services. Through the focused study of the Corriemoor water network, Professor Telukdarie and his team demonstrated the transformative potential of digital integration and data optimisation in urban water management.

The research provides valuable insights for the development of scalable solutions for Johannesburg and potentially other urban centres facing similar challenges. The use of multi-band SAR technologies, particularly the L-band, plays a crucial role in the comprehensive analysis of Johannesburg's water infrastructure. The L-band frequency was chosen for its superior depth of penetration, which enables effective subsurface imaging despite the presence of urban asphalt pavements. This capability is essential for accurately detecting underground anomalies, such as leaks. Moreover, employing a multi-band approach enables the differentiation between various types of materials and conditions within the pipeline network, enhancing the precision of leak identification. The research team also focuses on the nodes of the pipeline network, which are critical junction points where different segments of the water system intersect. By analysing these nodes, the team can identify potential vulnerabilities and high-risk areas where leaks are more likely to occur. The integration of shapefiles with SAR data allows for a detailed mapping of these nodes, facilitating targeted inspections and maintenance activities. Another key aspect of the project is the calculation of the moisture index, which provides insights into soil moisture levels around the pipelines. Variations in the moisture index can indicate the presence of leaks, as leaking water alters the surrounding soil moisture. The L-band SAR's ability to penetrate the ground and provide detailed subsurface images is instrumental in assessing these moisture variations. By combining moisture index data with SAR imaging, the team can detect and locate leaks with greater accuracy (Figure 5.1).

Strategic implementation of collaborative initiatives

At the heart of this initiative are robust university research partnerships, which serve as the backbone for innovative solutions. The University of Johannesburg, under Professor Telukdarie's leadership, collaborates with other academic institutions to conduct cutting-edge research in digital infrastructure and water management. These partnerships facilitate the exchange of knowledge, technologies, and best practices. The research frameworks developed in these academic settings are critical for analysing data, predicting future infrastructure needs, and developing scalable solutions. Stakeholder engagement is meticulously structured to ensure all voices are heard and considered. The project employs frameworks such as the stakeholder engagement framework (SEF) and the participatory planning and management framework (PPMF). These frameworks provide a structured approach for identifying key stakeholders, understanding their needs and concerns, and actively involving them in the decision-making process. The project ensures that the community's input is integrated into the planning and implementation phases through workshops, focus groups, and public consultations. The project also integrates the SDGs framework, specifically focusing on SDG 6: Clean water and sanitation. The initiative not only addresses immediate infrastructure issues but also contributes to long-term sustainability goals by aligning the project's objectives with the SDGs. This alignment ensures that the solutions developed are environmentally sustainable, socially inclusive, and economically viable.

Risk assessment in digital infrastructure: The integration of digital technologies necessitates robust risk assessment frameworks to safeguard the infrastructure's integrity and reliability. Professor Telukdarie's team conducts thorough evaluations of potential risks, including cyber threats, system vulnerabilities, and operational disruptions. Proactive risk management strategies are implemented to mitigate these risks, ensuring the digital infrastructure's security and resilience. This approach includes regular system audits, cybersecurity measures, and contingency planning to address potential failures and ensure uninterrupted water service. Additionally, the team employs General Data Protection Regulation (GDPR) compliance to protect personal data and maintain privacy standards, further strengthening the system's robustness against cyber threats.

Innovation in proactive infrastructure management: Innovation is central to proactive infrastructure management. Professor Telukdarie's approach includes the deployment of cutting-edge technologies for continuous monitoring and maintenance of the water infrastructure. By proactively identifying and addressing issues, the project minimises the impact of aging infrastructure and enhances the city's ability to provide a reliable water supply. This innovative approach not only addresses current inefficiencies but also sets the foundation for future advancements in urban infrastructure management. The initiative aims to reduce the city's water losses by at least 30% within three years, improve leak detection accuracy by 50%, and achieve cost savings of up to R1.5 billion annually.

Integrating advanced technologies: a structured approach to urban water management

Strategic deployment and continuous monitoring of water management technologies

The implementation of the initiative is structured into distinct phases, beginning with pilot testing in high-risk areas identified through initial data analysis. This critical first phase focuses on deploying satellite-based monitoring and SAR technologies in these targeted areas. Initial data collection and analysis are conducted to identify existing leaks and inefficiencies within the water distribution network. The performance of these technologies is meticulously evaluated and, based on these findings, the deployment strategy is refined and optimised for broader application. In the subsequent full-scale implementation phase, the deployment of monitoring technologies is expanded across the entire city. This expansion involves integrating data from various sources into a centralised management system, allowing for targeted maintenance efforts based on real-time data insights. Continuous monitoring and evaluation are integral components of this phase, ensuring that the system remains responsive and effective. Regular system audits and performance evaluations are conducted, with strategies and technologies adjusted as needed based on ongoing data analysis.

Stakeholder engagement is also a key aspect, with regular updates and feedback sessions ensuring that all parties are informed and involved in the process. The project faces several challenges, including technical integration issues, stakeholder resistance, and funding constraints. To mitigate these risks, comprehensive training programmes are provided to municipal staff and technology partners, ensuring the smooth integration and operation of the new technologies. Additionally, transparent communication strategies and regular stakeholder meetings are employed to address concerns and foster collaboration. Securing funding from multiple sources, including government grants, private sector investments, and international aid, is also crucial to ensuring the project's financial sustainability. The success of the initiative is measured against specific metrics, including reductions in non-revenue water (NRW), cost savings, and improvements in service reliability. The project aims to achieve a 30% reduction in water losses within three years, resulting in annual savings of up to R1.5 billion through reduced water losses and improved operational efficiency. Additionally, the initiative seeks to enhance the reliability of the water supply, as reflected in customer satisfaction surveys and service delivery metrics. Regular progress reports and performance evaluations ensure accountability and promote continuous improvement throughout the project's lifecycle.

Optimising data integration and environmental analysis for leak detection

The initiative also addresses the technical challenges of preprocessing satellite imagery to improve leak detection accuracy. Atmospheric influences on the imagery are mitigated through techniques such as atmospheric correction and radiometric calibration, which are crucial for accurate water leak detection (see Figure 5.2). The integration of satellite data with atmospheric datasets and detailed water pipeline network information is facilitated using MongoDB Atlas, ensuring efficient access and retrieval of satellite imagery while maintaining the integrity of the water network design. This holistic approach is further supported by overlaying water pipeline network data with satellite imagery (as shown in Figure 5.2, image 2 and image 3), incorporating atmospheric considerations to enhance detection accuracy. The careful selection of spectral bands, including thermal infrared and shortwave infrared, allows for the identification of temperature anomalies and spectral signatures that are indicative of leaks. Additionally, moisture index parameters, such as specialised indices tailored for hydrocarbon detection, play a vital role in identifying anomalous moisture patterns associated with leaks. Spatial and temporal analysis techniques, combined with ground truth and weather data, enable the project team to achieve a high level of accuracy in leak identification.

This advanced analytical capability allows pipeline operators to swiftly respond to detected incidents, safeguarding the environment and ensuring the continued reliability of the water supply system. The practical application of these technologies can be observed in the Landsat 8 image of the Corriemoor area, which serves as a key study site (Figure 5.2, image 3).

The image provides a detailed view of the landscape, including vegetation and built-up areas, offering valuable insights into the area's infrastructure. By overlaying water pipeline network data and topographical features onto this satellite imagery, the project team gains a comprehensive understanding of the physical infrastructure that delivers water to various locations. This spatial context is crucial for identifying potential leak points, assessing infrastructure vulnerabilities, and optimising maintenance and repair efforts. The use of contour lines, which represent elevation levels across the landscape, provides a three-dimensional perspective of the terrain (Figure 5.2, image 3). Understanding the topography is essential for analysing water flow patterns, identifying potential drainage areas, and assessing the impact of elevation changes on infrastructure design and performance. These insights are vital for ensuring the long-term sustainability and effectiveness of the water management system in Johannesburg.

Conclusion

The integration of satellite imagery and advanced data analytics, as explored in this chapter, marks a significant leap forward in addressing the critical challenge of water leak detection in urban settings. The collaboration between Johannesburg Water and the University of Johannesburg showcases the power of strategic partnerships and cutting-edge technologies in tackling systemic issues in water infrastructure. By leveraging technologies like SAR and satellite-based monitoring, the initiative not only improves operational efficiency but also paves the way for more sustainable urban water management. This project underscores the importance of data-driven decision-making in modern infrastructure management. Through continuous monitoring and real-time data optimisation, maintenance efforts can be precisely targeted, reducing water loss, lowering costs, and improving service delivery. The project's application of multi-band SAR technology, combined with geographic information system (GIS) tools and shapefile integration, demonstrates how innovative digital solutions can address long-standing challenges, offering a scalable model for other urban areas worldwide.

As the field of digital transformation continues to evolve, initiatives like this one reflect a forward-thinking approach to solving urban infrastructure problems. The ability to predict leaks, prevent costly damage, and optimise resource allocation is critical not just for Johannesburg, but for cities globally, where aging infrastructure threatens the sustainability of essential services. The proactive measures and collaborative framework established here set the stage for future advancements in infrastructure management, driven by continuous innovation and cross-disciplinary collaboration. Looking ahead, this initiative points to exciting possibilities in applying advanced technologies for urban resilience. As cities grow and climate challenges intensify, integrating satellite imagery with data analytics represents a powerful tool for cities to ensure water sustainability. Moreover, this research serves as a call to action for ongoing learning and innovation. The challenges of managing urban infrastructure will continue to evolve, and future leaders in this field must be equipped to adapt and innovate in response. The work done by Professor Telukdarie and his team provides a blueprint for digital transformation in water management. Their success offers valuable lessons on the need for constant adaptation, partnership, and innovation. The future of urban water systems will depend on the ability to apply these tools and methodologies, ensuring that cities like Johannesburg can not only manage today's challenges but also thrive in the future. Students and professionals in the field must remain committed to continuous learning and engage in forward-thinking research to stay at the cutting edge of digital transformation in infrastructure management.

Exhibits

Figure 5.1 Integration of L-band SAR data with pipeline network nodes and moisture index

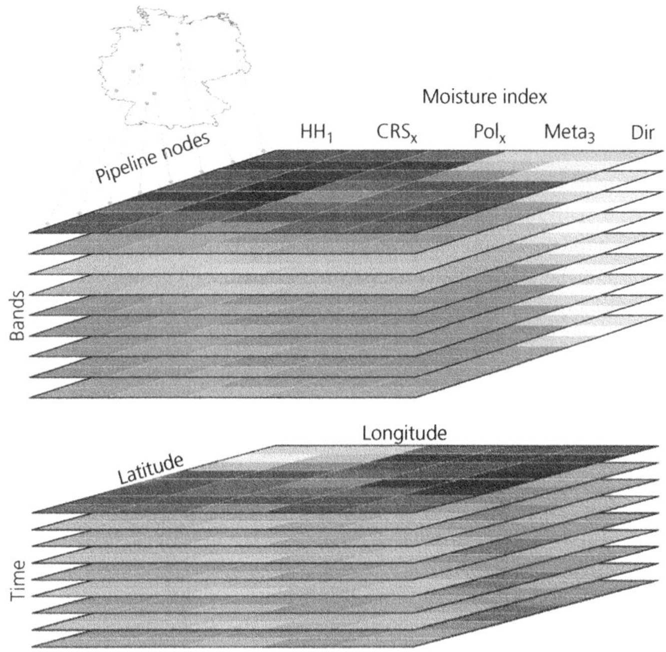

Figure 5.2 Landsat 8 Corriemoor area

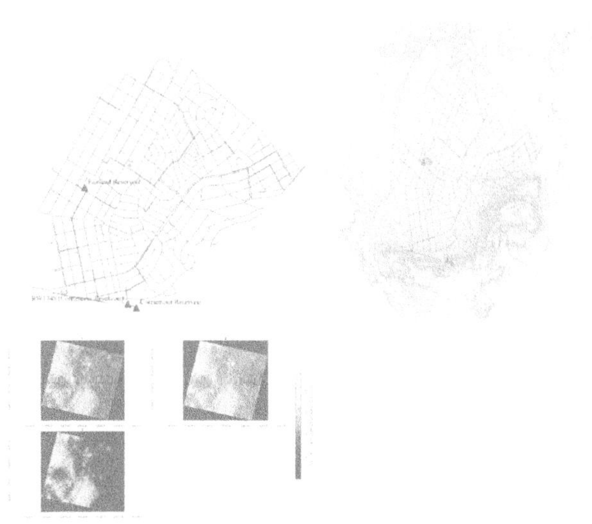

Satellite imagery for water leak detection: a joint research initiative by the city's Water Department and the University of Johannesburg

Teaching notes

Synopsis

This chapter explores the innovative use of satellite imagery and SAR technology to address the critical issue of water leak detection within Johannesburg's aging infrastructure. Faced with nearly 46.1% of the city's water being unaccounted for owing to leaks, the chapter outlines a strategic collaboration between the University of Johannesburg and the city's Water Department, led by Professor Arnesh Telukdarie. The primary objective is to demonstrate how integrating satellite-based monitoring and SAR technologies can significantly enhance leak detection precision, reduce water losses, and support sustainable urban infrastructure development. Through the deployment of GIS data, multi-band SAR imaging, and advanced analytics, the chapter reveals how these tools were applied to proactively manage Johannesburg's water infrastructure, particularly focusing on the Corriemoor water network. The findings underscore the importance of optimising data resolutions to support informed decision-making, showcasing the potential of interdisciplinary collaboration and digital innovation to transform traditional water management practices. This chapter not only provides a scalable model for other urban centres but also includes practical teaching objectives that equip students with the skills needed for future digital infrastructure projects, emphasising data optimisation, risk assessment, and collaboration. By addressing the complexities of water leak detection through digital integration, the chapter offers a blueprint for enhancing sustainability and efficiency in urban water systems.

Teaching objectives

Students will embark on an in-depth exploration of how digital technology intersects with urban infrastructure, with a focus on modernising systems to meet contemporary challenges. The teaching objectives are carefully structured to equip students with the essential skills and insights needed for this endeavour. Key objectives include the following.

- **Integrating advanced satellite imagery:** Students will delve into the integration of advanced satellite imagery within the context of digital transformation for urban infrastructure. The goal is to develop a comprehensive understanding of how digital technologies can address challenges like water leaks and aging infrastructure.
- **Optimising data for strategic decision-making:** The curriculum emphasises the importance of optimising spatial, spectral, temporal, and radiometric data resolutions to support effective decision-making. Students will acquire the skills to analyse and harness data for informed decisions in water management.

- **Understanding interdisciplinary collaboration:** Through a case study on the collaborative efforts between the City of Johannesburg and the University of Johannesburg, students will explore the strategic challenges and benefits associated with interdisciplinary collaboration in digital transformation projects.
- **Assessing risks in digital infrastructure:** The objective focuses on engaging students in risk assessment related to the implementation of digital solutions in aging water infrastructure. The goal is to foster a nuanced understanding of the complexities and potential risks involved in digital transformations.
- **Innovating in proactive infrastructure management:** The programme encourages students to explore innovative solutions in digital infrastructure management. By analysing how digital technologies enable proactive strategies, the curriculum emphasises a shift from reactive to proactive approaches in maintaining and upgrading urban infrastructure.

These teaching objectives equip postgraduate students with a comprehensive skill set to address the challenges of modernising urban systems. The exploration spans digital integration, focusing on advanced satellite imagery to tackle water leaks and aging infrastructure. Proficiency in data optimisation enables strategic decision-making in water management. The collaborative initiative between the City of Johannesburg and the University of Johannesburg highlights the importance of interdisciplinary collaboration in digital transformation. Emphasising risk assessment, the curriculum promotes a nuanced approach to decision-making in the implementation of digital solutions for aging infrastructure, ultimately encouraging innovative, proactive strategies in urban development.

Research methods

The research methodology employed in this chapter combines advanced digital technologies with a systematic approach to data collection and analysis to address the challenge of water leak detection in Johannesburg's aging infrastructure. The study began with the integration of GIS data provided by the city's Water Department, which served as a foundational geographic framework for the analysis. This data was then overlaid with satellite imagery and SAR data, specifically utilising an L-band SAR for its superior depth of penetration, essential for detecting subsurface anomalies in urban environments. The research team applied atmospheric correction and radiometric calibration techniques to enhance the accuracy of the satellite imagery, ensuring that environmental factors did not interfere with leak detection. A critical component of the methodology was the optimisation of spatial, spectral, temporal, and radiometric resolutions in the satellite data, allowing for precise identification of potential leaks. This optimisation was supported using advanced data analytics tools, which processed and analysed vast amounts of data to detect anomalies in water usage patterns. The integration of MongoDB Atlas facilitated efficient data management and retrieval, ensuring that the extensive datasets generated by the SAR and satellite imagery were readily accessible for analysis. Additionally, the team employed the sen2r package to streamline the preprocessing and analysis of Sentinel-2 satellite imagery, further enhancing the robustness of the research. Ground truthing was conducted to validate the findings, ensuring that the detected anomalies corresponded with actual leaks within the water network. The methodology also included a comprehensive risk assessment to evaluate potential challenges in implementing these digital solutions, ensuring that the infrastructure's integrity and security were maintained throughout the study. This multifaceted approach allowed for a thorough investigation of water leaks, providing actionable insights and a scalable model for future urban infrastructure projects.

Keywords and subject code

Satellite imagery, synthetic aperture radar, water leak detection, digital integration, data optimisation, urban infrastructure management.

Assignment questions

1. How does the integration of advanced satellite imagery contribute to the proactive management of water infrastructure, and what specific challenges does it address in the context of water leaks and aging infrastructure? Provide examples from the case study to support your answer.
2. Discuss the significance of optimising spatial, spectral, temporal, and radiometric resolutions in satellite data for effective decision-making in water management. How do these optimisations enhance the precision of water leak detection, and what role do they play in the overall success of the collaborative initiative?
3. Explore the strategic considerations and potential challenges associated with interdisciplinary collaboration, as highlighted in the collaborative effort between the City of Johannesburg and the University of Johannesburg. How can effective collaboration impact the success of digital transformation projects in urban infrastructure?
4. Assess the risks associated with the implementation of digital solutions in aging water infrastructure, as presented in the case study. Identify potential risks and complexities and propose strategies for mitigating these risks while ensuring the successful integration of digital technologies.
5. Analyse the research methodology employed in the case study, focusing on data collection and analysis. How do primary and secondary data sources contribute to the robustness of the research, and what role does technology, such as the use of MongoDB Atlas and the sen2r package, play in ensuring efficient data management and retrieval?

Additional materials

Bates BP, Kundzewicz ZW, Wu S and Palutikof JP (2008) Climate change and water. *Technical Paper of the Intergovernmental Panel on Climate Change, IPCC Secretariat*. IPCC Secretariat, Geneva, Switzerland.

Gens R (2010) Pixel-based and object-oriented image analysis for SAR data: the importance of spatial context. *Photogrammetric Engineering & Remote Sensing* **76(2)**: 191–200.

McGlade J (2012) The role of geospatial technologies in urban environmental management. *Journal of Urban Technology* **19(3)**: 39–56.

Richards JA (2013) *Remote Sensing Digital Image Analysis: An Introduction*. Springer, New York, NY, USA.

Schott JR (2007) *Remote Sensing: The Image Chain Approach*. Oxford University Press, Oxford, UK.
Ulaby FT, Moore RK and Fung AK (1981) *Microwave Remote Sensing: Active and Passive*. Artech House, Norwood, MA, USA.

Suggested core readings

Jensen JR (2007) *Remote Sensing of the Environment: An Earth Resource Perspective*. Pearson, Upper Saddle River, NJ, USA.
Parcak SH (2009) *Satellite Remote Sensing for Archaeology*. Routledge, London, UK.
Rahman MM, Hossain MS and Ahmed T (2019) Data-driven decision making in urban water management: the role of big data and analytics. *Journal of Environmental Management* **237**: 458–470.
Saatchi SS *et al.* (2011) Monitoring and modelling forest carbon stocks with satellite data: an integration of remote sensing, LIDAR, and field observations. *Global Change Biology* **17**(**2**): 831–846.
Singh A and Sharma P (2017) Application of synthetic aperture radar (SAR) data for urban flood mapping: a case study. *Natural Hazards* **89**(**1**): 497–511.
Weng Q (2010) *Remote Sensing and GIS Integration: Theories, Methods, and Applications*. McGraw-Hill, New York, NY, USA.

Teaching plan and timing

Introduction and case overview (10 min)

- Begin with a brief introduction to the case study, emphasising the challenges faced by the City of Johannesburg regarding water leaks and aging infrastructure.
- Discuss the significance of the collaborative initiative between the city's Water Department and the University of Johannesburg in addressing these challenges using advanced satellite imagery.

Setting the stage: digital integration and water management (15 min)

- Dive into the teaching objectives regarding digital integration in urban infrastructure.
- Discuss the role of advanced satellite imagery in the context of digital transformation for urban infrastructure, linking it to the specific challenges of water leaks and aging infrastructure outlined in the case study.
- Use a board plan to highlight key concepts and questions about digital integration.

Data optimisation and decision-making (20 min)

- Transition to the next teaching objective on data optimisation for decision-making.
- Explore the significance of optimising spatial, spectral, temporal, and radiometric resolutions in satellite data for effective decision-making in water management.
- Encourage students to relate these optimisations to specific examples from the case study.
- Facilitate a discussion around the board plan, focusing on student insights and critical thinking.

Interdisciplinary collaboration and strategic implementation (15 min)

- Move to the teaching objective of Understanding interdisciplinary collaboration.
- Explore the strategic considerations, challenges, and benefits associated with interdisciplinary collaboration, as exemplified in the collaborative effort between the City of Johannesburg and the University of Johannesburg.
- Use the board plan to map out key considerations and potential challenges.

Risk assessment and discussion (15 min)

- Transition to the teaching objective of risk assessment in digital infrastructure.
- Engage students in a risk assessment related to the implementation of digital solutions in aging water infrastructure, drawing insights from the case study.
- Encourage students to propose strategies for mitigating risks, fostering a nuanced approach to decision-making.
- Use the board plan to capture key risk factors and mitigation strategies.

Research methodology and technology integration (15 min)

- Shift to the research methods employed in the case study.
- Discuss a comprehensive approach to data collection and analysis, emphasising the role of primary and secondary data sources.
- Explore the integration of technology, such as MongoDB Atlas and the sen2r package, to ensure efficient data management and retrieval.
- Use the board plan to visually outline the key steps in the research methodology.

Synthesis and application of additional readings (10 min)

- Summarise key insights from the case study and connect them to the suggested core readings.
- Discuss how principles from *Remote Sensing of the Environment* (Jensen, 2007) and *Satellite Remote Sensing for Archaeology* (Parcak, 2009) align with or contribute to the strategies presented in the case study.
- Encourage students to draw parallels and critically analyse the cross-disciplinary applications.

Q&A, reflection, and assignment (10 min)

- Open the floor for students' questions and reflections.
- Assign the post-class assignment, emphasising independent research and critical analysis related to the case study.
- Conclude by highlighting the broader implications of the collaborative initiative for global water infrastructure management.

Analysis of assignment questions

Table 5.1 Analysis of assignment questions

Assignment questions	Conceptual framework/ model applied	Key learnings	Possible student responses	Author's own model/ framework
Integration of advanced satellite imagery for proactive management	TOE framework (technology, organisation, environment)	Interdisciplinary collaboration as a catalyst for innovation	Focus on strengths and opportunities, overlooking weaknesses	Satellite integration framework
Optimising spatial, spectral, temporal, and radiometric resolutions	DIKW pyramid (data, information, knowledge, wisdom)	Proactive water infrastructure management through technology integration	Emphasise resolution optimisation but overlook its role in overall success	Resolution optimisation pyramid
Strategic considerations and challenges of interdisciplinary collaboration	SWOT analysis	Innovative solutions through interdisciplinary collaboration	Prioritise strengths and opportunities, neglecting weaknesses	Interdisciplinary collaboration matrix
Risk assessment in digital infrastructure	Risk management framework	Nuanced decision-making in digital transformations	Focus on identifying risks, overlooking mitigation strategies	Risk assessment matrix
Research methodology and technology integration	Research onion model	Role of technology in robust data collection and analysis	Focus on technology integration, neglecting diverse data sources	Integrated research framework

Key learnings

Interdisciplinary collaboration as a catalyst for innovation

The case study underscores the transformative power of interdisciplinary collaboration between the City of Johannesburg's Water Department and the University of Johannesburg. The collaboration, driven by the convergence of expertise from water management and remote sensing, exemplifies how pooling diverse skills and knowledge can lead to innovative solutions. The key learning here is that addressing complex challenges, such as water leaks in aging infrastructure, often requires a collaborative, multidisciplinary approach. This aligns with the 'understanding interdisciplinary collaboration' teaching objective.

Proactive water infrastructure management through technology integration

A central theme in this case is the shift from reactive to proactive water infrastructure management, facilitated by the integration of advanced satellite imagery. The continuous monitoring enabled by high-resolution satellite data allows for early detection of water leaks, reducing response times and mitigating the overall impact on infrastructure. This highlights the importance of technology integration in achieving proactive and precise solutions. Students gain insights into the practical application of technology for infrastructure management, aligning with the objective of 'innovating in proactive infrastructure management'.

Risk assessment and nuanced decision-making in digital transformations

The case study introduces the concept of risk assessment within the context of implementing digital solutions for aging water infrastructure. It emphasises the complexities and potential risks associated with digital transformations. Students learn that while advanced technologies offer innovative solutions, a nuanced approach to decision-making involves identifying and mitigating risks. This aligns with the teaching objective of 'assessing risks in digital infrastructure', fostering an understanding of the broader implications of technology adoption.

Guidance for instructors

Encourage students to reflect on these key learnings and discuss how they align with the broader themes of digital transformation, collaboration, and risk management in the field of urban infrastructure.

Connect these learnings to real-world scenarios, prompting students to consider how similar approaches could be applied in different contexts or cities facing analogous challenges.

Foster a discussion on the ethical considerations related to technology integration in infrastructure management, encouraging students to critically evaluate the potential societal impacts of such initiatives.

emerald PUBLISHING ice Publishing

Prof. Arnesh Telukdarie
ISBN 978-1-83662-645-9
https://doi.org/10.1108/978-1-83662-642-820251006

Chapter 6
Digital twins in healthcare: a pathway to sustainable hospital optimisation

Arnesh Telukdarie and Joseph Mwanza

Introduction

Chipo,[1] a young mother from a remote village in Zimbabwe, had walked 6 km to reach Chiremba[2] District Hospital with her sick child. After waiting for hours in a crowded waiting room, she was devastated to learn that the medicine her child desperately needed was out of stock. This was not an isolated incident but a daily reality for many patients at Chiremba, a privately-owned level 1 district hospital situated in rural Zimbabwe. Such challenges not only compromised the well-being of individuals like Chipo and her child but also hindered the achievement of Sustainable Development Goal (SDG) 3, whose aim is to 'ensure healthy lives and promote well-being for all at all ages' (United Nations, 2015). The hospital's Chief Executive Officer (CEO), deeply moved by the struggles faced by patients like Chipo, was determined to find a solution that could address the root causes of these challenges. Chiremba, like many district hospitals in low- and middle-income countries (LMICs), faced a set of obstacles preventing it from delivering quality healthcare services to surrounding communities and upholding the principles of sustainable healthcare. Limited resources, including financial constraints and a shortage of skilled personnel, made it difficult to meet the growing healthcare needs of the population. This left the CEO confronted by the daunting task of improving efficiency and patient outcomes while working within these limitations. A sustainable solution was needed – one that could balance the economic realities of the hospital with the social imperative of providing equitable and accessible healthcare.

Driven by a profound desire to serve the community better and to ensure that neither poverty nor lack of resources hindered access to healthcare, the CEO embarked on a journey to explore innovative solutions. This journey, spanning a six-month period between January and June 2023, covered Chiremba's exploration, decision-making, and initial implementation of digital twin technology – a cutting-edge approach that creates a virtual replica of a physical system to optimise operations. This technological solution held the promise of transforming the hospital's operations, ultimately improving the lives of patients like Chipo and contributing to a more sustainable healthcare system. 'We wanted to serve all our surrounding communities better, ensuring we made a positive difference in their quality of lives', explained the CEO.

[1] Not her real name.
[2] The name of the hospital has been changed to mask its actual identity.

Company background and problem identification: description of the hospital

Founded in 1947 by Roman Catholic Jesuits, Chiremba District Hospital has grown over the past decades to become a vital primary healthcare provider within its rural district. The hospital, with its 98 beds and a dedicated team of seven resident doctors, 20 nurses, and support staff, serves a vast catchment area, treating patients primarily for infectious diseases, maternal and child health issues, and injuries. Chiremba offers essential outpatient, inpatient, emergency, and laboratory services. Patients travel from both nearby and distant villages, with some facing arduous journeys just to reach the hospital.

Identification of specific issues

Despite its critical role in the community, Chiremba had long struggled with operational inefficiencies that negatively impacted patient care. The COVID-19 pandemic further exacerbated these challenges, revealing the hospital's vulnerability to disruptions and highlighting the urgent need for improvement. One major issue was the excessively long patient wait times. From registration to consultation to receiving medication, patients could wait hours, sometimes an entire day, for treatment. These delays, often exceeding four hours even for urgent cases, caused frustration and sometimes led patients to leave without receiving the necessary care. Furthermore, the hospital grappled with mismatches between medicine supply and demand. Essential medications frequently ran out, while others expired on shelves owing to overstocking. This pointed to challenges in accurately forecasting demand and optimising procurement processes, resulting in both shortages and wastage of valuable resources, ultimately hindering the hospital's economic sustainability.

These operational inefficiencies had a direct and detrimental impact on patient outcomes. Delayed treatment potentially worsened existing conditions, and medication shortages forced patients to seek costly alternatives or forgo treatment altogether. The long distances travelled by many patients underscored the urgency of improving efficiency within the hospital to ensure timely and effective care for all who sought its services, thus upholding the principle of equitable healthcare access, a key component of social sustainability.

Exploring digital twin technology

Discovery and initial investigation

Chiremba's CEO, deeply committed to improving the hospital's services and sustainability, was determined to find innovative solutions despite the resource constraints. Traditional, fully-fledged enterprise resource planning (ERP) systems were out of reach owing to their complexity and cost, prompting the CEO to actively seek out alternative digital solutions. This led to the discovery of digital twin technology – a cutting-edge approach involving virtual replicas of physical systems or processes that leverage real-time data for operational optimisation (Deloitte, 2021). Recognising its potential to address the hospital's challenges cost-effectively and sustainably, the CEO delved deeper into researching this promising technology further. Inspiration came from the research work of the Johannesburg Business School's (JBS) 4IR Research Laboratory, whose extensive publications on fourth industrial revolution (4IR) technologies, including digital twins, showcased successful applications across a diverse range of industrial and commercial enterprises. Notably, their research highlighted the potential of digital twins to enhance healthcare systems in resource-limited environments through cost-effective means. The CEO was particularly encouraged by two case studies demonstrating improved hospital processes, resource allocation, and patient flow. While one case focused on vertical healthcare systems integration in a developed economy context, the other was on the development of an equipment maintenance optimiser for a developing economy. This latter initiative, which had been successfully implemented in Gauteng, South Africa, during the COVID-19 pandemic of 2020 and 2021, had helped improve availability of critical medical equipment dramatically (Mwanza *et al.*, 2023). These examples reinforced the CEO's belief that indeed a digital twin solution could be tailored to Chiremba's unique needs and constraints and implemented successfully.

Decision-making process

The CEO recognised that introducing a digital twin was a significant decision with potential risks and challenges. Consultations with the hospital's IT personnel confirmed the technical feasibility of integrating the technology without major infrastructural overhauls, although staff training would be required. To ensure a smooth transition, the CEO prioritised stakeholder engagement from the outset. Seminars were held to explain digital twin concepts to doctors, nurses, and administrative staff, providing a platform to gather feedback. The CEO also emphasised the potential for the digital twin to contribute to sustainable healthcare delivery by optimising resource use, reducing waste, and improving patient access to care (Lewis, 2024). Some staff expressed skepticism, raising concerns about potential job losses and the technology's suitability for a rural setting. However, the CEO emphasised that the digital twin was a tool to empower staff, freeing them from time-consuming manual processes and allowing them to focus on patient care (Jousset *et al.*, 2023).

Following extensive discussions with the Research Chair of the Johannesburg Business School, the CEO entered into a partnership agreement with the JBS team to co-create a custom digital twin solution tailored to the specific needs of Chiremba Hospital. This collaborative approach, informed by thorough research and stakeholder engagement, paved the way for implementing a digital twin that could potentially transform Chiremba's healthcare delivery and contribute to a more sustainable healthcare system (Vallée, 2023).

Implementation of the digital twin

The partnership between Chiremba Hospital and the JBS 4IR Research Laboratory propelled the digital twin implementation into action. With a well-defined project plan, strategic stakeholder engagement, and meticulous attention to detail, the hospital embarked on a journey to transform its operational landscape. The implementation steps followed by the hospital's management team are depicted in Figure 6.1. The diagram provides a visual representation of the sequence of steps involved in the implementation of digital twin technology in the hospital, showcasing the systematic approach and iterative nature of the model.

Planning and collaboration

The CEO, in collaboration with the JBS team, developed a comprehensive project plan, outlining key phases and milestones.

- **Data collection and mapping:** Data were meticulously gathered on existing patient flows, resource utilisation patterns, and medicine inventory levels. This provided a baseline for comparison and informed the customisation of the digital twin platform.

- **Digital twin development**: The JBS team tailored their digital twin platform to Chiremba's unique layout and processes. This involved creating a virtual replica of the hospital, incorporating real-time data feeds from various sources, allowing the hospital to simulate and analyse its operations in a dynamic virtual environment.
- **Staff training:** Recognising that the success of the implementation hinged on staff adoption, comprehensive training sessions were conducted. These sessions focused on data input, interpretation of the digital twin's analysis, and utilisation of the generated recommendations.
- **Pilot phase:** To mitigate risks and identify potential issues, a pilot implementation was initiated in a selected department. This allowed for testing and refinement of the system before a wider rollout.
- **Full rollout:** Following the successful pilot, the digital twin was gradually expanded throughout the hospital. Continuous monitoring and adjustments were made to optimise its performance and address any challenges encountered.

Stakeholder engagement and operational challenges

Recognising that successful implementation relied on stakeholder buy-in, the CEO maintained open communication channels throughout the process. Regular meetings with staff fostered a sense of ownership and provided opportunities to address concerns, gather feedback, and celebrate milestones. A patient education initiative was launched to explain the changes, emphasising the goal of reducing wait times and improving quality of care. The digital twin brought about transformative operational changes, enhancing the hospital's sustainability on multiple fronts.

Real-time patient tracking

By leveraging Bluetooth low energy (BLE) beacons embedded in patient bracelets and mesh beacons strategically placed throughout the hospital, the system tracked patients' movements throughout their hospital journey in real time. This provided a continuous stream of data on patient location, wait times at each stage (registration, consultation, lab, pharmacy, etc.), and overall time spent in the hospital. This highlighted bottlenecks and areas of delay, enabling staff to address issues proactively and streamline patient flow, leading to reduced wait times and improved patient satisfaction.

Optimised resource allocation

Analysing medicine stock patterns through the digital twin's predictive algorithms led to more accurate demand forecasting. This allowed for efficient inventory management, minimising stockouts of essential medications and reducing wastage owing to expiry. The system also enabled automated monitoring and control of storage conditions (temperature, humidity, light intensity) in real time, ensuring that medications remained chemically stable and effective throughout their shelf life.

Data-driven decision-making

The digital twin's real-time dashboards and analytics provided the CEO and managers with a comprehensive view of hospital operations. This data-driven approach allowed for prompt identification and swift resolution of bottlenecks, leading to more efficient patient scheduling, resource allocation, and staffing decisions. By visualising patient flow, identifying peak hours, and anticipating resource needs, the hospital could allocate staff and deploy resources more efficiently, leading to improved overall operational performance.

Challenges and mitigation strategies

The implementation was not without its challenges. Initial technical glitches caused some frustration, but these were swiftly resolved with the support of the JBS team. Additionally, changing long-standing work practices required ongoing support and encouragement, particularly for less tech-savvy staff. Targeted training programmes and one-on-one assistance helped address these concerns, ensuring that all staff members felt comfortable and confident using the new system. By proactively identifying and addressing these challenges, Chiremba Hospital successfully integrated the digital twin into its daily operations, thereby putting the hospital firmly on track towards achieving its goals of reduced wait times, optimised resource utilisation, and ultimately enhanced quality of care for patients.

Results and impact

Figure 6.2 illustrates the configuration and layout of the hospital digital twin implementation, including system architecture and users. The implementation of the digital twin at Chiremba Hospital yielded measurable improvements across key result areas, demonstrating the transformative power of data-driven solutions.

The BPM diagram in Figure 6.3 indicates the key events that were identified for incorporating into digital twin and simulation model. Event logs are created at start and end of each key event, in real time. These logs are process mined for data collation, process monitoring, advanced analytics and processing to generate data-driven decision-support. Figure 6.4 provides a detailed view of the various components making up the hospital digital twin. In addition, it also indicates the flow of operational data and of information for management and decision-support.

Quantitative and qualitative outcomes

As detailed in Table 6.1, the digital twin's impact was evident in the hospital's key areas and performance metrics.

Patient wait times

The implementation resulted in a significant reduction in average patient wait times. Before the digital twin, patients often waited for hours, with an average wait time of 120 min across all processes, from registration to discharge. After implementation, this wait time plummeted by over 46%, to just 65 min. This dramatic improvement meant that more patients could be treated within the same timeframe, effectively increasing the hospital's outpatient capacity without requiring additional physical resources or staff. This not only enhanced the hospital's operational efficiency but also contributed to social sustainability by ensuring equitable and timely access to healthcare for all patients.

Resource allocation

Just within the first year after implementation, the hospital observed a 14% reduction in medicine wastage owing to expiry, saving an estimated US $4 500 annually. The digital twin's predictive algorithms, which analysed historical consumption patterns and patient demographics, optimised stock levels, virtually eliminating stockouts of essential medications. This ensured patients consistently received necessary treatments, avoiding disruptions in care or the need to seek costly alternatives elsewhere.

Decision-making

Real-time insights provided by the digital twin revolutionised decision-making. For example, during peak hours, administrators could instantly identify bottlenecks in patient flow and reassign staff to alleviate congestion. This agility led to a smoother patient experience and reduced the burden on overworked staff. Furthermore, the data-driven approach optimised administrative staff scheduling, ensuring that registration and processing times were aligned with patient flow, effectively eliminating these steps as bottlenecks.

Stakeholder feedback

These quantitative gains were mirrored by positive qualitative feedback.

Staff

Nurses and doctors reported a significant decrease in administrative burdens, allowing them to dedicate more time to patient care. The initial apprehension surrounding the technology faded as staff experienced firsthand its positive impact on their day-to-day workflow.

Patients

Shorter wait times and a more consistent availability of medication led to noticeably improved patient satisfaction. One patient, Mrs. Bande,[3] shared, 'Before, I would spend an entire day at the hospital, even for a simple check-up. Now, it's much faster, and I can get back to my family sooner.' The CEO of Chiremba Hospital hailed the digital twin implementation as a positive transformation. 'The digital twin has been a game-changer for us, enabling us to make the most of our limited resources,' the CEO explains. 'It not only solved our most pressing challenges, but also empowered our staff and fundamentally changed how we manage our hospital. We have become more efficient, more responsive, and ultimately we are able to serve our surrounding communities better.' The medical staff echoed this sentiment, with one doctor stating, 'The digital twin has given us the tools to focus on what we do best – providing quality care to our patients.' Doctors and nurses were unanimous in acknowledging an improvement in their ability to deliver care owing to fewer stockouts and smoother patient flow. Even patients, who initially were simply grateful for shorter waits, started to appreciate that technology was playing a role in the betterment of their local hospital.

Discussion of multiple perspectives

The successful implementation of the digital twin at Chiremba Hospital was met with overall enthusiasm, but not without some initial apprehensions and varying viewpoints among different stakeholders. Diverse perspectives surrounded this significant technological shift.

Staff perspectives

While the digital twin implementation at Chiremba did not result in any job losses, concerns about technology potentially replacing human tasks remained a valid consideration.

[3] Not her real name.

Some staff members, particularly those less familiar with technology, voiced concerns about the potential for automation to replace human tasks. To proactively address this, the hospital leadership engaged in open dialogues with staff, assuring them that the technology was designed to augment, not replace, their roles. Furthermore, the hospital invested in upskilling programmes, providing staff with training on how to utilise the digital twin's insights to enhance their work. This focus on reskilling not only alleviated concerns about job security but also empowered staff to embrace the technology as a tool for improving patient care.

Patient perspectives

The digital twin implementation not only impacted hospital staff but also significantly influenced the experiences and perceptions of patients, with both positive and mixed reactions observed. While the majority of patients were delighted with the reduced wait times and improved access to medication, the introduction of the digital twin and its associated technologies was not without initial hesitations among some patients. Older patients, particularly those less familiar with technology, expressed apprehension about the tracking bracelets. Some were concerned about the discomfort of wearing them, while others worried about potential health effects. To address these concerns, the hospital staff engaged in proactive communication and education. They explained the purpose and functionality of the bracelets in simple terms, emphasising their role in improving patient flow and care delivery. Demonstrations were provided to show how the bracelets worked and to alleviate any fears about potential harm. Additionally, staff members were trained to provide personalised assistance to patients who needed help wearing or adjusting their bracelets. These efforts fostered a sense of trust and understanding, leading to wider acceptance of the tracking technology.

Beyond the bracelets, the hospital also took steps to ensure that the digital twin's user interface was intuitive and accessible to patients with varying levels of technological literacy. Clear instructions and user-friendly displays were implemented to guide patients through the registration process and provide updates on their wait times and appointments. These measures contributed to a smoother patient experience and enhanced the overall accessibility of the hospital's services.

Balancing efficiency with patient-centred care

While the digital twin promised to optimise hospital operations and improve efficiency, the leadership remained mindful of the potential for technology to overshadow individual patient needs. To strike a balance between efficiency and patient-centred care, the hospital established clear guidelines for using the digital twin's insights. These guidelines emphasised that the final decision regarding patient care rested with the healthcare professionals, not the algorithm. The digital twin was seen as a powerful tool to inform and support decision-making, not to replace the expertise and judgement of the medical staff.

Data privacy concerns and solutions

The hospital leadership understood the sensitivity of patient data, especially in a rural context where awareness of data protection might be limited. To ensure patient confidentiality and trust, Chiremba Hospital implemented a robust data protection plan, adhering to international best practices and local regulations. This plan outlined strict access controls, encryption protocols, and data anonymisation procedures to ensure that patient data remained confidential and secure. Additionally, the hospital conducted community outreach programmes to educate patients about their data rights and the measures taken to protect their privacy. This transparency fostered trust and encouraged patient participation in the digital twin initiative.

Ethical considerations

The optimisation of hospital processes through the digital twin raised ethical questions regarding the potential conflict between efficiency and individual patient needs. While the system aimed to improve overall care delivery, there was a risk that its recommendations might not always align with the nuanced needs of specific patients (Kouri *et al.*, 2022). To mitigate this, the hospital established clear guidelines for using the digital twin's insights, emphasising that the final decision regarding patient care rested with the healthcare professionals. The digital twin was positioned as a tool to inform decision-making, not dictate it, ensuring that patient-centred care remained the priority.

Cultural considerations

The implementation of the digital twin was carefully considered within the local cultural context. The hospital leadership recognised the importance of respecting and incorporating traditional health beliefs and practices prevalent in the community. In this context, it was noted that some community members held the belief that certain illnesses were caused by evil spirits, and often sought remedies from traditional healers, alongside modern medicine. The long wait times previously experienced at the hospital had discouraged some individuals from utilising its services, leading them to rely solely on traditional methods. However, the implementation of the digital twin and the resulting improvements in efficiency and patient flow helped to change this perception. As the hospital became more efficient, it regained the trust of the community, who appreciated that the technology did not dismiss their cultural beliefs, but rather complemented existing practices by providing faster and more accessible care. This cultural sensitivity was fostered through engagement with community leaders and traditional healers, who were consulted throughout the implementation process. Their insights helped shape patient education materials, which incorporated elements of traditional knowledge systems alongside explanations of the digital twin technology. This collaborative approach helped bridge the gap between modern medicine and local traditions, ensuring that the digital twin was not seen as a foreign imposition, but rather as a tool that could enhance the overall well-being of the community.

Conclusion

The implementation of a digital twin at Chiremba Hospital marked a pivotal moment in its history, transforming the way it delivered healthcare services. By harnessing the power of technology, the hospital was able to overcome long-standing challenges related to patient wait times, resource allocation, and operational efficiency. The success of the digital twin project was not merely a technological feat but a testament to the hospital's commitment to innovation, collaboration, and patient-centred care. The CEO's proactive approach in seeking solutions, the partnership with the Johannesburg Business School's 4IR Research Laboratory, and the active involvement of staff and community members throughout the implementation process all played a crucial role in the project's positive outcomes. As news of Chiremba's success spread, other healthcare facilities in the region began to take notice.

The hospital's experience sparked conversations about the potential for wider adoption of digital twin technology in similar settings across Zimbabwe and other LMICs. The CEO, reflecting on the journey, emphasised the importance of tailoring solutions to the unique needs and constraints of each context. 'While our experience with the digital twin has been transformative,' the CEO stated, 'it's important to remember that every hospital is different. What worked for us may need to be adapted for other settings.'

Key factors contributing to the project's success included the following.

- **Careful selection of partners:** Collaborating with the Johannesburg Business School's 4IR Research Laboratory ensured expertise and tailored solutions for Chiremba's unique context.
- **Focus on real operational problems:** The project prioritised addressing the most pressing issues faced by the hospital, ensuring a pragmatic and impactful implementation.
- **Continuous stakeholder engagement:** Involving staff, patients, and community leaders throughout the process fostered trust, ownership, and a higher likelihood of long-term sustainability.

Broader implications and future directions

Looking ahead, the CEO envisioned a future where Chiremba Hospital continues to explore innovative ways to leverage technology to improve healthcare delivery. The hospital planned to invest in further research to assess the long-term sustainability and scalability of the digital twin, as well as to investigate how the technology could be enhanced to meet the evolving needs of the community. The CEO also expressed a desire to share the hospital's learnings with other healthcare providers, contributing to a growing body of knowledge on how digital solutions can be harnessed to achieve the goals of SDG 3 in resource-limited contexts.

The Chiremba Hospital case study stands as a testament to the power of technological innovation to drive positive change in healthcare, even in the face of significant challenges. It underscores that when implemented strategically, collaboratively, and with cultural sensitivity, technology can be a powerful catalyst for achieving more accessible, efficient, and equitable healthcare delivery for all.

Exhibits

Figure 6.1 Digital twin implementation model

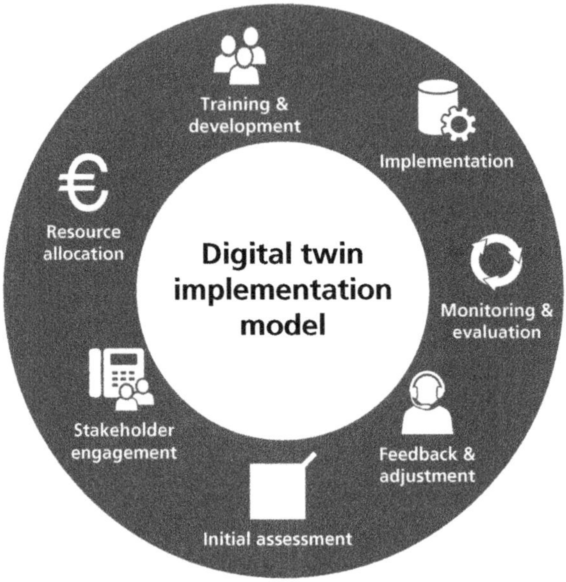

Figure 6.2 Digital twin configuration for hospital

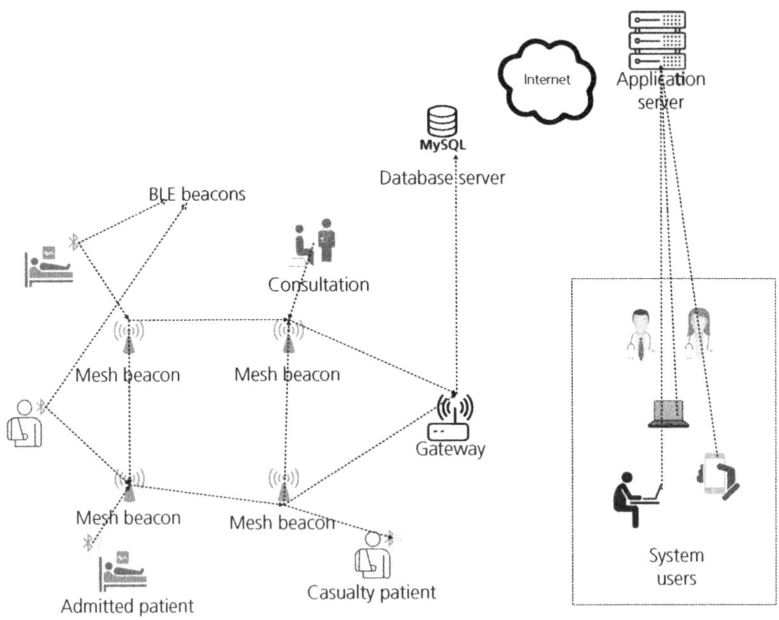

Figure 6.3 BPM key process events for hospital

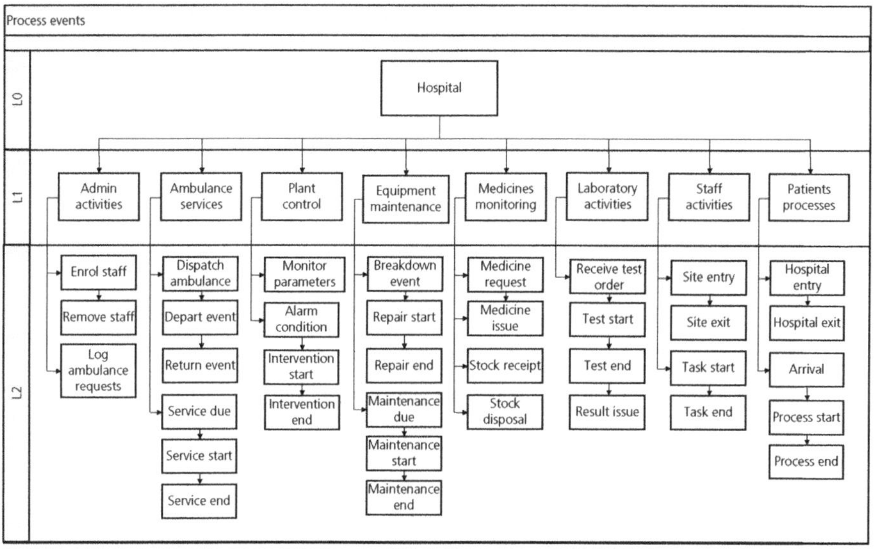

Figure 6.4 Hospital digital twin architecture

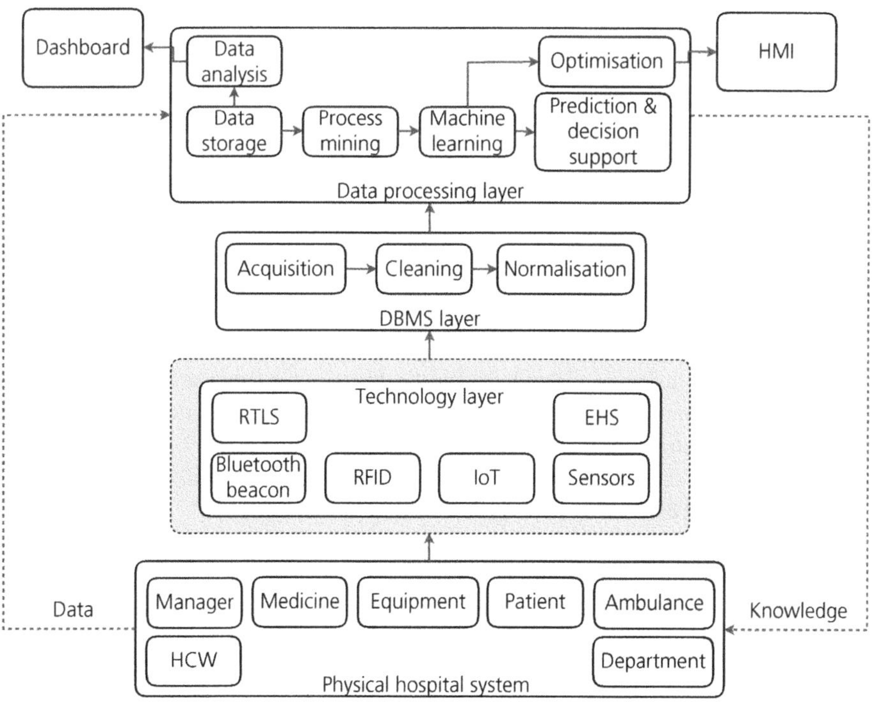

Table 6.1 Comparative data on Chiremba Hospital performance metrics

Metric	Pre-implementation	Post-implementation	Improvement
Average patient waiting time: min	122	65	47%
Resource utilisation rate: %	12.8%	17.4%	36%
Patient satisfaction level: %	37.0%	59.0%	60%
Operational cost/total revenue	TBC	TBC	–
Repeat cases: %	12.0%	4.1%	67%

Digital twins in healthcare: a pathway to sustainable hospital optimisation

Teaching notes

Synopsis

This case study presents an in-depth examination into the implementation of digital twin technology at Chiremba Hospital, a privately-owned level 1 facility in a rural district of Zimbabwe. The narrative centres around the hospital's CEO, who must tackle the task of increasing operational efficiency, reducing patient waiting times, and improving patient outcomes, all within a resource-constrained setting. As the story unfolds, the CEO confronts various challenges, including navigating the complexities of seamlessly integrating the digital twin into existing systems, managing stakeholder expectations, and working around financial constraints. This initiative represents a significant step towards technological innovation in a low- and middle-income country (LMIC), demonstrating the hospital's dedication to advancing patient care with cutting-edge, cost-effective solutions. The case delves into the strategic decision-making process, illustrating the complex dynamics of balancing technological advancement with practical healthcare delivery. It also explores the transformative impact of the digital twin on hospital operations, reducing patient waiting times, optimising resource allocation, and enhancing real-time decision-support, leading to a profound impact on the hospital's operational efficiency and quality of care. This narrative offers insights into the fusion of technology and healthcare management within resource-limited settings, underlining both the opportunities and challenges of implementing digital health solutions in LMICs.

Teaching objectives

This case study is primarily targeted at graduate-level students in healthcare management, business administration, or related fields with a specialisation in healthcare innovation or technology management. It is also highly relevant for executives and professionals participating in continuing education programmes focused on healthcare technology, emerging market strategies, digital transformation in healthcare, and leadership in health services. Aligned with Bloom's taxonomy, the teaching objectives are structured to guide students from understanding and analysis to application and evaluation, ensuring a comprehensive grasp of the complexities involved in adopting digital health technologies in LMICs. The case study blends theoretical knowledge with practical insights, preparing students for real-world application and decision-making in healthcare technology and management. The specific teaching objectives of the case study are as follows.

- Develop students' ability to evaluate the balance between technological advancement and practical healthcare management, leading to informed judgements about technology integration in similar settings. Students will critically analyse the challenges and strategies involved in implementing innovative healthcare technologies in resource-constrained environments.
- Enhance students' understanding of strategic decision-making processes in healthcare. Students will apply leadership skills in stakeholder management and resource allocation, essential for driving technological change in healthcare settings.
- Develop competencies in evaluating how technology can enhance operational efficiency, patient outcomes, and the overall sustainability of healthcare delivery. This will prepare students to make sound evidence-based recommendations in healthcare operations. Students should be able to assess the transformative impact of digital twin technology on healthcare delivery.
- Equip students with change management strategies for integrating new technologies into existing healthcare systems. They will learn to devise and implement training programmes for staff, facilitating smooth technological transitions in healthcare environments.
- Engage the students in the critical evaluation of digital health solutions' applicability in emerging markets. Students will be expected to assess scalability and replicability and formulate strategies for addressing healthcare challenges in diverse global contexts using digital health solutions.

Research methods

The case study was constructed using a mix of primary and secondary research methods. Primary research involved in-depth interviews with the hospital's CEO, key staff, and patients, providing insights into the decision-making process and operational impacts of implementing digital twin technology. Direct observations and analysis of hospital documents complemented these insights. Secondary research included a review of industry reports, academic journals, and news articles on digital twin technology in healthcare, focusing on its application in emerging markets. This approach ensured a balanced perspective, combining firsthand experiences with broader industry and academic context.

Keywords and subject code

Digital twin, healthcare management, technological innovation, lean healthcare, low- and middle- income countries, leadership in healthcare.

Assignment questions

1. Assess the CEO's decision-making process in implementing digital twin technology. Consider factors like stakeholder management, resource allocation, and anticipated operational challenges. Discuss how these factors influenced the final decision and the approach taken.
2. Analyse how the introduction of digital twin technology impacts the hospital's operational efficiency and quality of patient care. Explore how this technology addresses specific challenges in the healthcare sector of an LMIC such as Zimbabwe.
3. Critically assess the risks associated with implementing digital twin technology in a resource-constrained environment. Propose strategies for risk mitigation and discuss how these strategies could lead to successful implementation.
4. Discuss the challenges and strategies involved in integrating digital twin technology within the existing hospital systems. How should the hospital manage staff training and adaptation to ensure a smooth transition?
5. Evaluate the long-term sustainability and scalability of implementing digital twin technology in a healthcare setting such as a level 1 district hospital situated in a resource-limited environment. Consider the economic, social, and technological factors that might influence its ongoing success and potential expansion.

Additional materials

The following additional materials are suggested for students to access in preparation for the lecture. These materials enhance the learning experience and provide a comprehensive understanding of the case context.

YouTube clip: A detailed video explaining the concept of digital twin technology and its applications in healthcare. This clip will provide a foundational understanding of the technology central to the case study.

PixelPlex Inc. (Director) (2023) What is a digital twin in healthcare? https://www.youtube.com/watch?v=YMZV2y0NaiE (accessed 14/01/2025).

Interactive web resource: Access to an interactive website like Gapminder, which offers data visualisation tools. These tools can help students understand the healthcare challenges in emerging markets like Zimbabwe and the potential impact of technological solutions.

Health | Gapminder (n.d.) https://www.gapminder.org/tag/health/ (accessed 12/01/2024).

Case study simulation: A digital twin simulation module where students can engage in decision-making scenarios similar to those faced by the hospital's CEO. This interactive element will allow students to apply their theoretical knowledge in a practical context.

Suggested core readings

To further deepen instructors' and students' understanding of the topics related to the case, the following readings are recommended.

Bode M, Goodrich T, Kimeu M *et al.* (2020) Unlocking digital healthcare in lower- and middle-income countries. Mckinsey & Company. https://www.mckinsey.com/industries/healthcare/our-insights/unlocking-digital-healthcare-in-lower-and-middle-income-countries (accessed 12/01/2024).

Global Strategy on Digital Health 2020–2025 (2021) *World Health Organization.* https://www.who.int/publications-detail-redirect/9789240020924 (accessed 12/01/2024).

Laksono S and Darmawan ES (2021) The new leadership paradigm in digital health and its relations to hospital services. *Journal Ilmu Kesehatan Masyarakat* **12(2)**: article 2, 10.26553/jikm.2021.12.2.89-103.

Ngoc Dinh M, Nygate J, Hoang Minh Tu V and Thwaites CL (2020) New technologies to improve healthcare in low- and middle-income countries: Global Grand Challenges satellite event, Oxford University Clinical Research Unit, Ho Chi Minh City, 17–18 September 2019. *Wellcome Open Research* **5**: 142, 10.12688/wellcomeopenres.16008.2.

Shilpa KT (2022) Digital healthcare: current trends, challenges and future perspectives. In *Proceedings of the Future Technologies Conference (FTC) 2021, Volume 2. FTC 2021. Lecture Notes in Networks and Systems* (Arai K (ed.)). Springer, Cham, Switzerland, vol. 359, 10.1007/978-3-030-89880-9_48.

Stoumpos AI, Kitsios F and Talias MA (2023) Digital transformation in healthcare: technology acceptance and its applications. *International Journal of Environmental Research and Public Health* **20(4)**: 3407, 10.3390/ijerph20043407.

Teaching plan and timing

This teaching plan outlines a 90-min session structured to maximise student engagement and learning from the case study. Instructors are advised to give the case study to students for pre-reading before they attend the class. The case introduces complex concepts that may require background understanding, and additional reading materials provided in the teaching notes will support this preparatory work. The rationale for this approach is to optimise classroom time for discussion, analysis, and the application of concepts, rather than for reading the material. Pre-reading ensures that students arrive for the class with a foundational grasp of the case, prepared to go deeper into its analysis and to engage in meaningful discussions.

Introduction and context setting (15 min)

- Begin with a brief overview of the hospital's situation and the challenges faced in the healthcare sector in emerging markets. Introduce the central figure, the hospital's CEO, and the decision to implement digital twin technology. Engage students by asking a key question related to the case, such as: 'What are the primary challenges in implementing new technology in a resource-limited healthcare setting?' Collect responses to gauge initial understanding and perspectives.

Video presentation and discussion (10 min)

- Show a YouTube clip from the additional materials that explains digital twin technology in healthcare. Follow this with a short discussion on how such technology could be a game-changer in healthcare settings, particularly in emerging markets.

Analysis of the CEO's decision-making process (20 min)

- Delve into the first assignment question focusing on the CEO's strategic decision-making. Present a relevant decision-making framework from the teaching notes. Facilitate a class discussion on how the case details align with this framework and what other factors might influence decision-making in similar contexts.

Impact on operational efficiency and patient care (20 min)

- Address the second assignment question in a group activity. Students will work in small groups to analyse how digital twin technology affects operational efficiency and patient care, using real data and examples from the case. Select a few groups to present their findings, fostering a critical discussion on their analysis and viewpoints.

Risk assessment and interactive problem-solving (20 min)

- For the third and fourth assignment questions, engage students in a problem-solving exercise. They will assess potential risks and devise mitigation strategies. Encourage students to consider various angles – technological, financial, human resource – and present a holistic action plan. Use this segment to highlight the practical applications of theoretical knowledge.

Conclusion and key learnings (5 min)

- Conclude the session by summarising the key learning points. Discuss the actual outcomes of the case as stated in the postscript. Emphasise the importance of strategic decision-making, risk management, and the impact of technological innovations in healthcare, especially in emerging market conditions.

This teaching plan is designed to offer a balance of theoretical and practical learning, encouraging active participation, critical thinking, and application of concepts. Instructors can adjust the timing and focus of each section to suit their teaching style and the class's dynamics.

Analysis of assignment questions

Question 1: Evaluating strategic decision-making

Model answer: The CEO's approach in adopting digital twin technology reflects a complex strategic decision-making process. Key considerations include stakeholder management, resource allocation, and anticipating operational challenges. Utilising frameworks such as Kotter's eight-step change model (Kotter, 2012), the CEO's approach can be dissected into stages: establishing a sense of urgency, creating a guiding coalition, developing a vision and strategy, and communicating this vision. The case reveals that stakeholder management involved balancing the interests of healthcare professionals, patients, and financial backers. Resource allocation was particularly challenging, considering the limited funds and the need for a high return on investment (ROI). Anticipating operational challenges involved understanding the hospital's current limitations and the potential scalability of the technology. Students can be encouraged to analyse how different decision-making models apply to this scenario and how the CEO's approach aligns with theoretical best practices.

Possible student responses

- A student might argue that the CEO could have adopted a more participatory approach in decision-making, involving more staff members in the process. In response, the discussion can revolve around the pros and cons of participatory decision-making in high-stakes, technology-driven changes in healthcare settings.
- Others might question the prioritisation of digital twin technology over other pressing needs, such as additional staff or equipment. This could lead to a debate about resource allocation in resource-limited settings and the role of innovation in addressing systemic challenges.
- Students could also explore alternative decision-making models, such as the rational model or the incremental model, and discuss their applicability to this case.

Question 2: Impact on operational efficiency and patient care

Model answer: The implementation of digital twin technology in the hospital is expected to significantly enhance operational efficiency and patient care. This can be analysed through the lens of the technology acceptance model (Davis, 1989), which suggests that perceived usefulness and ease of use are fundamental in technology adoption. In this case, the digital twin technology directly addresses long wait times, a key issue in patient care, and enables more efficient resource allocation, a critical aspect of operational efficiency. Students can explore how the hospital's adoption of technology aligns with the model's parameters and the specific challenges it addresses in an LMIC like Zimbabwe.

Possible student responses

- Some students might argue that focusing on technology alone is insufficient and that addressing underlying systemic issues, such as staff shortages or inadequate infrastructure, is equally important. This could lead to a discussion about the role of technology as a tool for improvement rather than a panacea.
- Others might question the long-term sustainability of the technology, especially given the hospital's limited resources. This could spark a debate about the need for ongoing investment and support to ensure the continued success of the digital twin.
- Students might also suggest alternative technologies that could have been adopted. The discussion can then focus on comparing the benefits and drawbacks of different technologies in such a healthcare setting.

Question 3: Risk assessment and mitigation strategies

Model answer: The risks associated with implementing digital twin technology in a resource-constrained environment include technical challenges, staff resistance, and potential financial overruns. Risk mitigation strategies involve comprehensive training for staff to ensure smooth technology integration and adoption, phased implementation to monitor and manage financial outflows, and technical partnerships to address potential technical hurdles. This analysis can be grounded in Lewin's change management model (Lewin, 1947), focusing on unfreezing current processes, making changes, and then refreezing to solidify these changes. Students can be encouraged to assess these risks and strategies critically, considering the unique context of a resource-limited healthcare setting.

Possible student responses

- Students could propose a more proactive approach to risk management, such as conducting a thorough risk assessment before implementation and developing contingency plans for potential setbacks.
- Others might suggest involving external stakeholders, such as community leaders or technology experts, in the risk assessment and mitigation process.
- Students might propose alternative mitigation strategies, such as exploring the role that government policy and regulations could play in mitigating risks associated with adopting digital health technologies in LMICs, or they could propose seeking more substantial government support or international aid as alternative mitigatory strategies. Discussions can then delve into the feasibility and implications of these alternatives.

Question 4: Technology integration and staff adaptation

Model answer: Integrating digital twin technology into the existing hospital systems presents several challenges, primarily in staff training and adaptation. The technology-organisation-environment (TOE) framework (Tornatzky and Fleischer, 1990) can be used to analyse these challenges. Technologically, the hospital needs to ensure compatibility with existing systems and ease of use. Organisationally, there is a need for effective change management, as identified by Kotter's model, to encourage staff buy-in and minimise resistance. Environmentally, considering the external factors like healthcare regulations and market readiness for such technology is crucial. Staff training must be comprehensive, focusing not just on the technical aspects but also on the value addition of this technology to their daily operations and patient care. Students can discuss different training and adaptation strategies, considering the hospital's context and resources.

Possible student responses

- A student might propose a more gradual, phased approach to technology integration to mitigate resistance. The discussion could then focus on the trade-offs between a phased approach and a more rapid implementation.
- Students might suggest alternative approaches to staff training, such as mentorship programmes or ongoing support from the technology provider.
- They could also discuss the importance of creating a culture of innovation within the hospital to encourage staff to embrace new technologies.

Question 5: Long-term sustainability and scalability

Model answer: Assessing the long-term sustainability and scalability of digital twin technology in the hospital involves examining economic, social, and technological factors. Economically, the ROI and funding models need to be sustainable. Socially, the technology's impact on patient care quality and staff satisfaction are vital. Technologically, the adaptability of the digital twin system to future healthcare needs and its compatibility with emerging technologies are crucial. The diffusion of innovations theory (Rogers, 1962) can be applied here to understand how this innovation might be accepted and scaled in the healthcare sector. Students should analyse the potential for this technology to be replicated in similar healthcare settings, considering the successes and challenges faced by the district hospital.

Possible student responses

- Students might question the feasibility of replicating such a technology in other resource-limited settings. This can lead to a discussion on the variables that affect the scalability of healthcare innovations in different environments.
- Students might question the feasibility of replicating the model in other resource-limited settings, considering local infrastructure, funding, and political challenges.

- They could propose alternative funding models, such as public–private partnerships or international aid, to ensure sustainability.
- Students could also discuss the importance of local ownership and capacity building to ensure the long-term success of digital health solutions.

Key learnings

The Chiremba Hospital case study narrates the journey of a rural Zimbabwean hospital facing operational challenges, including long patient wait times, inefficient resource allocation, and the need to improve patient outcomes. It focuses on the CEO's decision to implement a digital twin solution, detailing the exploration, implementation, and initial results of this initiative. The case highlights the challenges and opportunities of adopting innovative technology in a resource-limited setting and the impact on patient care, resource management, and stakeholder engagement. Specifically, the case study offers several key learnings that are vital for students and practitioners in the field of healthcare management and technology.

- **Strategic decision-making in resource-limited environments:** This case discusses the complexities of decision-making in resource-constrained healthcare settings. It emphasises the need to balance stakeholder interests, resource allocation, and operational challenges effectively. Decision-making models like Kotter's eight-step change model offer valuable frameworks for navigating these decisions.
- **Impact of technological innovation on healthcare delivery:** This case provides a clear example of how technological innovations, like digital twin technology, can transform healthcare delivery. It shows the potential for such technologies to enhance operational efficiency and improve patient care, particularly in LMICs facing unique challenges.
- **Change management and technology adoption:** The case underlines the critical role of change management in successfully adopting new technologies. It highlights the importance of aligning technological solutions with organisational goals, staff training, and adaptation. The TOE framework is particularly useful in guiding this process.

Postscript

The implementation of digital twin technology at the district hospital has led to significant advancements. Operational efficiency has markedly improved, with reduced patient waiting times and better resource allocation, thanks to the technology's capability for real-time operational tracking and management. This has resulted in a more efficient healthcare delivery system. The hospital's CEO reports increased patient satisfaction and decreased operational bottlenecks. Staff members, initially hesitant, adapted to the new technology following comprehensive training and support. Economically, the hospital has achieved cost savings through improved allocation of its limited resources. The technology's impact has extended beyond the hospital, providing a broader understanding of digital health solutions' potential to improve healthcare delivery in resource-limited settings. The technology's success has garnered interest from other regional healthcare providers, sparking discussions about its scalability and replication in similar settings.

REFERENCES

Davis FD (1989) Perceived usefulness, perceived ease of use, and user acceptance of information technology. *MIS Quarterly* **13(3)**: 319–340, 10.2307/249008.

Deloitte (2021) Digital Twins – Bridging the Physical and Digital. https://www2.deloitte.com/us/en/insights/focus/tech-trends/2020/digital-twin-applications-bridging-the-physical-and-digital.html (accessed 12/07/2024).

Jousset O, Kimeu M, Sforza G *et al.* (2023) *How Digital Tools Could Boost Efficiency in African Health Systems.* Mckinsey & Company. https://www.mckinsey.com/industries/healthcare/our-insights/how-digital-tools-could-boost-efficiency-in-african-health-systems (accessed 12/07/2024).

Kotter JP (2012) *Leading Change.* Harvard Business Review Press, Boston, MA, USA.

Kouri P, Kaija-Kortelainen M, Kennedy M and Turjamaa R (2022) *Ethical Considerations in Digital Health.* Routledge, Abingdon, UK.

Lewin K (1947) Frontiers in group dynamics: concept, method and reality in social science; social equilibria and social change. *Human Relations* **1(1)**: 5–41.

Lewis K (2024) Digital Twin Technology in Healthcare: Current & Future Uses. https://www.linkedin.com/pulse/digital-twin-technology-healthcare-current-future-uses-kevin-lewis-dwkcc?trk=public_post (accessed 12/07/2024).

Mwanza J, Telukdarie A and Igusa T (2023) Impact of Industry 4.0 on healthcare systems of low- and middle-income countries: a systematic review. *Health and Technology* **13(1)**: 35–52, doi: 10.1007/s12553-022-00714-2.

Rogers EM (1962) *Diffusion of Innovations.* Free Press, New York, NY, USA.

Tornatzky LG and Fleischer M (1990) *The Processes of Technological Innovation.* Lexington Books, Lanham, MD, USA.

United Nations (2015) *Transforming Our World: The 2030 Agenda for Sustainable Development.* United Nations Department of Economic and Social Affairs. https://sdgs.un.org/2030agenda (accessed 14/01/2025).

Vallée A (2023) Digital twin for healthcare systems. *Frontiers in Digital Health* **5**: 1253050, 10.3389/fdgth.2023.1253050.

Winter T (2023) Digital twins in healthcare: a holistic approach to predictive human maintenance. *TechUK.* https://www.techuk.org/resource/digital-twins-in-healthcare-a-holistic-approach-to-predictive-human-maintenance.html (accessed 14/01/2025).

World Health Organization (2019) WHO Guideline: Recommendations on Digital Interventions for Health System Strengthening. Geneva: Licence: CC BY-NC-SA 3.0 IGO. https://iris.who.int/bitstream/handle/10665/311941/9789241550505-eng.pdf?ua=1 (accessed 14/01/2025).

THEME 2

SUSTAINABLE BUSINESS PRACTICES

Prof. Arnesh Telukdarie
ISBN 978-1-83662-645-9
https://doi.org/10.1108/978-1-83662-642-820251007

Chapter 7
Business process for digital sustainability

Arnesh Telukdarie, Megashnee Munsamy, Regionald Mongwe, Christian Tshukudu and Tatenda Katsumbe

Introduction

The CEO of the Chemical Industries Education and Training Authority (CHIETA), a South African services sector organisation, led the organisation to a pivotal moment in its organisational evolution in January 2022. This pivotal moment entailed spearheading a digital transformation roadmap to inform digital business status, in line with the CHIETA business strategies for the current five-year cycle. These strategies are critical to the CHIETA's business landscape at the strategic and operational levels and entail innovation, information communication and technology (ICT), risk management, monitoring and evaluation (M&E), digital transformation (DT), post-learnership absorption, regional outreach, sector skills planning, and stakeholder management. Through this digital transformation roadmap, the CHIETA seeks to catalyse the transition of the organisation towards being a digital and intuitive enterprise that fully leverages the technological merits of Industry 4.0. The full leverage of Industry 4.0's technological merits, informed by the envisaged digital transformation roadmap, further positions the CHIETA to align with Sustainable Development Goal (SDG) 8, which is premised on promoting sustained, inclusive, and sustainable economic growth, full and productive employment, and decent work for all. This emanates from the CHIETA's pivotal role within the chemical industry sector in facilitating skills development and addressing skills needs. Addressing skills unleashes other co-benefits appended to the SDGs holistically, such as closing the unemployment gap, addressing the gender parity gap, South Africa's overall economic growth, and regional outreach through global partnerships. Additionally, this aligns with SDG 9, premised on promoting inclusive, sustainable industrialisation and the fostering of innovation and infrastructure. According to best practices, a crucial step in formulating a digital business strategy for an organisation is to fully comprehend the current (as-is) state of the organisation with the intent to formulate and model the envisaged (to-be) state, aligned with best practices and business objectives. This is done by hierarchical business process mapping with the associated role players. The CHIETA CEO mandated the Johannesburg Business School (JBS) research team, under the leadership of Professor Telukdarie, to map the CHIETA activities for digital interventions and identify areas for improvement.

Background profile of the CHIETA

The South African Skills Development Act 97 of 1998 formed the CHIETA as a statutory body to facilitate skills development, specifically targeting the chemical industry sector. The CHIETA facilitates skills development by identifying and addressing skills needs through diverse initiatives, including collaborative projects with the South African chemical industry sector, higher education institutions (HEIs), and consultants. The CHIETA comprises multiple business functional units, such as finance, corporate services, governance and risk, research monitoring and evaluation, and grants. Each of these functional disciplines comprises unique yet interdependent processes. These functional units are supported by various systems and tools in their execution. Full systems integration is essential for the CHIETA to effectively capitalise on business operational efficiency, thus necessitating the need for digitalisation of the various standard service functions. This augments the CHIETA's endeavours to facilitate advanced functions such as agile enterprises, data analysis, automated reporting, and artificial intelligence (AI).

Informed by the need to optimise business operational efficiency through opportunities presented by the fourth industrial revolution, the CHIETA CEO embarked on a mission to fully digitalise the CHIETA's business operations by 2030. As a result, the CHIETA collaborated with Professor Telukdarie to design the CHIETA's digitalisation strategy. The merits of JBS as a leading research institution on digital transformation motivated this collaboration. Furthermore, this collaboration illustrated both the CHIETA and JBS's commitment to sustainability and the betterment of the broader chemical industry sector. Spearheaded by Professor Telukdarie (Professor of Digital Business, JBS), the JBS team sought to design and document the digital transformation roadmap for the CHIETA accordingly. To propose a global best practice-based digital transformation roadmap for the CHIETA, a fundamental initial activity entailed the mapping of the current (as-is) CHIETA business processes. The business process mapping was executed for purposes of defining the as-is digital maturity state of the CHIETA. This allows for the development of a data-driven digital strategy roadmap for successful digital transformation towards the CHIETA's future state. This case study outlines the business process mapping of the as-is digital maturity state of the CHIETA using a structured, hierarchical, role-based methodology.

Mapping the CHIETA as-is business processes

The initial step in the development of a digital business strategy is mapping the as-is business processes for the CHIETA. This informs the analysis of intra- and inter-dependencies among business processes in order to identify areas for optimisation and bottlenecks. However, it is essential to mention that this analysis is not the scope of this case study. The focus of this case study is solely to illustrate the steps taken to map the as-is business processes of the CHIETA. Figure 7.1 illustrates the consultative and collaborative approach adopted for hierarchical business process mapping. The business process mapping was based on BPMN 2.0, a widely adopted business process mapping notation (Tsagkani and Tsalgatidou, 2022).

Each step of the approach followed will be illustrated in detail. The first step entailed engaging with the various business functional units at the CHIETA to comprehend each department's objectives and output. Documentation in the form of standard operating procedures (SOPs) or process flow maps illustrating the current business activities was also requested from the respective CHIETA business process owners to gain a full understanding of the execution procedures of the various business activities. Herein, business process owners are described as the personnel responsible for the business processes. Following an understanding of the as-is process, the processes were mapped on Microsoft Visio for business process mapping.

The different role players identified per business process were also mapped against their activities. After mapping the as-is business processes, the processes were shared with the respective CHIETA business process owners for a review. This was an iterative process to ensure that the mapped processes are a true representation of the CHIETA activities. The final mapped as-is processes were then shared with the respective CHIETA business departments for confirmation of the validity of the processes, serving as the final step of the business process mapping approach. The business process mapping is outlined in detail below.

Mapping the as-is CHIETA processes

The business process hierarchy enables the graphical representation of an organisation's business processes and the alignment of business activity tasks to business goals. It illustrates the breakdown of complex business processes, sub-processes, and process segments, each containing detailed activities (Holotiuk and Beimborn, 2017). A business is categorised into hierarchies, as per the hierarchical framework in Figure 7.2, structured from business organisation to business units (P_n) to business areas per business unit (F_n).

For this project, a four-level hierarchy was adopted to illustrate the CHIETA's process hierarchy as follows:

- Level 0: Business function
- Level 1: Sub-function
- Level 2: Business area per business function
- Level 3: Process area per business area
- Level 4: Business process activity.

Based on the adopted four-level hierarchy, the identified CHIETA business functions were: (1) human resources (HR); (2) finance; (3) governance and risk; (4) grants; (5) education and training quality assurance (ETQA); (6) research and skills planning unit (RSPU); (7) information communication and technology (ICT); (8) marketing; and (9) facilities management. Figure 7.3 illustrates the CHIETA's business processes overview map, which shows the level 0–level 2 mapped processes of the CHIETA.

The next activity entailed breaking down the identified level 0 (business function) into respective level 2–level 4 requirements. This was done through iterative meetings with CHIETA process owners and the JBS team between April 2023 and February 2024. The JBS team and the CHIETA business process owners held follow-up meetings after mapping the as-is business processes to confirm that the mapped processes accurately represented the as-is processes. It is crucial to keep the following in mind.

- The mapping and follow-up meetings were iterative, until the mapped as-is processes represented the activities of the CHIETA department.
- Despite the identification of ICT, marketing, governance and risk, and facilities management during initial engagements (Figure 7.3), these processes did not map to level 3 or 4.
- Whether a process ended at level 3 or 4 depended on the specific complexity of the business function.
- Not all level 4 activities are illustrated herein, owing to their complex level of detail.

Validation of the mapped as-is CHIETA processes

The respective business owners validated the as-is processes, establishing an accurate baseline. This was a crucial step, as the as-is processes were the baseline for the development of the CHIETA digital strategy.

Examples of the validated as-is processes

Examples of the validated as-is processes are illustrated below in the form of respective overview maps for the identified business functions, showing the level 0–3 details. Furthermore, an example of level 4 activities is illustrated, using induction to portray the interdependencies between activities and the respective role players. Figure 7.4 illustrates the overview map for human resources (HR), showing level 0–level 3 mapped processes in a hierarchical structure. The following key points are highlighted.

- Level 0 demarcates human resources at the business function level.
- Level 1 demarcates the delineation of the business function strategy in line with the overall business strategy.
- Level 2 demarcates the sub-functions of human resources, such as recruitment & onboarding, employee performance, payroll, and employee relations.
- Level 3 demarcates the business areas per business function, with each sub-function (level 2) mapped accordingly to the respective business areas per business function (level 3). For example, recruitment, induction, and termination are the business areas per business function falling under recruitment & onboarding.

Figure 7.5 illustrates the mapped level 4 activities, using induction as an example. Induction is demarcated at level 3 under the human resources business function (Figure 7.4). The level 4 activities of induction are illustrated in sequence and mapped against the respective role players such as new employee, HR business partner, and induction team. There are interdependencies between the different role players, indicated by the connecting arrows (Figure 7.5), which show information exchange between the different role players according to their respective levels of authority.

Figure 7.6 illustrates the grants overview map for grants, showing the level 0–level 3 mapped processes, in a hierarchical structure.

Figure 7.7 illustrates the mapped level 4 activities, using discretionary grants (DG) – advertisement as an example. DG – advertisement is demarcated at level 3 under the grants function (Figure 7.6). The level 4 activities of DG – advertisement are illustrated in sequence, mapped against the respective role players such as grants and projects manager, and executive manager – MANCO. There are interdependencies between the different role players, indicated by the connecting arrows (Figure 7.7), showing information exchange between the different role players according to their respective levels of authority.

Lessons learned

The sequential steps adopted to map the CHIETA business processes are illustrated with examples of graphical presentations. Thus, it is crucial to highlight the lessons learnt in this case study, inclusive of the importance of adequate knowledge gathering through preliminary engagements, informed by the relevant business process owners; the adoption of the BPMN 2.0 as a structural configuration through which the business processes of an organisation can be mapped; the importance of iterative engagements with updates for purposes of ensuring accurate mapping of the business process of an organisation; and the importance of validation by respective business process owners to endorse the mapped as-is processes and facilitate documentation. These lessons are further expanded, as shall be illustrated in the ensuing paragraphs below, and summarised in Figure 7.8.

- Through the preliminary engagements held between the JBS team and the diverse CHIETA business process owners, and enabled by the BPMN 2.0 notation, it was learnt that a business organisation can be structured according to its constitutive business functional units. According to the BPMN 2.0 hierarchical notation, these business functional elements are denoted at level 0, within this hierarchical structural manner of arrangement, with examples such as human resources, grants and finance, as illustrated in the business process overview map (Figure 7.3).
- Furthermore, through follow-up engagements between the JBS team and the CHIETA business process owners, it was learnt that each business functional unit can be further disaggregated into the constitutive sub-functions denoted at level 1, business areas per business function denoted at level 2, process areas per business area denoted at level 3, and business process activities denoted at level 4, respectively. Accordingly, Figures 7.4–7.7 are herein displayed as examples, to illustrate the pathways taken to disaggregate business functional units into business process activities, respectively.
- It was further learnt that the mapping of business processes is an iterative process, as exemplified by the numerous back-and-forth engagements between the JBS team and the CHIETA team, with updates when and as necessary, to ensure the accurate capture of the as-is business processes of the CHIETA. This accurate capture of the as-is business processes is essential for follow up steps such as simulation and analysis to identify gaps, which in turn inform the development of the digital business strategy pathway, accordingly, to address these gaps. It is crucial to mention that the simulation, analysis, and development of the digital business strategy pathway were not within the scope of this case study, hence they were not illustrated within the case study, accordingly.
- Finally, it was also learnt that the validation of the business process by the CHIETA business process owners was essential for purposes of endorsing the as-is mapped processes, and to enable documentation of the mapped processes, which in turn is essential for the follow up steps – that is, the aforementioned analysis, simulation, and digital strategy formulation.

Overall, collaboration is an essential component of business process mapping, as signified by the numerous iterative engagements between the JBS team and the CHIETA business process owners towards successfully delivering the mapped as-is CHIETA business processes.

Exhibits

Figure 7.1 As-is process mapping including validation

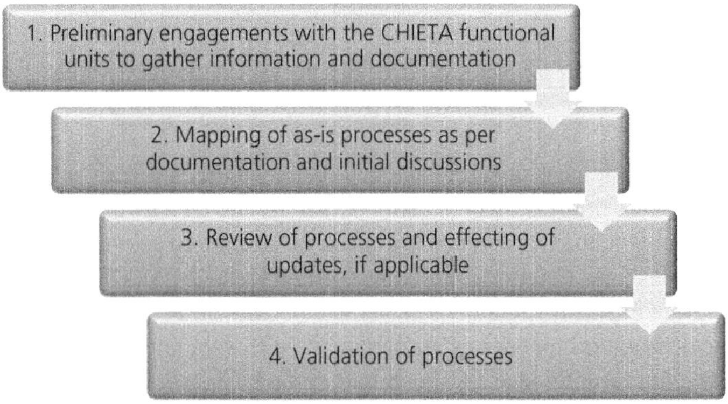

Figure 7.2 Business process hierarchy according to Harmon (2019)

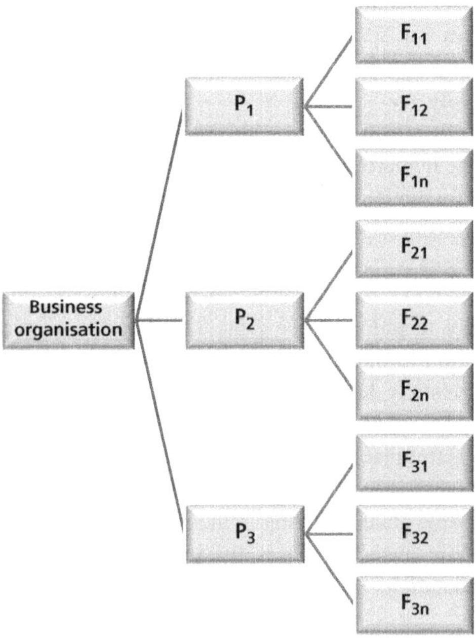

Figure 7.3 The CHIETA business processes overview map illustrating the level 0–level 2 details

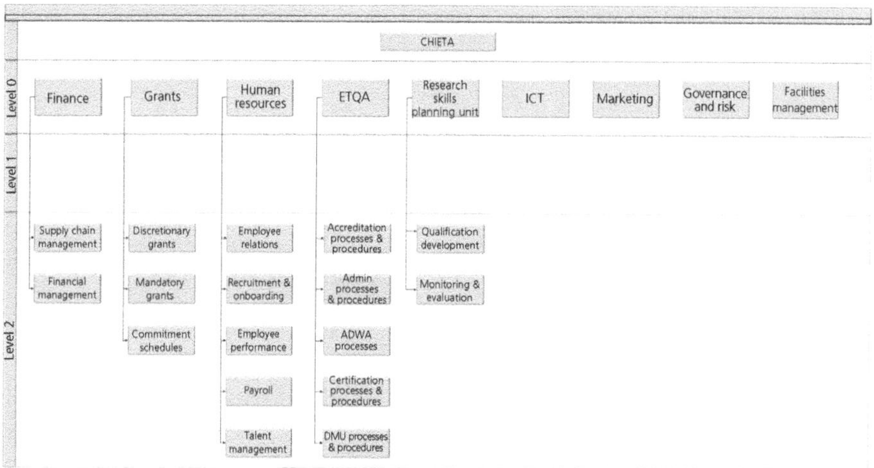

Figure 7.4 Human resources overview map illustrating level 0–level 3 details

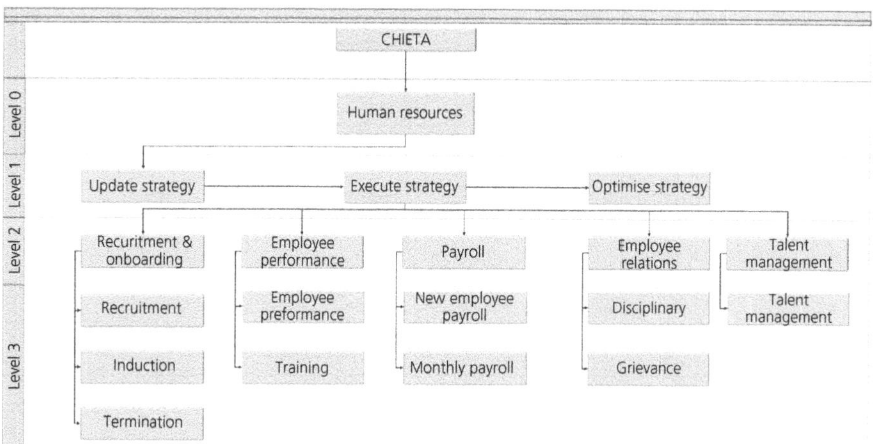

Figure 7.5 Illustration of the level 4 mapped activities, using induction as an example

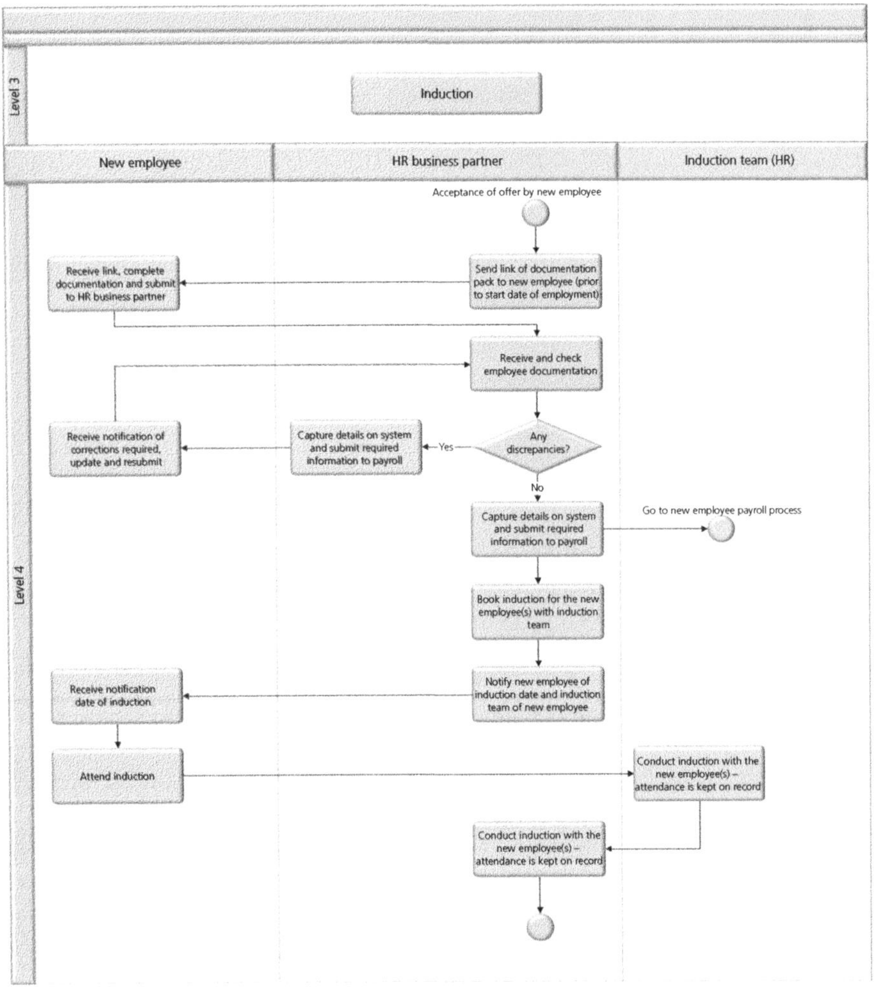

Figure 7.6 Grants overview map illustrating level 0–level 3

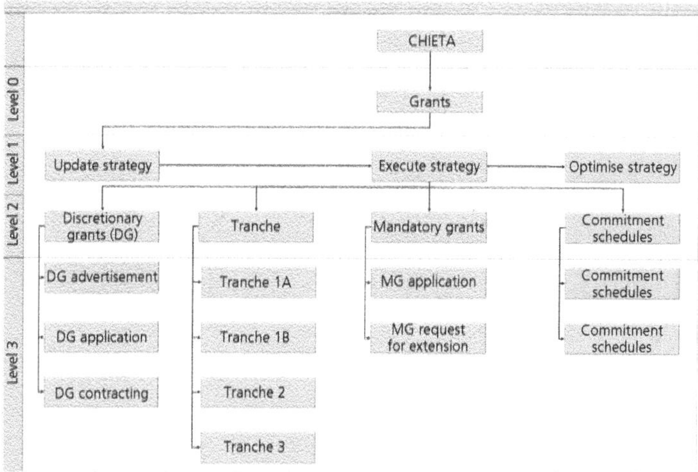

Figure 7.7 Illustration of the level 4 mapped activities, using DG advertisement as an example

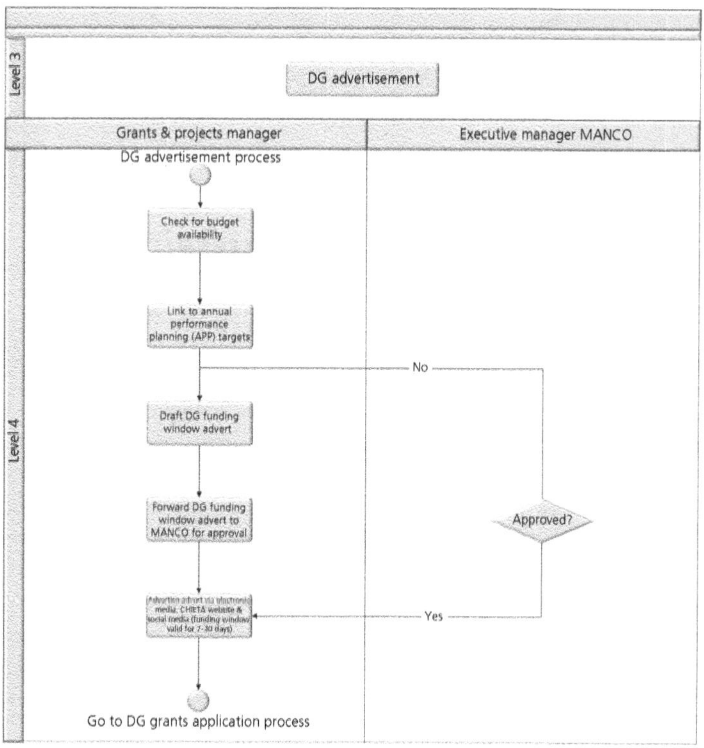

Figure 7.8 Collaboration as the key commonality in the lessons learnt

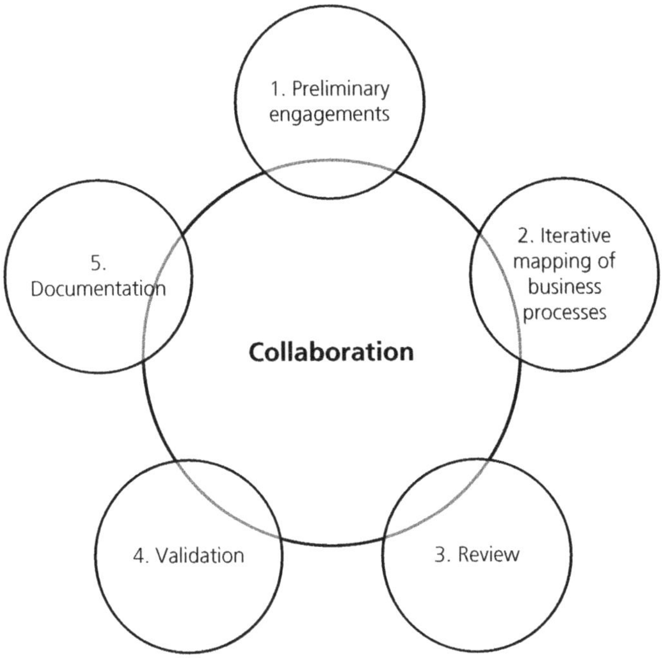

Note: The diagram is author generated, emanating from the lessons learnt in this case study.

Business process for digital sustainability

Teaching notes

Synopsis

The ongoing wave of digital transformation ignites complexities within business realms, propelling businesses to innovate their business process models towards gaining a competitive advantage. As one of the 21 South African Sector Education and Training Authority (SETA) organisations, the Chemical Industries Education and Training Authority (CHIETA) is on a digital transformation journey, emanating from the CHIETA business strategy of full digitalisation by 2030. The CHIETA appointed Professor Telukdarie, Research Chair of Digital Business at the University Johannesburg (UJ), to spearhead the development of a digital business strategy aimed at achieving an integrated and intuitive business. The initial task undertaken by the Chair was to scope and map the current CHIETA business process activities using a structured, hierarchical methodology from April 2023 to February 2024. Business process mapping was executed to comprehend the as-is state of these activities. Professor Telukdarie and his team engaged with various CHIETA business functions to map all processes, including identifying manual or system-based processes, and interdependencies. This initial task of business process mapping informs the context-based and data-driven development of a digital business strategy for the migration of the as-is to the to-be for the CHIETA. This case study seeks to outline Professor Telukdarie and his team's steps in mapping the as-is business processes for the CHIETA.

Teaching objectives

This case study is intended for use in undergraduate, MBA, and executive education courses in Business Strategy Management and Digital Business Transformation. The nature of the teaching objectives suggests this case study is more appropriate for undergraduate programmes or introductory strategy courses in postgraduate programmes. The case study has two teaching objectives, which are as follows.

- Business process mapping: To understand the systemic steps taken to map the business processes of an organisation using a collaborative and iterative approach with the respective process owners.
- Business organisations as an ecosystem: To understand the interlinkages between the activities per business functional unit, inclusive of interdependencies between role players. A business is an ecosystem, with various business functions, internal and external, constituting the systems of the business ecosystem.

Research methods

A consultative and collaborative approach based on hierarchical business process mapping and incorporating the BPMN 2.0 notation is adopted. This case study draws on the insights gained from information gathering engagements with the various business process owners at the CHIETA. The aim was to understand the current state (as-is) of their business processes and activities, identify key role players, and identify which activities are manual or system-based. After mapping the current business processes, the process owners validated them for quality and accuracy. The CHIETA business process owners then signed off the as-is processes. Understanding this as-is state is essential to inform the future state (to-be) of the organisation, but the to-be state mapping is not part of the scope of this case study. It is essential to note that business process mapping is only the initial step in the development of a digital transformation pathway for an organisation, with other steps not documented within the scope of this case study.

Keywords and subject code

Business process, business process mapping, business strategy, SETA, digital business strategy, digital transformation, digital maturity.

Assignment questions

1. What are the essential steps taken to map the business processes of an organisation that are essential in informing the strategic digital transformation path of a business?
2. In groups of three, think of a hypothetical business of your own, and demonstrate the usage of the business process hierarchy illustrated in question 1 to map the business processes of your hypothetical organisation.

The questions seek to test the ability of students to understand that the initial step in the digital transformation path of an organisation entails understanding the organisation's as-is business processes. Furthermore, the aim is to assess students' comprehension of the sequential steps involved in mapping an organisation's business processes. These steps include iterative collaborative engagements to comprehend business activities, the capture and documentation of as-is processes, verification with business process owners, and finally, the validation of the business processes. Furthermore, the questions seek to test the ability of students to identify the role players in business processes, and finally to test students' ability to identify intradependencies within business processes as well as interdependencies across business processes, with a particular emphasis on role players and business process activities.

Additional materials

Harmon P (2019) Chapter 8 – Understanding and scoping process problems. In *A Business Process Management Guide for Managers and Process Professionals,* 4th edn. (Kaufmann M (ed.)). Morgan Kaufmann, Cambridge, MA, USA, pp. 179–202.

Holotiuk F and Beimborn D (2017) Critical success factors of digital business strategy. In *13th International Conference on Wirtschaftsinformatik, St. Gallen, Switzerland*, pp. 991–1005.

Kıyıklık A, Kuşakcı AO and Mbowe B (2022) A digital transformation maturity model for the airline industry with a self-assessment tool. *Decision Analytics Journal* **3**: 1–13.

Sishi M and Telukdarie A (2020) Implementation of Industry 4.0 technologies in the mining industry case study. *International Journal of Mining and Mineral Engineering* **11(1)**: 1–22.

Tsagkani C and Tsalgatidou A (2022) Process model abstraction for rapid comprehension of complex business processes. *Information Systems* **103**: 101818.

Wang W, Chen T, Indulska M, Sadiq S and Weber B (2022) Business process and rule integration approaches – an empirical analysis of model understanding. *Information Systems* **104:** 101901.

Wurm B, Grisold T, Mendling J and vom Bracke J (2020) Business management and routine dynamics. In *Handbook of Routine Dynamics* (Feldman MS, Pentland PT, D'Adderio L *et al.* (eds)). Cambridge University Press, Cambridge, UK.

Teaching plan and timing

For a class of 90 min, the researcher recommends the following teaching plan.

Introduction (5 min)

- The instructor provides a summary of the case.

Assignment question 1 (30 min)

- What are the essential steps taken to map the business processes of an organisation that are essential in informing the strategic digital transformation path of a business? The lecturer will outline steps taken to map the CHIETA business processes, with emphasis on the four-level hierarchical notation, identification of the CHIETA business processes for mapping, and the associated business process owners. The lecturer will explain how each business process was partitioned into four hierarchical levels and how the business process activities were populated as per the four-level hierarchy, inclusive of the identified role players. The lecturer must emphasise that the mapping procedure was an iterative collaborative task between the JBS team and the CHIETA business process owners. Emphasis must be given to the importance of verification of each captured business process as a quality check and validation by the CHIETA business process owners as a means to sign off the accuracy of the processes thereof.

Conclusion (5 min)

- Wrap up the case and summarise key learnings.

Assignment question 2 (30 min)

- In groups of three, think of a hypothetical business of your own, and demonstrate the usage of the business process hierarchy illustrated in question 1 to map the business processes of your hypothetical organisation. The lecturer will prompt the students to group themselves into groups of three and think of a hypothetical business organisation of their own. The students are furthermore required to demonstrate the utilisation of the business process hierarchy to map the business processes of their organisation. The lecturer will then ask a volunteering group to outline their mapped business processes.

Conclusion (20 min)

- The lecturer will give feedback on the outlined business processes by the volunteering group and wrap up the lecture.

Analysis of assignment questions

1. What are the essential steps taken to map the business processes of an organisation that are essential in informing the strategic digital transformation path of a business?

[Teaching suggestions] The value proposition of this question is for learners to appreciate that developing a successful digital business strategy for an organisation stems from understanding its current (as-is) status. Learners must understand that the initiation point of an organisation towards a digital business strategy lies in the initial comprehensive understanding of its as-is business process model. The analysis of the to-be (not covered in this case study), in turn, informs the development of a suitable digital strategy.

A business process hierarchical configuration is illustrated in Figure 7.2, whereby this hierarchy can be arranged from business organisation level to the business units of an organisation (P_n) and respective business areas per business unit (F_n). Using this hierarchical configuration as best practice, a modification of this hierarchy to suit the capture of the business processes for the CHIETA was done as follows.

- Level 0: Business function
- Level 1: Sub-function
- Level 2: Business area per business function
- Level 3: Process area per business area
- Level 4: Business process activity.

An adequate conceptual understanding of the business's as-is status serves as the foundation for the optimal development of a business strategy. Adopting a typical business process model framework like the one above and identifying each of the level 0–4 inputs can enable this as-is status. The JBS team then proposed a systematic approach to mapping the CHIETA's business processes using the aforementioned modified framework, which entails mapping the business processes of each department from level 0 to level 4. The JBS team and the CHIETA business process owners, who were significant contributors to the digital transformation of organisations, underpinned this approach and method with extensive knowledge.

Mapping the as-is business processes

Preliminary information gathering was done by reviewing the best global practices as well as the CHIETA website in order to gain a background understanding of typical business processes for mapping. This was then followed by initial engagements with the stakeholders responsible for the various areas of the business. These stakeholders are also referred to as the business process owners and include finance, human resources, grants, research and skills planning, governance and They initiated the business process mapping and agreed on an initial business process overview for each business process that informs the processes for mapping from level 0 to level 4 as per the modified adopted business process hierarchy. Consultative, iterative engagements were conducted with the business process owners to map and graphically present their as-is business process on Microsoft Visio. These as-is mapping engagements also enabled the identification of currently installed systems (if available), business operations, functions, processes, and resources (inclusive of role players within the business process). Sharing of the processes was done with the business process owners prior to holding verification meetings, which serve as quality audit checks and guarantee that the processes were mapped as a true as-is replica of the CHIETA business processes. Whenever updates were needed, the business process owners communicated them to the JBS team. The JBS team then made the necessary changes and shared the revised version with the process owners for validation. This process was repeated iteratively until the updates were fully approved by the process owners.

Documentation: as-is and to-be statuses

After validating the as-is mapped business processes, the JBS team documented a comprehensive report and shared it with the CHIETA CEO, including the graphically captured processes. This then concluded the initial phase of developing a digital business strategy – that is, mapping the business processes, which in turn informs a business's current digital maturity state.

The instructor can use a typical business process, such as HR, as an example to demonstrate the adoption of the business process hierarchy for mapping HR.

2. In groups of three, think of a hypothetical business of your own and demonstrate how to use the business process hierarchy illustrated in question 1 to map the business processes of your hypothetical organisation.

The value proposition of this question is for learners to demonstrate their understanding of and utilisation of the business process hierarchy to map the business processes of their hypothetical organisation.

The lecturer will allow the students to group themselves into groups of three and map the business processes of their hypothetical organisation. The lecturer can walk around the classroom and observe as the students conduct the exercise.

The lecturer will then ask a volunteer group to present their findings to the class. The lecturer will, in turn, give feedback and comments.

Key learnings

Collaboration is an essential component of business process mapping, as signified by the numerous iterative engagements between the JBS team and CHIETA business process owners towards successfully delivering the mapped as-is CHIETA business processes.

REFERENCES

Harmon P (2019) Chapter 8 – Understanding and scoping process problems. In *A Business Process Management Guide for Managers and Process Professionals,* 4th edn. (Kaufmann M (ed.)). Morgan Kaufmann, Cambridge, MA, USA, pp. 179–202.

Holotiuk F and Beimborn D (2017) Critical Success Factors of Digital Business Strategy. In *13th International Conference on Wirtschaftsinformatik, St. Gallen, Switzerland*, pp. 991–1005.

Tsagkani C and Tsalgatidou A (2022) Process model abstraction for rapid comprehension of complex business processes. *Information Systems* **103**: 101818.

Prof. Arnesh Telukdarie
ISBN 978-1-83662-645-9
https://doi.org/10.1108/978-1-83662-642-820251008

Chapter 8
Strategies for sustainable digitalisation: advancing environmental, social, governance metrics

Arnesh Telukdarie, Megashnee Munsamy and Mahure Hlobisile Natasia

Introduction

Professor Telukdarie launched a strategic project in January 2022, designed to enhance the sustainability performance of companies. Professor Telukdarie is a highly esteemed researcher, philanthropist, and professor hailing from the University of Johannesburg in South Africa. He has been a prolific contributor to academia, penning a variety of articles and books that serve as practical guides for digitalisation in a range of industries, including healthcare, manufacturing, education, mining, and digitalisation for rural development. As the year progressed, Professor Telukdarie geared up to tackle the new challenge of integrating digitalisation into the realm of environmental, social, and governance (ESG). ESG is a concept that addresses the social, environmental, and governance effects produced by companies while striving for profits. It includes a range of criteria that investors and other stakeholders use to evaluate a company's performance and operations in these three crucial areas (Li *et al.*, 2021). Regulatory bodies such as the Sustainable Accounting Standards Board (SASB) and the International Integrated Reporting Council (IIRC) were created to promote and ensure sustainable performance among companies. These organisations set standards and regulations that guide ESG reporting, which allows companies to report on their ESG practices (Li *et al.*, 2021). Despite the growing adoption of ESG reporting, traditional ESG rating methods have received widespread criticism for their lack of transparency, comparability, and consistency in measures, making it difficult for investors to make informed decisions and for companies to improve their ESG performance. As a result, various organisations and researchers, such as Professor Telukdarie, have been exploring innovative ways to enhance ESG reporting. Professor Telukdarie has successfully leveraged digitalisation in other fields and aims to do the same for ESG reporting. With his track record of success, it raised the question of whether he would succeed in integrating digitalisation into the complex world of ESG.

The intricacies of ESG: context and challenges

The concept of ESG factors has its roots in the ethical movements of the 1960s and 1970s. These movements raised awareness about the adverse impact of corporate activities on society and the environment, prompting a shift towards considering non-financial aspects in investment decisions (Dye *et al.*, 2021). The United Nations (UN) has played a significant role in promoting and advancing the concept of ESG. In 2004, the UN formally introduced the UN Global Compact, which encouraged businesses to embrace sustainable and socially responsible policies, thereby emphasising the integration of ESG factors into their decision-making processes. Moreover, the UN has been instrumental in developing the sustainable development goals (SDGs) as a universal framework to address global challenges, such as poverty eradication, environmental conservation, and promoting peace and prosperity by 2030. The SDGs and ESG share the common goal of fostering sustainable development and ethical business practices. ESG rating agencies and investors utilise ESG metrics to assess the sustainability performance of companies. For instance, SDGs such as climate action (SDG 13) and life on land (SDG 15) directly target environmental challenges. The E in ESG focuses on evaluating companies' commitment to environmental protection in areas such as climate action, biodiversity preservation, and mitigating land pollution.

Today, ESG has become a key consideration for investors, who increasingly view it as a critical component of long-term value creation and risk management. With this increased focus on ESG, the need for standardising reporting and accountability practices has become crucial (Eccles and Stroehle, 2020; Godelnik, 2021; Sachini *et al.*, 2021). This has led to the establishment of various regulatory bodies, each playing a crucial role in shaping the guidelines for ESG reporting (Sachini *et al.*, 2021). Figure 8.1 presents an overview of various ESG regulatory bodies and their respective introduction years. These regulatory bodies aim to promote ESG practices in businesses and ensure their accountability for their impact on the environment and society. The Global Reporting Initiative (GRI) was established in 1997 to promote transparency in corporate sustainability reporting. The Carbon Disclosure Project (CDP) was introduced in 2000 to encourage companies to disclose their carbon emissions and climate change-related risks (Godelnik, 2021). The Sustainable Accounting Standards Board (SASB) was established in 2011 to develop industry-specific sustainability accounting standards. The Task Force on Climate-Related Financial Disclosures (TCFD) was introduced in 2015 to guide climate-related financial risks and opportunities. The European Union's Non-Financial Reporting Directive (NFRD) was introduced in 2014, mandating large companies to disclose non-financial information, including ESG performance (Godelnik, 2021). The Network for Greening the Financial System (NGFS) was established in 2017 to promote the integration of climate-related risks into financial management and decision-making (Godelnik, 2021). Finally, the International Sustainability Standards Board (ISSB) emerged as a key standard-setting body in the ESG field, intending to develop globally accepted sustainability reporting standards (Godelnik, 2021). However, despite the presence of these regulatory bodies, issues in ESG reporting persist (Eccles and Stroehle, 2020; Sachini *et al.*, 2021).

Pitfalls of ESG reporting: a critical perspective

There are several limitations to the current ESG reporting methods that must be considered. One of the main concerns is that ESG rating agencies often rely on information provided by companies themselves to develop the ratings, which may not be entirely accurate (Dye *et al.*, 2021). This can lead to biased ratings that do not accurately reflect a company's sustainability practices. For example, a company may report positive ESG practices without providing concrete evidence to support its claims. Moreover, different ESG rating agencies use different methodologies in evaluating the sustainability performance of firms, which can lead to inconsistent ratings for the same company (Dye *et al.*, 2021). For instance, a company that receives a high ESG rating from one agency may receive a lower rating from another agency. This inconsistency can be confusing for investors and may result in incorrect assessments of a company's sustainability.

ESG ratings have been subjected to criticism for their lack of transparency. Rating agencies do not disclose the methodologies or data sources they use. This makes it difficult for investors to evaluate the ratings and make informed investment decisions (Godelnik, 2021). In some cases, a company may receive a high ESG rating without investors knowing how the rating agency arrived at that conclusion. Additionally, ESG ratings tend to focus on a company's current sustainability practices rather than its ability to adapt and improve over time (Eccles and Stroehle, 2020). This can lead to a focus on short-term sustainability goals rather than long-term sustainability strategies. For example, a company may receive a high ESG rating for implementing a single sustainability initiative without considering the company's overall sustainability goals and strategies (Kotsantonis and Serafeim, 2019). This has led to a loss of trust in ESG reporting, as illustrated in Figure 8.2, where renowned business magnate Warren Buffett states, 'ESG is a fraud, and companies should refrain from investing in it'. Therefore, it is crucial to address the current limitations of ESG reporting and ensure that companies report their sustainability issues transparently and accurately. This will ensure the preservation of the culture of responsible and sustainable business practices and build a sustainable future for all stakeholders.

Harnessing digital tools for enhanced ESG reporting: a path forward

Professor Telukdarie, from the University of Johannesburg, recognised the current challenges in ESG reporting and took it on himself to leverage digitalisation to improve the process. The University of Johannesburg, as a leading institution in the fourth industrial revolution (4IR), has always been at the forefront of embracing sustainability issues. To achieve this ambitious goal, Professor Telukdarie, together with his research team, set to develop innovative methods to improve ESG reporting through digitalisation. The primary objective of the project was to improve the comparability and transparency of ESG reporting, which is crucial for the sustainable growth of businesses and the economy. To put this vision into motion, Professor Telukdarie detailed the methodology that would serve as a roadmap to develop a comprehensive and standardised approach to ESG metrics that considers all companies from different industries. The methodology involved a comprehensive approach that included conducting a comprehensive literature review to gain insight into the current ESG landscape, developing a consolidated ESG matrix, and digitalising the data collection using artificial intelligence (AI)-based tools. The goal was to create a reporting framework that would be both effective and practical, which could be implemented by companies across various sectors. Figure 8.3 indicates the methodology comprising three main components: a literature review, the development of a consolidated ESG matrix, and the digitalisation of ESG data.

ESG data consolidation and digitalisation procedure

Step one: The first step involved conducting a comprehensive literature review to gain an understanding of the current ESG landscape (Telukdarie *et al.*, 2024).

Step two: The second step involved developing a consolidated ESG matrix. This included categorising and consolidating ESG metrics into a single approach based on the three main components of ESG, namely environmental, social, and governance. The consolidation process involved defining a single unit of measure for each ESG metric to ensure that all metrics are standardised and easily comparable (Telukdarie *et al.*, 2024).

Step three: To make the consolidated ESG matrix more precise and efficient, the third step involved digitalising ESG data collection. This step entailed identifying applicable and available data sources and applying them to determine AI-based tools for automatic data collection (Telukdarie *et al.*, 2024).

In summary, the steps in developing a novel methodology recommended by Professor Telukdarie and his research team aimed to provide a roadmap to develop a comprehensive and standardised approach to ESG metrics that considers all companies from different industries. It involved conducting a comprehensive literature review to gain insight into the current ESG landscape, developing a consolidated ESG matrix, and digitalising the data collection using AI-based tools.

Mapping process: key results in digitalising ESG metrics

The research team made three significant discoveries, which are outlined below.

Literature review

The research team conducted a comprehensive literature review on ESG and found that ESG reporting faces several key challenges, including a lack of transparency, comparability, standardisation, and resources. While many recommendations to address these challenges were noted, only a few considered digitalisation as a means of improving ESG reporting. The team also identified several theories related to sustainability, including triple bottom line, stakeholder theory, and natural capitalism theories, which provided a strong theoretical underpinning for the methodology developed to develop a consolidated ESG matrix with automated inputs per company.

Consolidated matrix

In the second step, ESG data were collected from various sources such as corporate reports, news articles, social media posts, and government websites. To consolidate the ESG indicators, data were initially collected from various ESG rating agencies, including Bloomberg, Sustainalytics, MSCI, Moody's, S&P Global, Refinitiv, and FTSE Russell.

All ESG indicators and key performance indicators (KPIs) were collected from these rating agencies for each and consolidated, as indicated in Table 8.1. The researchers established a standardised unit of measure for each metric to facilitate easy comparisons across different measuring or reporting methods. A single unit of measurement, such as standardising greenhouse gas emissions to metric tons of CO_2 equivalent, simplifies comparisons and consolidations of ESG metrics across different industries and companies.

Digitalisation

After conducting extensive research, Professor Telukdarie and his research team found that natural language processing (NLP) could be leveraged in five different contexts to gather and analyse ESG data using distinct sets of rules and techniques. NLP is an AI technique that examines unstructured data, such as news articles and social media posts, to identify ESG-related keywords, themes, and sentiments. In summary, these five contexts detailed below were the outcome of this initial research.

Context 1: Open web data. This involves defining a list of relevant KPIs for ESG assessment and developing rules for web scraping relevant websites for ESG data. NLP techniques can be used to extract ESG data from the scraped websites, and the extracted data are then filtered to remove irrelevant information. Finally, the extracted data for each KPI are consolidated, as indicated in Table 8.2.

Context 2: Company websites and documents. The second context develops rules for web scraping company websites for ESG data; NLP techniques are used to extract ESG data from company websites and documents. The extracted data are then filtered to remove irrelevant information and consolidated for each KPI. The data from both contexts 1 and 2 are then combined to create a comprehensive dataset for ESG assessment (Table 8.2 and Figure 8.4).

Context 3: Social media + other. The third context develops rules for collecting ESG data from social media platforms and other sources; NLP techniques are used to extract ESG data from them. The extracted data are filtered to remove irrelevant information and consolidated for each KPI. The data from all three contexts are then combined to create a comprehensive dataset for ESG assessment (Figure 8.5).

Context 4: Best practice agencies. In the fourth context, relevant best practice agencies for ESG assessment are identified, and rules are developed for collecting ESG data from them. NLP techniques are used to extract ESG data from the collected reports, and the extracted data are filtered to remove irrelevant information. The extracted data for each KPI is consolidated, and the data from all four contexts are combined to create a comprehensive dataset for ESG assessment.

Context 5: Consolidation. In the final context, a multi-AI model is used to consolidate the data from all contexts. The accuracy of the model is verified using a rule-based approach and benchmarking against industry standards. A basic integrated comprehensive performance indicator (ICPI) methodology is used to track the ESG ratings of the companies over a specified period. AI-generated data update the ESG matrix, and an evolutionary and long-term approach ensures its relevance over time (Figure 8.6). The process involves developing rules for web scraping and using NLP techniques to extract ESG data from multiple sources such as the open web, company websites and documents, social media, and best practice agencies. The extracted data are then filtered to remove irrelevant information and consolidated for each KPI. Finally, a multi-AI model is used to consolidate the data from all contexts, and the accuracy of the model is verified using a rule-based approach and benchmarking against industry standards. The ICPI methodology is used to track the ESG ratings of companies over time, and the ESG matrix is updated using AI-generated data.

Conclusion: wrapping up the initial phase of the project

Professor Telukdarie and his research team made interesting discoveries about the potential of digitalisation to enhance ESG reporting for companies. The team discovered that it was possible to develop a standardised ESG matrix and leverage digitalisation for efficient and comprehensive ESG data collection from diverse industries using NLP. The team also learned about AI's ability to automatically update ESG data and update ESG metrics with new KPIs as they emerge, making it all-encompassing and relevant to all sectors using AI. Despite not being able to fully implement a software platform, they made significant progress towards developing a robust, standardised ESG matrix applicable to various industries.

The team was excited about the potential benefits of leveraging digital tools to enhance ESG reporting as they completed the first stage of the project. Looking ahead to the new year, Professor Telukdarie and his team intend to build on their current methodology and explore ways to enhance it further. They were committed to working tirelessly to implement their findings and continue contributing to the growing body of research on ESG reporting. With their dedication and expertise, they hope to make significant strides in improving sustainability reporting practices across industries.

Exhibits

Figure 8.1 ESG regulatory bodies

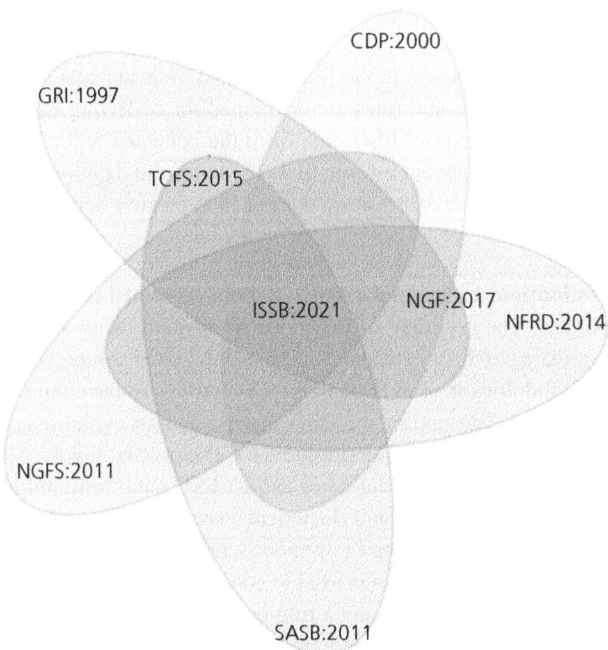

Figure 8.2 Screenshot of Google results indicating the controversies in ESG reporting (Google, 2024)

Figure 8.3 Digitalisation of ESG metrics (Telukdarie *et al.*, 2024)

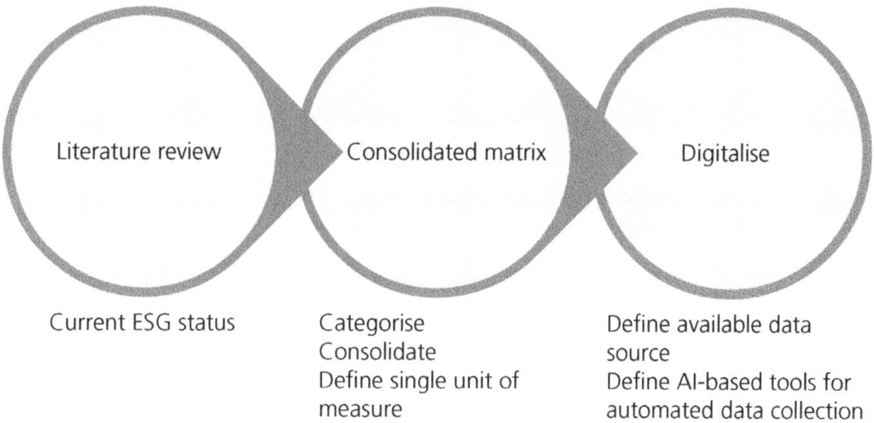

Figure 8.4 Results of scraped environmental KPIs from company documents

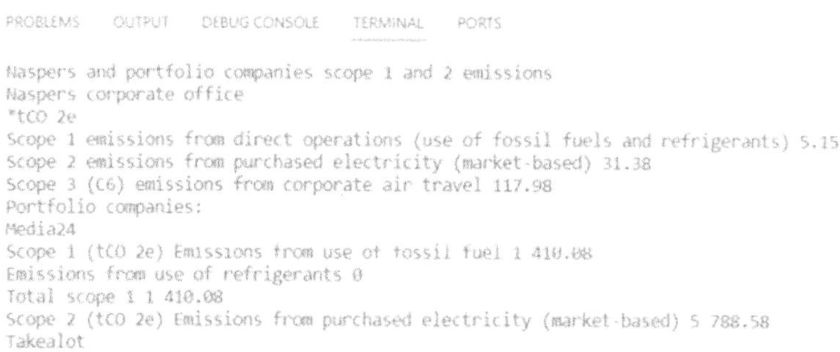

Figure 8.5 Company X scraping social media posts

web-scraper-start-url	News_blog
https://www.naspers.com/news-insights/portfolio-updates/2023/naspers-takealot-and-media24-join-the-south-african-plastics-pact-to-reduce-plastic-waste	Naspers, Takealot and Media24 join the South African Plastics Pact to reduce plastic waste across value chain
https://www.naspers.com/news-insights/portfolio-updates/2023/naspers-takealot-and-media24-join-the-south-african-plastics-pact-to-reduce-plastic-waste	Naspers, and its subsidiary businesses, Takealot and Media24, have joined the South African Plastics Pact, an industry collaboration of packaging producers, brand owners, retailers, recyclers, governments and NGOs to tackle plastic packaging waste by creating a circular economy for plastics
https://www.naspers.com/news-insights/portfolio-updates/2023/naspers-takealot-and-media24-join-the-south-african-plastics-pact-to-reduce-plastic-waste	Naspers, and its subsidiary businesses, Takealot and Media24, have joined the South African Plastics Pact, an industry collaboration of packaging producers, brand owners, retailers, recyclers, governments and NGOs to tackle plastic packaging waste by creating a circular economy for plastics
https://www.naspers.com/news-insights/portfolio-updates/2023/naspers-takealot-and-media24-join-the-south-african-plastics-pact-to-reduce-plastic-waste	The SA Plastics Pact was launched in 2020 as part of The Ellen MacArthur Foundation's New Plastics Economy initiative. It is managed by green economy non-profit, GreenCape and currently has 44 members representing key organisations across the plastics value chain, including Unilever, Woolworths, Clicks, Food Lovers Market, Spar and Pick n Pay among others
https://www.naspers.com/news-insights/portfolio-updates/2023/naspers-takealot-and-media24-join-the-south-african-plastics-pact-to-reduce-plastic-waste	was launched in 2020 as part of The Ellen MacArthur Foundation's New Plastics Economy initiative. It is managed by green economy non-profit, GreenCape and currently has 44 members representing key organisations across the plastics value chain, including Unilever, Woolworths, Clicks, Food Lovers Market, Spar and Pick n Pay, among others
https://www.naspers.com/news-insights/portfolio-updates/2023/naspers-	The SA Plastics Pact aims to transform the country's plastic packaging use by meeting four ambitious targets

Figure 8.6 Consolidated ESG data from companies A, B and X

CompanyName	KPICategory	KPIName	KPIValue
Company A	Governance	Number of serious cybersecurity incidents	1
Company A	Governance	DX certification	Certified company
Company A	Governance	Number of violations of antitrust laws	0
Company A	Governance	Number of violations of anti-bribery laws	0
Company A	Governance	Response to human rights risks in the supply ch...	Held meeting with suppliers
Company A	Governance	External awards from customers	8
Company A	Governance	Compliance with timely and appropriate disclosure	100%
Company B	Environmental KPIs	Total waste generated (metric tons)	108,282
Company B	Environmental KPIs	Waste generated which is recycled (metric tons)	7,932
Company B	Environmental KPIs	Waste generated which is reused (metric tons)	92,760
Company B	Environmental KPIs	Total waste intensity (metric tons/$1 million rev...	24.20
Company B	Environmental KPIs	Waste recycled intensity (metric tons/$1 million ...	1.80
Company B	Environmental KPIs	Hazardous waste (metric tons)	6,950
Company B	Environmental KPIs	Hazardous waste incinerated (metric tons)	1,295
Company B	Environmental KPIs	Hazardous waste landfilled (metric tons)	520
Company B	Environmental KPIs	Hazardous waste other disposal method (metric...	112
Company B	Environmental KPIs	Hazardous waste intensity (metric tons/$1 millio...	1.58
Company X	Scope 1 emissions	Scope 1 emissions from direct operations (use o...	5.15

Table 8.1 Social indicators and KPIs from ESG rating agencies

Social performance category	Description	Key performance indicator	Unit
Employees and diversity	Breakdown of employees per type of contract	Permanent employees	%
		Temporary employees	%
		Apprenticeship/internship	%
	Breakdown of employees per age	Employees under the age of 30	%
		Employees between 30 and 50	%
		Employees above 50	%
	Breakdown of employees per employment type	Full-time employees	%
		Part-time employees	%
	Breakdown of employees per gender and per category	Male employees	%
		Male middle management	%
		Male company management	%
		Thereof male group management	%
		Female employees	%
		Female middle management	%
		Female company management	%
		Thereof female group management	%
	New employee hires and employee turnover	Recruitment rate	Rate
		Employee voluntary turnover rate	Rate
		Employee turnover rate	Rate

Table 8.2 Context 1: open web scraping results for company website: https://www.toyota-boshoku.com/global/sustainability/management/kpi/

KPI evaluation items	KPI values
Governance	
Number of environmental abnormalities and complaints	1
Number of serious cybersecurity incidents	1
DX certification	Certified company
Number of violations of anti-trust laws	0
Number of violations of anti-bribery laws	0
Response to human rights risks in the supply chain (development of human rights due diligence)	Held meetings with suppliers
External awards from customers	8
Compliance with timely and appropriate disclosure	100%

Strategies for sustainable digitalisation: advancing environmental, social, governance metrics

Teaching notes

Synopsis

This case study revolves around Professor Arnesh Telukdarie, a distinguished professor specialising in digital transformation, Industry 4.0, and sustainable business practices at the University of Johannesburg in South Africa. In January 2022, Professor Telukdarie recognised a need to enhance the methods of measuring the sustainability performance of companies. He identified several inadequacies in traditional ESG (environmental, social, and governance) reporting techniques, such as a lack of transparency, comparability, and consistency. Professor Telukdarie embarked on a journey to develop and digitalise a consolidated ESG matrix that would undergo continuous updates to address these issues. This approach provided a standardised and reliable way to measure and interpret ESG performance, which can aid investors and other stakeholders in making informed decisions. The case study emphasises the importance of incorporating digitalisation as well as transparency, comparability, and consistency in ESG reporting. As a result, the case is highly relevant for students studying sustainable business practices, environmental management, digital innovation, the 4IR, and the changing landscape of ESG reporting.

Teaching objectives

This case study is intended for master's and Master of Business Administration students keen on digital business, sustainability management, environmental studies, and ESG practices. It is a valuable resource for courses that cover environmental management and economics, sustainable business practices, digital business, and ESG reporting. It is also in line with academic areas that examine the intersection of technology, sustainability, and corporate responsibility. Furthermore, this resource will benefit professionals interested in understanding the complexities of incorporating digitalisation into ESG in today's business landscape. The case study provides insights into responsible business practices and how digitalisation can enhance ESG reporting. Successful students will be able to do the following.

- Critique current ESG methods used to measure sustainability performance in companies.
- Examine the role of digitalisation in advancing and enhancing ESG reporting.

Research methods

This case study was developed using both primary and secondary sources of data to collect comprehensive and in-depth information. The primary data were collected through participant observation, which involved immersing the researcher in the context of the study for a period of seven months. This approach enabled the researcher to actively engage with and observe participants while also being a part of the activities or events being investigated. Participant observation is a well-established research technique that allows researchers to gain firsthand insights into behaviours, interactions, and dynamics within the research setting. In addition to the primary data, the research also utilised secondary data from various sources, such as literature, company reports, and websites. This supplemented the primary data and provided additional context and background information that was relevant to the case study. Pseudonyms were used instead of real names to maintain confidentiality and anonymity in this study.

Keywords and subject code

ESG, digitalisation, sustainability, artificial intelligence, corporate social responsibility, innovation, sustainable business practices.

Assignment questions

1. What are the current limitations of traditional ESG rating methods? Critically analyse issues related to transparency, comparability, and consistency.
2. What role can digitalisation play in enhancing ESG reporting? Assess the benefits and costs associated with leveraging these technologies for ESG reporting.

Additional materials

YouTube. Warren Buffett, we'll never waste time and money on ESG reporting. https://www.youtube.com/watch?v=Mx6ePidFBZo (accessed 27/08/2024).

YouTube. How ESG metrics work and why all investors should care. https://www.youtube.com/watch?v=4LPRQaG83Ls (accessed 27/08/2024).

YouTube. The Difficult Truth about ESG Investing with Aswath Damodaran. https://www.youtube.com/watch?v=8os1cmVXW0c (accessed 27/08/2024).

YouTube. Innovation showcase: The digital transformation of ESG management. https://www.youtube.com/watch?v=REWVBGeEQyA (accessed 27/08/2024).

Suggested core readings

Balakrishnan C and Rahman H (2022) Nexus of digitalization, social, and governance factors. In *Digitalization and the Future of Financial Services: Innovation and Impact of Digital Finance* (Vukovic DB, Maiti M and Grigorieva EM (eds)). Springer, Cham, Switzerland, pp. 181–200.

Clements R (2021) Why comparability is a greater problem than greenwashing in ESG ETFs. *William & Mary Business Law Review* **13(2)**: 441–485.

Cort T and Esty D (2020) ESG standards: looming challenges and pathways forward. *Organization & Environment* **33(4):** 491–510.

Joubrel M and Maksimovich E (2023) ESG data and scores. In *Valuation and Sustainability: A Guide to Include Environmental, Social, and Governance Data in Business Valuation* (Maiti M, Vukovic DB and Grigorieva EM (eds)). Springer, Cham, Switzerland, pp. 67–98.

Kumar S (2023) A review ESG performance as a measure of stakeholders theory. *Academy of Marketing Studies Journal* **27(S3)**.

Moro-Visconti R (ed.) (2022) Digitalization and ESG-driven valuation. In *The Valuation of Digital Intangibles: Technology, Marketing, and the Metaverse*. Springer, Cham, Switzerland, pp. 685–764.

Rezaee Z (2017) Corporate sustainability: theoretical and integrated strategic imperative and pragmatic approach. *The Journal of Business Inquiry* **16(1)**.

Zenkina I (2023) Ensuring the transparency of ESG reporting based on the development of its standardization. In *E3S Web of Conferences*, vol. 371. EDP Sciences, Paris, France.

Teaching plan and timing

Introduction (15 min)

- To start the session off on the right foot, the facilitator can pose a thought-provoking question that goes something like this: 'Do you think that current ESG ratings accurately reflect businesses' sustainable practices?'
- To make the class more engaging, the facilitator can conduct a quick anonymous poll and encourage students to share their opinions on the matter, along with their reasons why.

Contextualising the study (5 min)

- The facilitator can enhance the teaching and learning process by sharing a relevant YouTube clip or material that illustrates the current ESG challenges and demands.

Discussion (25 min)

- To begin the discussion session, the instructor must present Assignment Question 1, which focuses on evaluating the drawbacks of traditional ESG ratings and finding solutions to rectify the issues of transparency, comparability, and consistency. The instructor should then encourage the students to brainstorm alternative strategies that could effectively address the problems. The instructor should emphasise the importance of critical thinking while proposing alternative solutions.

Small group activity (35 min)

- Introduce the topic of Assignment Question 2, which focuses on how digitalisation can enhance learning. The facilitator should then form small groups and assign each group the task of creating an action plan. After this, randomly select two groups to present their plans to the class.

Conclusion and key learning points (5 min)

- The facilitator should present a summary of the main takeaways following the class discussion and plans.
- The facilitator should emphasise that ESG challenges are complex and require innovative solutions to address them effectively.
- Finally, the importance of incorporating digitalisation into advanced ESG rating techniques should be highlighted.

Analysis of assignment questions

Question 1. Assess the limitations of traditional ESG ratings.

Possible model answers:

Lack of standardised, globally accepted ESG frameworks: The current ESG rating methods face several challenges and limitations making it difficult to evaluate companies' ESG performance effectively and reliably. One of the significant issues is the lack of a globally accepted standardised framework for ESG ratings (Iamandi *et al.*, 2019). The absence of such a framework makes it difficult to compare ratings across regions and countries, as each region or country has its own unique challenges and ESG priorities (Iamandi *et al.*, 2019). Moreover, there are significant differences in ESG laws and regulations across regions and countries that further complicate the process (Gernego *et al.*, 2024).

Even within industries, there are different ESG KPIs that are given different priorities, making it challenging to establish a globally accepted and reliable ESG rating system. For instance, companies in the energy sector may prioritise greenhouse gas emissions, while those in the financial sector may prioritise diversity and inclusion. As a result, there is a need for a more comprehensive and standardised approach to ESG rating that takes into account the unique characteristics of each industry, region, and country (Matakanye *et al.*, 2021).

Dynamic nature of KPIs

ESG KPIs play a vital role in measuring and reporting a company's sustainability performance. However, the landscape of ESG KPIs is constantly evolving and requires a standardised, dynamic framework to keep pace with rapidly changing trends and external factors such as technological advancements, shifting consumer preferences, and emerging industries. As new trends emerge, industries evolve, and priorities shift, ESG KPIs need to accurately measure and reflect current priorities. This not only requires a standardised framework that is suited for all industries but also a dynamic ESG framework that can continuously update industry KPIs reflecting current trends.

Unfortunately, the current ESG reporting methods have not been able to address this challenge effectively. Therefore, it is crucial for companies to adopt a dynamic and standardised ESG reporting framework that can keep up with the constantly evolving sustainability landscape and accurately measure and report their performance.

Transparency, comparability, and consistency

An alternative way to answer this question is to say that the current ESG rating methods suffer from a lack of transparency, comparability, and consistency (Cardoni *et al.*, 2019). The absence of standardised metrics or definitions for ESG factors makes it difficult to compare companies. Additionally, data on a company's ESG practices can be challenging to access, leading to a lack of transparency (Cardoni *et al.*, 2019). Moreover, there is no consensus on what constitutes ESG, which can cause confusion and conflicting information. All these factors make it difficult for investors to navigate the ESG landscape and exacerbate the problem (Cardoni *et al.*, 2019).

Question 2. Digitalisation strategies

Possible model answers: Technology can be utilised to enhance ESG reporting.

Digital platforms can enforce standardised data formats and structures for ESG reporting, ensuring that data are consistent across different organisations. This makes it easier to compare and analyse the data.

Automated data collection through digital tools can be used to gather ESG data from various sources, including IoT devices, databases, and external sources, ensuring that the data are up-to-date and accurate.

Blockchain technology can be leveraged to create transparent and immutable records of ESG data, which ensures that data cannot be tampered with. Furthermore, machine learning algorithms can analyse large volumes of ESG data to identify trends, correlations, and anomalies, helping companies improve their ESG performances.

Key learnings

The case study presented emphasises the critical intersection of digitalisation and addressing the challenges associated with ESG reporting. The key takeaway from this study is the recognition of the need to synergise these two aspects for developing innovative and effective solutions to measure and report ESG performance. The students are made aware of the limitations of traditional ratings and understand the importance of considering economic, social, and environmental dimensions collectively. This holistic perspective aligns with the broader goals of sustainable business practices, which aim to create long-term value for all stakeholders. The students are also taught to navigate ambiguity, apply relevant frameworks, and make informed decisions that align with organisational goals and stakeholder expectations. By doing so, they are equipped to meet the challenges of today's business environment and contribute to the development of sustainable solutions that benefit everyone.

REFERENCES

Cardoni A, Kiseleva E and Terzani S (2019) Evaluating the intra-industry comparability of sustainability reports: the case of the oil and gas industry. *Sustainability* **11(4)**: 1093 –1105.

Dye J, McKinnon M and Van der Byl C (2021) Green gaps: firm ESG disclosure and financial institutions' reporting requirements. *Journal of Sustainability Research* **3(1)**.

Eccles R and Stroehle J (2020) Why corporate sustainability reporting needs to move beyond incrementalism. *Harvard Business Review* **98(1)**: 28–30.

Gernego I, Petrenko L, Lavrynenko V and Dyba M (2024) Trends and focuses for ESG investment support: a cross-country comparative analysis. *European Journal of Sustainable Development* **13(1)**: 167.

Godelnik R (2021) *Rethinking Corporate Sustainability in the Era of Climate Crisis: A Strategic Design Approach*. Springer Nature, Berlin, Germany.

Google (2023) Screenshot of search results for ' Warren Buffett on ESG '. https://www.google.com/search?q=warren+buffett+on+ESG&sca_esv=5a7521aecf6aaae0&rlz=1C1JJTC_enZA1075ZA1075&sxsrf=ADLYWILU0kaPKcHq19IVmWoR9UzKAIqVg%3A1736218503211&ei=h5d8Z8_JDJ6cigPtKqRmQs&ved=0ahUKEwiPqYv5zeKKAxUezgIHHTRVJLMQ4dUDCBA&uact=5&oq=warren+buffett+on+ESG&gs_lp=Egxnd3Mtd2l6LXNlcnAiFXdhcnJlbiBidWZmZXR0IG9uIEVTRzILEAAYgAQYkQIYigUyCxAAGIAEGIYDGIoFMgsQABiABBiGAxiKBTILEAAYgAQYhgMYigUyBRAAGO8FMgUQABjvBUjMFFDxA1jBC3ABeAGQAQCYAbACoAGrD6oBBTItNi4xuAEDyAEAAEBmAICoAKnAsICChAAGLADGNYEGEeYAwCIBgGQBgKSBwUxLjAuMaAHgxY&sclient=gws-wiz-serp (accessed: 22/08/2024).

Iamandi IE, Constantin LG, Munteanu SM and Cernat-Gruici B (2019) Mapping the ESG behavior of European companies. a holistic Kohonen approach. *Sustainability* **11(12)**: 3276.

Kotsantonis S and Serafeim G (2019) Four things no one will tell you about ESG data. *Journal of Applied Corporate Finance* **31(2)**: 50–58.

Li TT, Wang K, Sueyoshi T and Wang DD (2021) ESG: research progress and future prospects. *Sustainability* **13(21)**: 11663.

Matakanye RM, van der Poll HM and Muchara B (2021) Do companies in different industries respond differently to stakeholders' pressures when prioritising environmental, social and governance sustainability performance? *Sustainability* **13(21)**: 12022.

Sachini E, Tsagarakis K, Kokkinaki A and Vlachos P (2021) Environmental, social, and governance performance and financial performance: a systematic review of the literature. *Journal of Cleaner Production* **279(1)**: 123703.

Telukdarie A, Mahure H and Sishi M (2024) The digitization of ESG matrix. *Procedia Computer Science* **239**: 808–815.

Prof. Arnesh Telukdarie
ISBN 978-1-83662-645-9
https://doi.org/10.1108/978-1-83662-642-820251009

Chapter 9
The use of robotic process automation and artificial intelligence in designing a digital system landscape: a sustainable approach

Arnesh Telukdarie and Makoni Logistic

Introduction to the CHIETA and the design of a digital landscape initiative

The Chemical Industries Education and Training Authority (CHIETA) reached a pivotal moment in its organisational evolution and embarked on a bold digital transformation journey, driven by the imperative to adapt to the demands of the modern business environment and the need to achieve sustainability. According to the 2021–2022 Annual Report (CHIETA, 2022), the chemical sector contributes 3% to the total GDP and 23% to the total manufacturing gross value added (GVA) of South Africa. Given this significant impact, the CHIETA believes it is in its best interest to support the digital optimisation of small and medium-sized enterprises (SMEs) in the chemical sector. This initiative aims to enhance the sustainability of these SMEs and ensure their continued positive economic contribution over the long term. According to the CHIETA, digital enablement is one of the key initiatives that may increase capacity and outputs while playing a significant role in addressing the challenges that businesses face today and, most importantly, being a key driver and enabler of sustainable economies. Accordingly, in January 2023, the CHIETA decided to initiate a project building on business process optimisation and systems enablement to close the gap in limited digital optimisation. This programme, which was set to run for the year, focused on the digital strategy associated with the system and functional landscape that can deliver maximum Sector Education and Training Authority (SETA) value for today and tomorrow. To make this project work, the CHIETA approached it from a business, function, and IT perspective. This project was implemented with a focus on supporting sustainable development through digital optimisation.

The CHIETA's digital transformation and SDG conversations

Over the years that it has been in existence, the CHIETA has been one of the key SETAs in implementing transformation and economic growth initiatives through a number of education and training programmes for many marginalised communities and SMEs in South Africa. The current journey of designing a digital landscape is one of these many initiatives, which was developed to bridge the gap between service delivery in the chemical sector and current market trends that are tied to the flourishing digitalisation era. By so doing, the role of the CHIETA in achieving the sustainable development goals (SDGs) from a developmental perspective is undeniable. The CHIETA's projects may not directly address all of the SDGs, but they certainly support a number of them, especially those that deal with partnerships, industry, education, employment, and innovation. The CHIETA's emphasis on partnerships, industry development, employment, and education makes a big difference in the country's overall progress toward sustainable development and aligns with a number of SDGs. Table 9.1 presents the SDGs that are aligned with the CHIETA's current project of designing a digital system landscape.The SDGs provide a framework for tackling global issues and attaining sustainable development in a number of fields. SDGs 4, 8, 9, 11, 12, 13, and 17 relate to the design of the digital landscape in the chemical sector in several ways. SDG 4 emphasises the significance of universal access to high-quality education. Digital technologies are essential for giving experts and individuals in the chemical industry access to education and training. Creating digital learning resources and platforms with the chemical industry's demands in mind can improve worker proficiency, advance safety procedures, and stimulate creativity. This is related to SDG 8, which is about ensuring that everyone has access to decent work and sustained, inclusive, and sustainable economic growth. Digitalisation in the chemical industry has the potential to generate new employment possibilities, especially in cybersecurity, artificial intelligence (AI), and data analytics. The sector can promote innovation, productivity, infrastructure (SDG 9), and job creation while guaranteeing decent working conditions for employees by investing in digital infrastructure and skills development. SDG 9 also emphasises the significance of developing innovative practices, encouraging inclusive and sustainable industrialisation, and creating resilient infrastructure.

Automation, the Internet of Things (IoT), and machine learning are examples of digital technologies that are revolutionising the chemical industry by streamlining production procedures, increasing resource efficiency, and lessening environmental impact. The industry may reduce its environmental impact and increase its competitiveness by investing in digital infrastructure and innovation ecosystems. SDG 11 will help support sustainable cities and communities. By facilitating intelligent and effective management of resources, waste, and emissions, digitalisation in the chemical industry can help create sustainable communities. For instance, data analytics can optimise transportation routes and logistics, lowering traffic and emissions in urban areas, while digital sensors and monitoring systems can assist in identifying and mitigating pollution issues.

In addition, SDG 12 emphasises the need to guarantee sustainable patterns of production and consumption. Digital technologies enable businesses to monitor and optimise resource use, reduce waste, guarantee product safety, and comply with regulations, thereby promoting increased openness and traceability in the chemical supply chain. By adopting digital solutions for a circular economy and sustainable production, the industry may lessen its environmental impact and help achieve SDG 12 ambitions. This is also attributable to the conscientious actions taken in support of global climate action (SDG 13), which calls for the mitigation of climate change and its effects. The chemical industry may benefit greatly from digitalisation as it moves toward a resilient and low-carbon economy. Digital modelling and simulation technologies, for example, can maximise process efficiency and energy usage, while proactive adaptation efforts can be enabled by predictive analytics, which can also foresee dangers associated with climate change. The industry may help achieve SDG 13 objectives by utilising digital technology to lower greenhouse gas emissions and boost climate resilience.

Overall, SDG 17 highlights how crucial it is to fortify international collaborations in order to accomplish the SDGs. To fully realise the promise of digitalisation for sustainable development in the chemical industry, cooperation between governmental bodies, industry players, academic institutions, and civil society is vital. Partnerships have the potential to push collective action towards attaining the SDGs and expedite the adoption of digital technologies through the promotion of knowledge exchange, best practice sharing, and open innovation ecosystems. The collaboration between the CHIETA and the University of Johannesburg (UJ) in the project of designing a digital landscape is typically a promotion of the sustainability partnership goals (SDG 17). The fact that SDGs 4, 8, 9, 11, 12, 13, and 17 are in line with how the chemical industry is developing its digital landscape highlights how revolutionary digitalisation can be in promoting sustainability, innovation, and equitable growth in the sector. The chemical sector can address global concerns through collaboration and the use of digital technology. Despite the above, a thorough understanding of global best practices is imperative to effectively implement digital landscape design for sustainability in developing nations. These best practices offer invaluable insights and guidance on tried-and-true tactics, technologies, and policies that can be customised and adjusted to meet unique local contexts and challenges.

The CHIETA's global best practices approach to digital system landscape and sustainability

As the chemical industry's preferred partner, the CHIETA has positioned itself as a leader in education and training, providing innovative solutions for the creation of long-term sustainable values. Similar to other SETAs in the country, the CHIETA provides short-term, medium-term, and long-term strategies for the chemical industry, integrating these strategies across various sectors, institutions, and individuals within South Africa's chemical industry. In this case, the CHIETA sought to develop optimum enablement in order to maximise delivery with the adoption of the fourth industrial revolution (4IR) by SMEs. The CHIETA designed this enablement process as a solution to the current limited efficiency and lack of skills and expertise that are limiting SME growth and sustainability in the chemical sector, as identified and documented in its Annual Report 2020/2021 (CHIETA, 2021). The enablement included robotic process automation (RPA), integration, and the adoption of data analytics. RPA refers to the automation of service tasks that replicate human work (Ribeiro *et al.*, 2021). It represents an emerging solution for automating workflow processes, such as data processing. RPA has transformed workflow automation by combining the power of automation with the precision of technology, offering numerous benefits to businesses across various industries. RPA bots are highly accurate, minimising errors that can occur owing to manual data entry or repetitive tasks (Sharma *et al.*, 2022). This accuracy is crucial in industries like finance and healthcare, where precision is paramount. RPA eliminates the need for human intervention in repetitive tasks, reducing labour costs.

The efforts of the CHIETA in implementing RPA and AI in designing a digital system landscape are based on lessons learned from global best practices. In this case, a detailed literature study was conducted to provide insights into the global best practices associated with typical service-oriented enterprises, such as the CHIETA. Global best practices provide insights into the overall ecosystem and individual system constructs. One of the major best practices applicable to the CHIETA was a reflection on enterprise architecture (EA). EA is an approach to managing organisational complexity by aligning business analysis, governance, and IT (Řepa and Svatoš, 2019). EA is difficult to implement owing to complex maintenance, comprehensive coverage of all organisational areas, and conflicting frameworks and business needs. Given the complexities of implementing EA in a developing country like South Africa, the CHIETA deemed it crucial to reflect on this. Comprehending the technological practices of organisations, which are analogous to EA best practices (EABPs), is essential for successful implementation. Enterprise resource planning (ERP) is also crucial because it is the backbone of business intelligence (BI) and provides an integrated analysis of processes involved, from suppliers to customers, into an integrated system with shared data and visibility (Freitas Junior *et al.*, 2019). Cloud-based ERP systems are rapidly growing and significantly impacting current business models. Essential factors for ERP implementation include business process alignment with specific vendors' standard ERP systems; meeting customer and stakeholder needs; lowering maintenance costs; flexible reporting systems for ERP; and aligning IT with business strategy. Systems integration is an essential component of ERP.

Systems integration (IT/software integration) is the process of joining software and hardware modules into one cohesive infrastructure. Vertical, horizontal, data, and physical systems integration are all types of systems integration. The advantages of systems integration are increased productivity, accurate and trustworthy data, faster decision-making, and cost-effectiveness (Freitas Junior *et al.*, 2019). Typical best practices include modernising legacy systems by connecting to data warehouses and transportation management systems (TMS); and establishing newer information systems and technology solutions. Business-to-business (B2B) integration automates transactions and document exchanges across companies, for example, by connecting a retailer's purchasing system to a supplier's ERP system. Other important best practices include the following.

- **Business intelligence (BI):** BI is the ability of an enterprise to meaningfully leverage available data. Categories include competitive intelligence, consumer intelligence, industry intelligence, commodity intelligence, and operational intelligence.
- **Role-based visualisation:** Conventional architectures have slow analytical processing capacity and swamp IT with persistent ad hoc analyses and one-off reports, thus frustrating decision-makers. Visual analytics enable an interactive user experience that combines powerful in-memory technology, an easy-to-use interface, and advanced data visualisation.
- **Services sector systems:** Manufacturing operations management refers to workflow control to produce desired products and includes production operations, inventory operations/internal logistics, maintenance operations, and quality operations (Erasmus *et al.*, 2020).
- **Customer relationship management (CRM):** CRM is a 'cross-functional strategic approach aimed at formulating improved shareholder value via customer relationship development' (Al-Homery *et al.*).
- **Human resources (HR):** Employees' innovative work behaviours (IWBs) are critical to the functioning and long-term survival of organisations. Regrettably, a knowledge gap

exists around multilevel HR processes that motivate employees to display IWBs, such as leadership styles, work group characteristics, personality traits, job characteristics, and job demand.

- **Finance:** Organisational financial transformation systems strive to drive business growth through efficient transaction processing and reporting. The CHIETA viewed global best practices as crucial in providing valuable insights into critical areas of digital system landscape design. These insights would address the identified issue of limited digital optimisation and foster sustainable economic growth in South Africa by leveraging the chemical sector's contributions. From a practical standpoint, the CHIETA and the chemical industry can apply global digitalisation best practices to create a digital system landscape using RPA and AI. This approach encourages the execution of various strategies, ultimately leading to the successful attainment of sustainable objectives. Such strategies include, but are not limited to, data analytics, cloud computing, automation, and robotics. To make the project a success, the CHIETA, in alignment with SDG 17, had to partner with experts in the field. The CHIETA approached UJ's team for assistance in implementing the solutions.

The CHIETA and the University of Johannesburg collaboration: a digital system landscape design

As a driving force behind industry-related projects, UJ ensures close adherence to the working world and fosters innovation through applied research. UJ is poised to become the epicentre for critical intellectual inquiry across Africa. In order to accelerate the transition to 4IR and sustainability, UJ has committed to turning a number of departments and institutes into flagship organisations. It is for this reason that the CHIETA stakeholders then approached UJ, a seasoned university specialising in digital transformation, to provide support in implementing the design of a digital system landscape using RPA and AI. Given the history of UJ as pioneers of 4IR implementation in organisations and the institution's recent achievements in global participation in 4IR projects, ranking the institution among the top on the continent, the CHIETA saw it as the best decision to partner with the institution for the success of the digital system landscape. The UJ team was tasked with providing a critical analysis of the current CHIETA system, reviewing international best practices, and developing a sustainable strategy for digital transformation. The UJ team, in response to the task, hoped to create a digital system landscape for the CHIETA based on the availability of global systems, capital availability, and a strategic implementation plan. As a result, the UJ team conducted a comprehensive review of the CHIETA with supporting information in the form of processes, with a focus on gap analysis, needs analysis, and digitalisation. The purpose was to provide a digital enablement strategy for the CHIETA. The UJ team adopted a three-step research approach in the process (Figure 9.1).

The team conducted a detailed literature review to provide insights into the global best practices associated with service-oriented enterprises, such as the CHIETA. The analysis was divided into three parts:

- summary and overall digital strategy
- literature in support of the work presented in this study
- current CHIETA systems review and analysis.

The above was considered and unpacked by the team during the research process. A digital strategy was only the first step, with subsequent steps including business prioritisation, digital design, market scans, and then a full implementation strategy. This is shown in Figure 9.2.

The UJ team, regarding the processes taken (shown in Figure 9.2), undertook practical considerations. These included cost implications, skills, and timelines. Given that the CHIETA has defined a detailed ambition for digital optimisation with multiple focus areas, including operationalisation, optimisation, digitalisation, and intuitive operations, the UJ team focused this research on supporting the CHIETA in this ambitious strategy. In order to achieve digital migration, some fundamentals were considered by the team. The strategy was reflective of a comprehensive business change that would have a significant impact on business operations if implemented at once. The team therefore proposed two suggestions. First, the CHIETA was advised to develop an implementation strategy based on the key (backbone functionality) systems. This was deemed necessary to orient around the data warehouse and integration functions, followed by the functions that would deliver the most significant value to the business. Second, a recommendation was given to have the digital strategies grounded, with IT theories as the most appropriate grounding. The key IT approach adopted by the team was converting business activities into a structure that could interface with the development team.

The team in this business case converted all business activities into business processes. The processes were tiered into levels zero to three for ease of understanding. Digital maturity has been quantified through business process mapping and IT landscape analysis. The integrated data layer is the most important component of an integrated digital business. The integrated data layer serves as the central data layer, accommodating all functions within a single structured data warehouse. This serves as the foundation for the BI layer and, by default, AI can only function when all data are visible. The functions above were brought together by the UJ team into a data layer or enterprise data layer. Following this, the CHIETA was advised to do the following.

- Develop an integrated data warehouse, preferably cloud based.
- Develop an enterprise data design.
- Ensure functional data integration.
- Define and manage services for data integration.

The data and BI layers are synchronous and are highly interdependent. To achieve the above, an enterprise structure or template was designed, and the processes were developed. This was done in conjunction with standard operating procedures (SOP) updating and the repository. Process mapping is a science that consists of business processes or activities that begin at level zero (business function) and explode down to level seven (activity level). A significant amount of work was done in capturing and managing business processes. All IT vendors used business process maps to track the current status of activities. Additional activities implemented in the project include the following.

- Appoint a change management team.
- Develop a change strategy.
- Carry out change workshops.
- Develop a communications strategy and plan.
- Develop a skills plan (HR).
- Align skills training to systems prioritisation.
- UJ: develop as-is against to-be with engagement of all functions.
- Develop a request for quote (RFQ) to go to market for all priority solutions.

Project outcomes

The University of Johannesburg, Johannesburg Business School (UJ–JBS) team's effective accomplishment in designing the system landscape embodies a significant shift in cultivating operational efficiency and advancing the SDGs. In relation to SDG 4, the implementation of the digital landscape will ensure that there is heightened access to education and training tools and resources, endorsing skills and capacity building among chemical sector employees and the South African community as a whole. In addition to this, the enhanced data management abilities and restructured processes help achieve SDG 8 by promoting innovation, creating employment, and facilitating comprehensive economic growth and sustainability in the chemical industry. The CHIETA–JBS project also contributes to SDG 9 by implementing digital tools and technologies and encouraging innovation, thereby advancing a sustainable industry and enhancing infrastructure development.

In addition, SDG 11, which emphasises robust infrastructure and sustainable cities, is aligned with the digital landscape system's emphasis on resource optimisation and the alleviation of environmental impacts. SDG 12 is reinforced by the system design through the restructured patterns of consumption and production. SDG 12 fosters resource competence and waste minimisation, ensuring the implementation of the CHIETA's projects with a cautious approach to environmental well-being. The digital landscape system also assists in achieving SDG 13 by utilising data analytics and modelling to help predict and reduce climate change risks and dangers.

The CHIETA–JBS collaboration successfully achieved SDG 17, which underscores the significance of stakeholder collaboration and collective measures to achieve SDGs. The system largely enabled industry–academic collaboration and knowledge sharing in this project. The CHIETA–JBS system landscape design serves as a typical example of how innovation and digital technologies can be used to achieve the SDGs, benefitting various organisations as well as societies on a global scale.

Conclusion

Following the project's successful completion, the two stakeholders – the CHIETA and UJ–JBS team – engaged in conversations to determine the next steps. Architecture determinations can be complex for business representatives, but the ownership of architecture resides in IT, with functional determinations residing in business. The UJ–JBS team examined the existing portfolio and conducted a thorough review of recent papers in the services sector. The team collaborated with the current software developers, conducting an analysis of their current capabilities and designing the next version of software. The UJ–JBS team identified operations and functional opportunities that should be considered to improve current CHIETA operations and strategically address future systems that would catapult the CHIETA into the digital age. Automation and system-centric business enablement are the focus. To enhance the operations of the chemical sector through this design, the executive and board of the CHIETA have decided to prioritise the following considerations.

The refinement of the current service model: A first-to-third-line information technology infrastructure library (ITIL)-aligned set of contracts together with a structured service level agreement (SLA) has to be developed and agreed. This would reduce the risk of current and future service management, as well as the transition thereto. The above must include a license model for all software. Some of the current databases pose a risk, as the CHIETA does not host or have access to data. Operational data storage must be reconfigured with the use of CHIETA SharePoint. This comes with an active directory (AD) role-based access and management. In the current configuration, the database is handled by the software developer. The database services must be ring-fenced and separated by an SLA. The CHIETA's ICT department must have the capacity to support a first-to-third-line structure, monitor utilisation, conduct upgrades, manage ICT capacities, manage security, monitor SLAs, update and manage services, enhance reporting and overall operations, and ensure quality management of the system and processes. The ICT team must also manage AD.

Reconsidering the system's landscape: Best practice dictates the modularisation of enterprise functions, but current and forecast systems do not include functions like HR and finance. The system incorporates the CRM's key function as a prominent module. With the above in place, the reporting database should be singular, and the entire database must be available for analytics. The CHIETA team must be configured to view all data and reports, as per roles, or have the ability to build new reports.

Systems functionality: The current and forecast systems require some key functionality. This includes a fully enabled CRM that integrates into the CHIETA's processes. This includes all applications, levies, services, and other workflows.

- Integration: this must be focused on including SharePoint integration, workflow integration, reporting integration, integration with the Department of Higher Education and Training (DHET), integration with SARS, CIPLA, and Home Affairs. This would provide for enhanced authentication and acceleration of workflows, especially on the administration side.
- Services app: A learner and service provider app must be considered. This would provide a secure application for comprehensive monitoring and management of both learners and service providers. This can be integrated with BI and automated reporting.
- Business intelligence: a BI layer must be constructed. A reporting layer must be developed with CHIETA ICT to support it.

The major lesson learned from the UJ–JBS team and the CHETA collaboration was that the digital systems landscape is the key to a sustainable enterprise. The CHIETA aims to optimise the chemical industry for a profitable and sustainable future, and this collaboration brings us one step closer to achieving this goal and the overall achievement of the SDGs. It is believed that integrated digital systems can help close associated gaps and promote the sustainable efficiency of the businesses involved. CHIETA-related businesses, especially SMEs, require integrated digital systems that can enable them to adopt a variety of mechanisms and solutions to improve their efficiency and competitiveness. Currently, an integrated digital system is lacking in SMEs found in developing countries, especially in South Africa. The JBS team's integrated approach, with its potential for long-term sustainability and growth, holds significant value for the CHIETA.

Exhibits

Figure 9.1 Research approach adopted by the UJ team

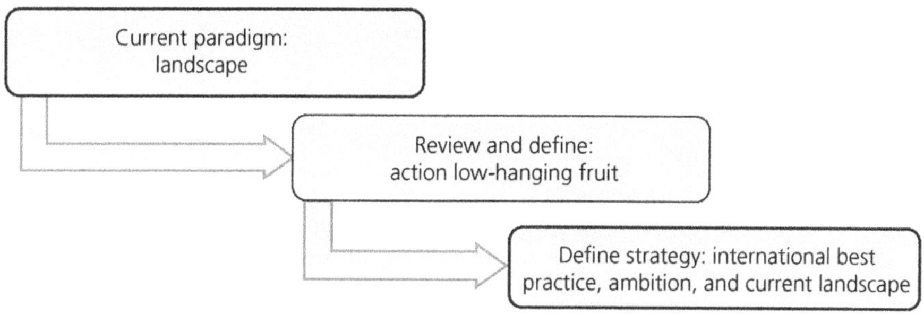

Figure 9.2 Research implementation steps taken by the UJ team

Figure 9.3 SDGs aligned to the CHIETA project

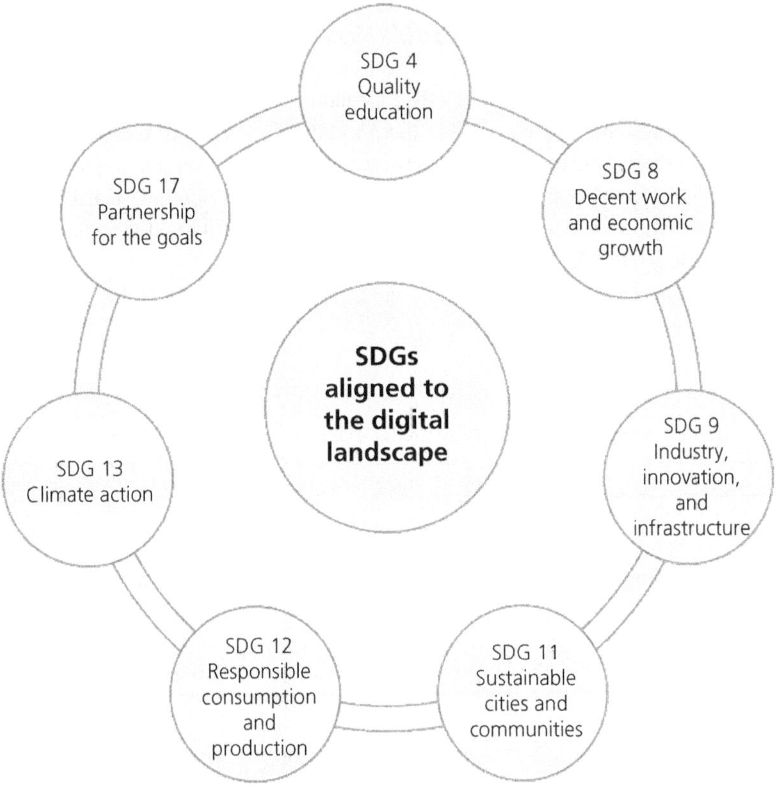

Table 9.1 Alignment of the CHIETA's project with the SDGs (continued on next page)

SDG	Description	Alignment to the CHIETA's digital landscape project
4	Quality education	By giving access to online learning platforms, individualised learning experiences, and educational resources, the integration of RPA and AI technologies in digital systems can improve educational opportunities and help ensure that all students receive a high-quality education.
8	Decent work and economic growth	Organisations can align with SDG 8 by generating more job opportunities, improving labour productivity, and stimulating economic growth through the use of RPA and AI to streamline processes and increase productivity.

Table 9.1 Continued

SDG	Description	Alignment to the CHIETA's digital landscape project
9	Industry, innovation, and infrastructure	With a particular focus on inclusive and sustainable industrialisation, this goal fosters technological innovation. This objective is advanced by the employment of RPA and AI in digital system architecture, which promotes innovation and increases infrastructure efficiency.
11	Sustainable cities and communities	RPA and AI integration in digital system design can enhance resource management, service delivery, and urban efficiency – all of which support resilient communities and sustainable urbanisation.
12	Responsible consumption and production	By encouraging sustainable consumption patterns and effective resource usage, RPA and AI technologies can optimise processes and reduce resource consumption, which is in line with SDG 12's aims.
13	Climate action	By optimising energy usage, eliminating waste, and promoting sustainable behaviours, digital transformation enabled by RPA and AI can assist enterprises in lowering their carbon footprint and thus lessen the effects of climate change.
17	Partnerships for the goals	Achieving the SDGs requires cooperation between governments, corporations, and technology providers to use RPA and AI technologies in digital system architecture. SDG 17's core tenets are cooperation and partnership.

The use of robotic process automation and artificial intelligence in designing a digital system landscape: a sustainable approach

Teaching notes

Synopsis

This case follows the Chemical Industries Education and Training Authority's (CHIETA)'s efforts to develop optimal enablement, aiming to maximise delivery through the adoption of 4IR. This includes robotic process automation, integration, the adoption of data analytics, and AI. To achieve this goal, the CHIETA partnered with the University of Johannesburg (UJ), a leading institution in digital transformation, to facilitate the design of a digital system landscape. This collaboration had a one-year lifespan in 2023. The major challenge faced by the CHIETA, in this case, was limited solutions for digital transformation, which the UJ–JBS team hoped to overcome through the digital system landscape design using RPA and AI. A critical review of global best practices in architecture development, focusing on the services environment, together with an analysis of current CHIETA systems and the development of a proposed digital strategy, was done. These teaching notes provide a pedagogy on the digital system landscape using the case of the CHIETA's pursuit to implement the systems design through RPA and AI. The case study falls under the academic sub-fields of digitalisation, sustainability, and 4IR.

Teaching objectives

This case study is useful for students pursuing business management, entrepreneurship, and related skills. Given that digital technologies and digitalisation are currently the focus of discussions and developments around business growth, competitiveness, and sustainability, this case study fulfills the purpose of providing insights on digital systems and maturity analysis to students in their respective relevant disciplines. Furthermore, this case study exposes students to knowledge on how futuristic projects such as the CHIETA's project of designing a digital landscape can be aligned to the SDGs and how to provide digital solutions to solve global problems and achieve the stated goals, as indicated in Figure 9.3.

Students will be exposed to the knowledge that designing a digital landscape assists in accomplishing the above-mentioned SDGs through the enhancement of productivity, business transparency, and the provision of access in various sectors, as well as the promotion of sustainable economic growth and social integration. Furthermore, the case study aims to educate students on how such a design enables data-driven policymaking and collaboration, leading to more organised resource distribution and focused partnerships to address the global issues the SDGs aim to tackle. As such, it is anticipated that, on completion of reading this case study, the students will be exposed to the following.

- Knowledge and understanding of the significance of the CHIETA and the chemical industry to the sustainability of the South African economy.
- Understanding of the challenges that are associated with a limited, or lack of, implementation of best practices digital systems in today's and future business.
- Evaluate the importance of the design of digital systems integration through RPA and AI on sustainable growth of economies.
- Understanding of the advantages of integrated digital systems in today's business environment, with an emphasis on sustainability principles and how they affect social and environmental aspects.

Research methods

This case study sought to focus on the digital strategy associated with the system and functional landscape that can deliver maximum SETA value for today and tomorrow.

The research is approached from a business, functional, and IT perspective. The methodology employed by the UJ team in this case study was based on a three-step process.

- **Summary and overall digital strategy:** this step included an outlook on the current paradigm on digital system landscapes.
- **Literature in support of the work presented in this study:** this step included a review of international best practices and a strategy for digital transformation.
- **Current CHIETA systems review and analysis:** this included a situational analysis of the current CHIETA systems to understand the current gaps and limitations on digital optimisation. This helped in providing a comprehensive review of the CHIETA with supporting information in the form of processes.

The three-step process helped the UJ team gain insights into the current paradigm and the CHIETA's current system, as well as provide a critique relative to best practices for this case. The team further identified and added new, technologically aligned methods to enhance the overall system's opportunities. The UJ team implemented various components to realise the integrated architectural strategy.

- **Change management:** Case studies have proven that the key limitation in digital transformation is change management. The CHIETA was advised to commence the change cycle immediately.
- **Business case:** The CHIETA was advised to convert the strategy into a full business case, including costing, an implementation plan, a change strategy, a skills strategy, an investment plan, cost recovery, and impact and business model strategy.
- **Technical team:** The migration to digital would require a strong internal team from IT, and business, together with external consultants. The CHIETA was advised to develop a strategy and empower a technical team.
- **Skills:** The reskilling and upskilling of employees is a key aspect of digital transformation. Linked to this strategy is the decoupling of non-value-add activities and localised development.
- **System reproducibility:** A key aspect of the development of digital solutions centres around reproducibility. With the architecture provided, the design is for a multisite model. ERP and data warehouses can be cloud services, along with other enterprise systems.
- **KPIs:** The ability to manage a business through key performance indicators (KPIs) is essential. The ability to develop trends for any KPI or combination of KPIs is critical for optimisation and transition management.
- **Process mapping and standard operating procedures:** Institutional memory is a key consideration for modern companies migrating into digital space. Traditionally, SOPs and, more recently, business processes capture this. The UJ team emphasised the importance of securing institutional memory in order to mitigate resource transitions out of the organisation.

Keywords and subject code

Digital optimisation, digital systems landscape, systems integration, robotic automation process, artificial intelligence.

Assignment questions

1. Discuss the significance of the CHIETA for the South African economy.
2. What are the challenges associated with a limited implementation of best practices digital systems in today's business?
3. Evaluate the importance of RPA on the designing of digital systems integration.
4. What are the benefits associated with integrated digital systems in today's business practice?

Additional materials

The following references and materials are recommended for further information regarding this case study.

Chingapi A and Steyn AA (2022) SMEs in South Africa: the era of adopting mobile payment solutions. *Lecture Notes in Networks and Systems* **235**: 429–447, 10.1007/978-981-16-2377-6_41.

Disse S and Sommer C (2020) *Digitalisation and its impact on SME finance in Sub-Saharan Africa: Reviewing the hype and actual developments*, Discussion paper 4/2020. Deutsches Institut für Entwicklungspolitik, Bonn, Germany.

Eniola AA and Entebang H (2015) SME firm performance-financial innovation and challenges. *Procedia – Social and Behavioral Sciences* **195**: 334–342, 10.1016/j.sbspro.2015.06.361.

Kennedy J and Hyland P (2003) A comparison of manufacturing technology adoption in SMEs and large companies. *16th Annual Conference of Small Enterprise Association of Australia and New Zealand*. University of Ballarat, Ballarat, Australia, pp. 1–10.

Sanga B and Aziakpono M (2022) The impact of technological innovations on financial deepening: implications for SME financing in Africa. *African Development Review* **34(4)**: 429–442, 10.1111/1467-8268.12668.

Sen D, Ozturk M and Vayvay O (2016) An overview of big data for growth in SMEs. *Procedia – Social and Behavioral Sciences* **235(October)**: 159–167, 10.1016/j.sbspro.2016.11.011.

Stoldt J, Trapp TU, Toussaint S *et al.* (2018) Planning for digitalisation in SMEs using tools of the digital factory. *Procedia CIRP* **72**: 179–184, 10.1016/j.procir.2018.03.100.

Suggested core readings

Bhattacharyya S, Banerjee JS and De D (eds) (2023) *Confluence of Artificial Intelligence and Robotic Process Automation*. Springer Nature, Berlin, Germany. https://www.chieta.org.za/resource-center/annual-reports/ (accessed 15 October 2023).

Moderno OBDS, Braz AC and Nascimento PTDS (2024) Robotic process automation and artificial intelligence capabilities driving digital strategy: a resource-based view. *Business Process Management Journal* **30(1)**: 105–134.

Ray JK, Sultana R, Bera R, Sil S and Alfred QM (2023) A comprehensive review on artificial intelligence (AI) and robotic process automation (RPA) for the development of smart cities. In *Confluence of Artificial Intelligence and Robotic Process Automation* (Bhattacharyya S, Banerjee JS and De D (eds)). Springer Nature, Berlin, Germany, pp. 289–311.

Teaching plan and timing

For a class of 90 min, the following teaching plan is recommended.

Introduction (5 min)

- The instructor should provide a summary of the case.

Teaching the assignment questions (80 min)

- This should be structured in the following manner.

Assignment question 1 (20 min)

- Discuss the significance of the CHIETA to the South African economy.

Assignment question 2 (20 min)

- What challenges are associated with the limited implementation of best practices in digital systems in today's business?

Assignment question 3 (20 min)

- Evaluate the importance of RPA in the design of digital systems integration.

Assignment question 4 (20 min)

- What are the benefits of integrating digital systems in today's business practice?

Conclusion (5 min)

- Wrap up the case and summarise the key learnings.

Analysis of assignment questions

Question 1. Discuss the significance of the CHIETA to the South African economy.

The instructor can ask the learners to evaluate the role of the CHIETA in promoting growth in the South African economy. A possible answer could be as follows.

The organisation has positioned itself as a leader in education and training, providing innovative solutions for the creation of long-term sustainable values as the chemical industry's preferred partner (CHIETA, 2021). Similar to other SETAs in the country, the CHIETA provides short-, medium-, and long-term strategies for the chemical industry, integrating these strategies across various sectors, institutions, and individuals within South Africa's chemical industry.

Question 2. What are the challenges associated with a limited implementation of best practices in digital systems in today's business?

The instructor can ask the learners to identify associated challenges that may arise from a limited ability to comprehensively enable best practices within the chemical sector.

Before getting into the discussion, it is crucial to enlighten the learners on what best practices mean. A best practice is a method or technique that has been generally accepted as superior to other known alternatives because it often produces results that are superior to those achieved by other means or because it has become a standard way of doing things. In today's business environment, digital technologies and digital maturity dictate the pace of current best practices in conducting business. In the chemical sector, best practices are the ones that promote increased efficiency, growth of the sector, and eventually a meaningful contribution to economic development and growth. As the world continues to evolve, new technologies are invented to ensure a positive outcome in the production of goods and services. Only those businesses that are quick to adapt to these changes will see much-needed growth. Moreover, the convenience of purchasing and consuming goods and products attracts consumers. Businesses that fail to make it easier for consumers to access their offerings tend to fail. Therefore, a limit on enabling new technologies may result in the CHIETA businesses' slow growth and transformation, as they will be considered unattractive and lack the skill to adapt to environmental changes. The above situation will result in the sector not being able to fulfill its role in economic growth and transformation, especially in developing economies. This will also affect the sector's role in overall economic sustainability.

Question 3. Evaluate the importance of RPA on the designing of digital systems integration.

The instructor can ask the learners to identify the importance of RPA in the design of digital systems integration.

RPA refers to the automation of service tasks that replicate human work. It represents an emerging solution for automating workflow processes, such as data processing. RPA has transformed workflow automation by combining the power of automation with the precision of technology, offering numerous benefits to businesses across various industries. RPA bots are highly accurate, minimising errors that can occur owing to manual data entry or repetitive tasks. This accuracy is crucial in industries like finance and healthcare, where precision is paramount. RPA eliminates the need for human intervention in repetitive tasks, reducing labour costs.

Question 4. What are the benefits associated with integrated digital systems in today's business practice?

The instructor can ask the learners to identify the importance of designing digital systems integration.

Systems integration (IT/software integration) is the process of joining software and hardware modules into one cohesive infrastructure. Vertical, horizontal, data, and physical systems integration are all types of systems integration. The advantages of systems integration are increased productivity, accurate and trustworthy data, faster decision-making, and cost-effectiveness.

Key learnings

The instructor can conclude the lesson by highlighting the importance of the sector's digital enablement. It is also crucial to provide a pedagogy that emphasises the significance of a comprehensive systems and maturity analysis, as well as the associated benefits of doing so for the sector. For example, increased proficiency and business functionality, increased competition, and ultimately, economic growth and transformation are some of the key outcomes of digital enablement.

REFERENCES

Al-Homery HA, Ashari H and Ahmad A (2023) Customer relationship management: a literature review approach. *International Journal of Global Optimization and its Application* **2(1)**: 20–38, 10.56225/ijgoia.v2i1.160.

CHIETA (2021) 2020/21 Integrated Annual Report. https://www.chieta.org.za/wp-content/uploads/2022/02/CHIETA-ANNUAL-REPORT-2020_2021.pdf (accessed 14/01/2025).

CHIETA (2022) 2021/22 Integrated Annual Report. https://www.chieta.org.za/wp-content/uploads/2022/11/CHIETA-ANNUAL-REPORT-2021_2022.pdf (accessed 14/01/2025).

Freitas Junior JC da S, Cabral PMF and Brinkhues RA (2020) Digital transformation: the gap between digital leadership and business performance. In *Proceedings of the Information Systems in Latin America (ISLA 2020)*.

Řepa V and Svatoš O (2019) Model consistency as a tool for digital business architecture verification. *Procedia Computer Science* **159**: 2144–2153.

Ribeiro DMN *et al.* (2021) Digital sustainability: How information and communication technologies (ICTs) support sustainable development goals (SDGs) assessment in municipalities. *Digital Policy, Regulation and Governance* **23(3)**: 229–247.

Sharma S, Kataria A and Sandhu JK (2022) Applications, tools and technologies of robotic process automation in various industries. In *2022 International Conference on Decision Aid Sciences and Applications (DASA)*. IEEE, pp. 1067–1072.

THEME 3

SUSTAINABLE URBAN AND RURAL DEVELOPMENT

Prof. Arnesh Telukdarie
ISBN 978-1-83662-645-9
https://doi.org/10.1108/978-1-83662-642-820251010

Chapter 10
Enhancing sustainable rural development through collaborative innovation: a case study

Arnesh Telukdarie and Thabile Dube

Introduction

The International Labour Organization (ILO, 2022) indicated that there is a lot of significance attached to effective rural development strategies, considering that almost half of the world's population resides in rural areas. These areas are responsible for a large number of people who live below the poverty line, need food security, and have no jobs (ILO, 2022). However, traditional investments in rural communities meant to promote rural development have generally been unsuccessful, and the Chemical Industries Education and Training Authority (CHIETA) management of directors understood this. Besides, they also realised that these programmes were not addressing all the challenges faced by people at the grassroots level. In 2023, the CHIETA management resolved to move from mere talk to action aimed at changing their approach towards rural development. Through conversations with the University of Johannesburg (UJ), the CHIETA management conducted extensive research on understanding rural communities' unique needs and factors blocking their progress, acknowledging the limitations of conventional strategies and the transformative possibilities of rurality. The initiative spearheaded by these two organisations had the potential to create change in those disadvantaged regions. The combined effort emanated from directors who dreamed about transforming these areas into environmentally sustainable economies, aligning with broader global aspirations for a greener and more equitable future (Stanef-Puică *et al.*, 2019). As such, sustainability lies at the heart of this aspiring journey, which goes beyond the preservation of natural resources, encompassing the promotion of social equity and economic resilience (Hariram *et al.*, 2023; Mensah, 2019). Through empowering local communities, preserving cultural heritage, and fostering innovation aligned with environmental principles, the study aims to holistically address the challenges faced by rural areas. The green journey began between the two organisations.

The CHIETA industry

The CHIETA was set up in terms of The Skills Development Act (1998) as a key player that was to address skills shortages and ensure learning opportunities in the field of chemical industries. The mission of the CHIETA is to facilitate skills development, education, and training through innovative solutions for sustainable livelihoods. Considering that the chemical industry is one of the vital sectors in South Africa's economy, there has been an increasing need for highly competent employees capable of handling such skills within the industry. Consequently, the CHIETA took on itself a number of strategies aimed at empowering individuals. Through its strategy interventions and collaboration efforts, the CHIETA has triggered a change in mindset in job creation, fostering an environment conducive to sustainable growth and success within the chemical sector. As the chemical industry landscape continues to change, so does the need for creativity and quality, hence the CHIETA's commitment to innovation and excellence, which has seen it remain relevant, playing a critical role in steering sector development and promoting socio-economic well-being in South Africa.

Breaking the chains: the UJ and the CHIETA forge a path to rural prosperity

Despite their well-known potential, there are significant disparities in rural areas. A remarkable portion of the rural population face perennial hardships such as poverty, healthcare access limitations, education gaps, and lack of basic living conditions (Shrestha, 2021). It is therefore important to have comprehensive strategies for rural development because these challenges not only affect individuals' well-being but also hinder overall contributions by rural areas towards global socio-economic aims. That is what has made it necessary for the UJ and the CHIETA to conduct this study. The main aim of this study is to investigate how systems and subsystems interact with each other in rural areas. This initiative seeks to develop an all-inclusive system model to help uncover the complex determinants that inform rural development and sustainability. In addition to identifying these systems, this model will prioritise them strategically, providing a guide on how interventions should be targeted and how resources should be allocated. This cooperation does not merely encompass theoretical frameworks but instead focuses on their implementation in order to address the challenges encountered by populations in outlying regions.

Review of rural community systems literature

The rural ecosystem involves interconnected socio-economic systems and infrastructure that affect rural life. The so-called socio-economic system, which includes education, healthcare, and agriculture, is challenged by resource scarcity and the poor state of infrastructure (Vorobyov and Bugai, 2019). This becomes a problem for rural schools trying to change the situation and improve education. Meanwhile, the healthcare sector has its issues characterised by such hurdles as transport challenges and lack of practitioners besides limited access to modern equipment, all hampering the smooth delivery of health services. As a pillar for income and resource stability, agriculture faces uncertainty regarding meeting the growing global food demand. Integrating sustainability in the sector could solve the challenges and bring about change. This is because sustainable development in education, healthcare, and agriculture involves promoting practices that ensure long-term resource availability and resilience (Mensah, 2019). For instance, sustainable infrastructure can provide better learning environments, while implementing sustainable agricultural practices can improve food security and environmental health (Mensah, 2019; Wu *et al.*, 2019). Furthermore, providing the backbone of the socio-economic framework is infrastructure made up of transportation, energy/electricity, and water/sanitation (Wu *et al.*, 2019). Nonetheless, this facet has not been without problems either. Transport inefficiencies make service provision difficult, resulting in a lack of mobility among people living in rural areas, thus increasing their economic disadvantages. Electricity disparities within rural regions impact many aspects of life, such as lighting in homes or functions related to farming (Wu *et al.*, 2019).

The UJ and the CHIETA approach to rural system modelling

A comprehensive exploration into the complexities of rural systems and a strategic plan for rural communities was undertaken by the university and the CHIETA management. The partners' joint effort came about because they acknowledged that there was potential in rural areas and wanted to transform them using new approaches. The CHIETA understood these challenges well enough to design its own approach to their resolution, which considered specific problems affecting education, healthcare, agriculture, transportation, energy and water supply issues, and sanitation. To ensure thorough research efforts on this project, the institutions made use of already existing studies; thus, they used academic knowledge from UJ. Identifying rural systems involved several activities. As an institution for higher learning, UJ contributed scientific research methodologies into this process with the quantitative analysis approach and data modelling, among others. Although the main aim was to identify these rural systems, it also sought to understand their interconnections with each other, thereby acknowledging that there are several relationships between various components. This comprehension formed the basis for coming up with a complete system model. The model did not confine itself to just informing what went wrong but instead sorted out systemic priorities.

Results from the word2vec analysis

The search for better knowledge, skills, and capacities for effective rural development at the UJ started with the development and undertaking of very detailed research – aimed at studying and understanding the systems and subsystems that are the building blocks of the rural ecosystem. The ultimate goal here was to craft a neat system model that was not too complicated to understand, making it easy to get a grasp of the complexities of the whole situation while also assisting with the long-term sustainability of the priorities. This study is distinguished by the fact that the system, the goal, and the subsystems are assessed quantitatively and compared with each other. The journey got a new start with the adoption of the most sophisticated techniques, word2vec being the most useful since it helps in the analytics of the micro-affinities that can be identified in the data. UJ leveraged modern machine learning to discover relationships and identicalities of elements of the rural ecosystem that had never been found before. The effect resulted in heat maps. This kind of relation graph shows where the connections of strong and positive correlations are located (represented by marks that are dark shades – see Figures 10.1–10.8), where the moderate correlations are located (represented by marks that are light shades), and where no correlation is found (represented by marks of white shades). Following on from the heatmaps, UJ ensured that it carried out a robust regression analysis to dig even deeper into the linkages between the various subsystems. As expected, coordinates, ranging from -1 to 1, were calculated for every two subsystems. The most positive correlation is the correlation coefficient of 1, which means that there is a perfect positive correlation; the opposite is the absolute value of -1, which signifies a perfect negative correlation, and the coefficient 0 shows no correlation.

The data in Figures 10.1–10.8 formed the background, and UJ attempted to paint a colourful picture of rural interconnectedness, which may have been easily noted at a glance.

Analysing correlation coefficients

UJ examined the data and the system's ID parameters; these included rural demographics, agriculture, education, electricity/energy, healthcare system, science and technology, transport, and water. Table 10.1 shows the resulting analysis.

Through systemic analysis, a successful rural development plan consists of several facets, including policies that enable the farming systems to recognise the cross-component interactions. The collaboration between the UJ and the CHIETA management provides a foundation for a more precise, strategic, and productive approach to rural development, which is a path that relates to the desires and aspirations of the inhabitants of these rural areas.

Conclusion

With the fall of the curtain on UJ's insightful fieldwork into the underbelly of rural dynamics, the regression coefficients are more than just numbers; they are an outline of an impeccable roadmap that can shape the future of rural development. The application of new approaches and examination generated means not only academia but also policymakers and stakeholders can adopt these approaches, thus improving life in rural areas. The correlation coefficient derived from the rural systems now brings in a comprehensive picture of the sophisticated web that defines rural development. The intervention developed by UJ does not just tell the stories of problems but also gives decision-makers a direction, a roadmap, and insight into strategies they can use to make a difference. Furthermore, sustainability is integral to this roadmap, ensuring that development is not only effective but also enduring. Also, the same correlations match the CHIETA's vision, thus enabling it to adopt a strategic tool to shift rural areas into more environmentally friendly and fairer futures. UJ is not only about theoretical knowledge, but rather, it is a tool for change. It drives the CHIETA management to an outstanding rural community transformation that is based on a problem-focused approach and rationality. When UJ's story concludes, it is not the end; instead, this is the story of hope, empowerment, and the future of sustainable rural practice, which is highly dependent on connecting and resilience. The UJ team will strive to achieve the goals of the CHIETA board of directors, which is a step towards the development of a system model that provides a strategic approach. The story about rural development is just at its beginning.

Exhibits

Figure 10.1 Demographic systems

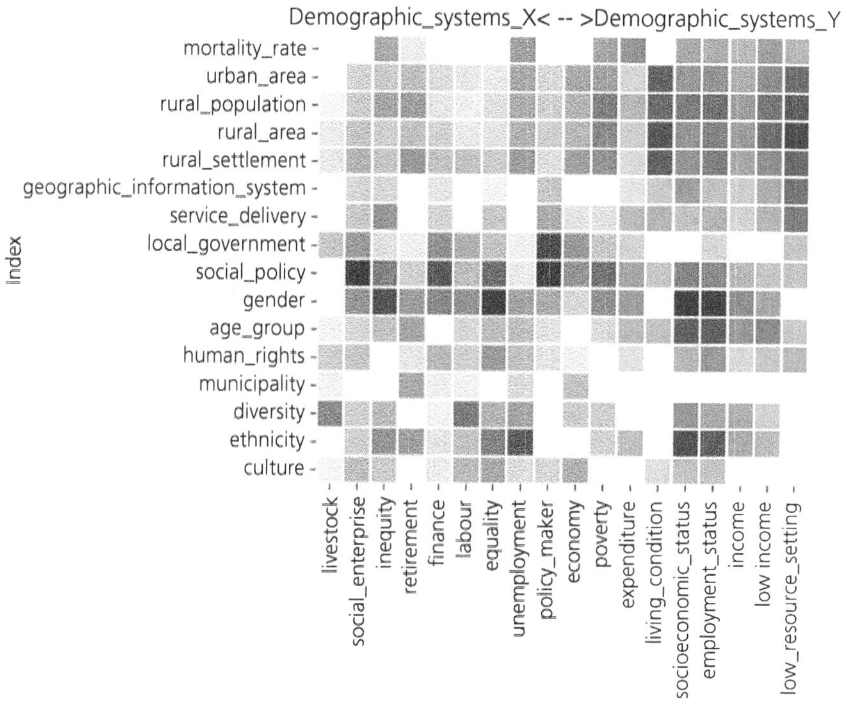

Figure 10.2 Agriculture systems

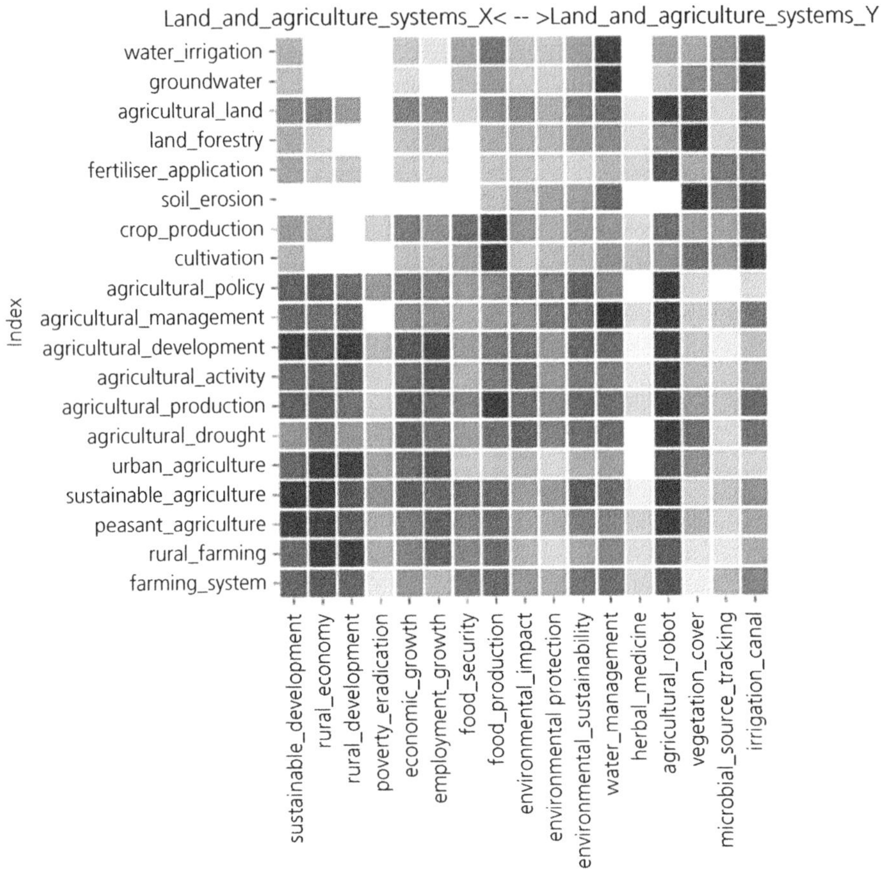

Figure 10.3 Education systems

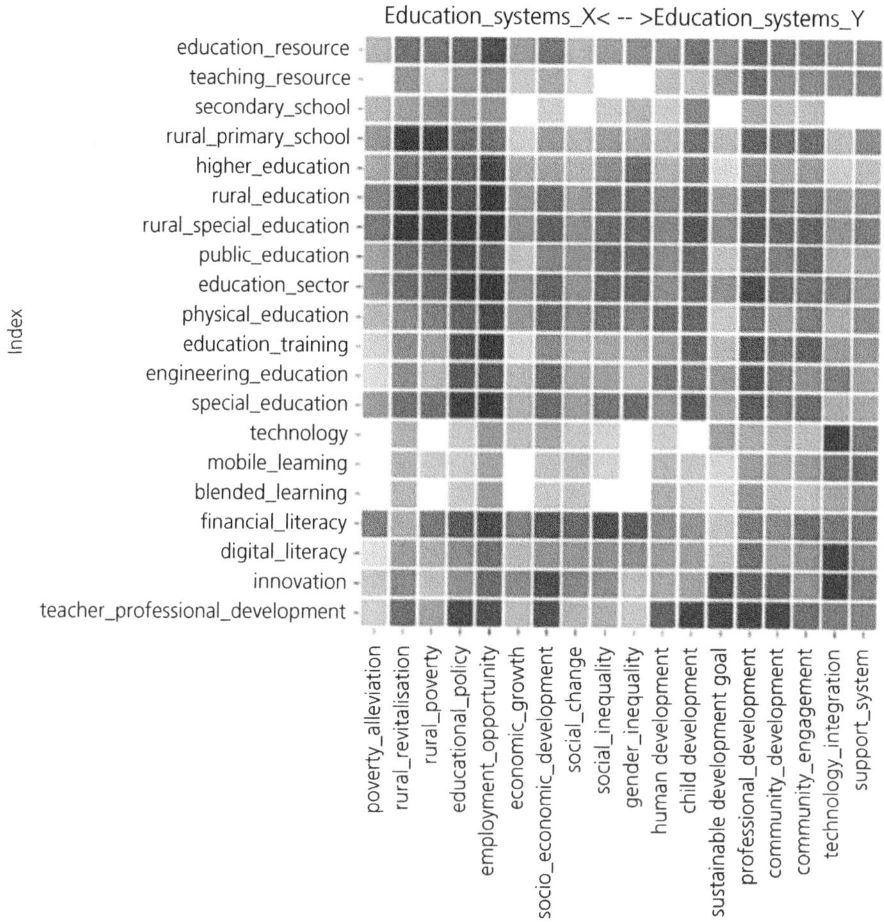

Figure 10.4 Electricity/energy systems

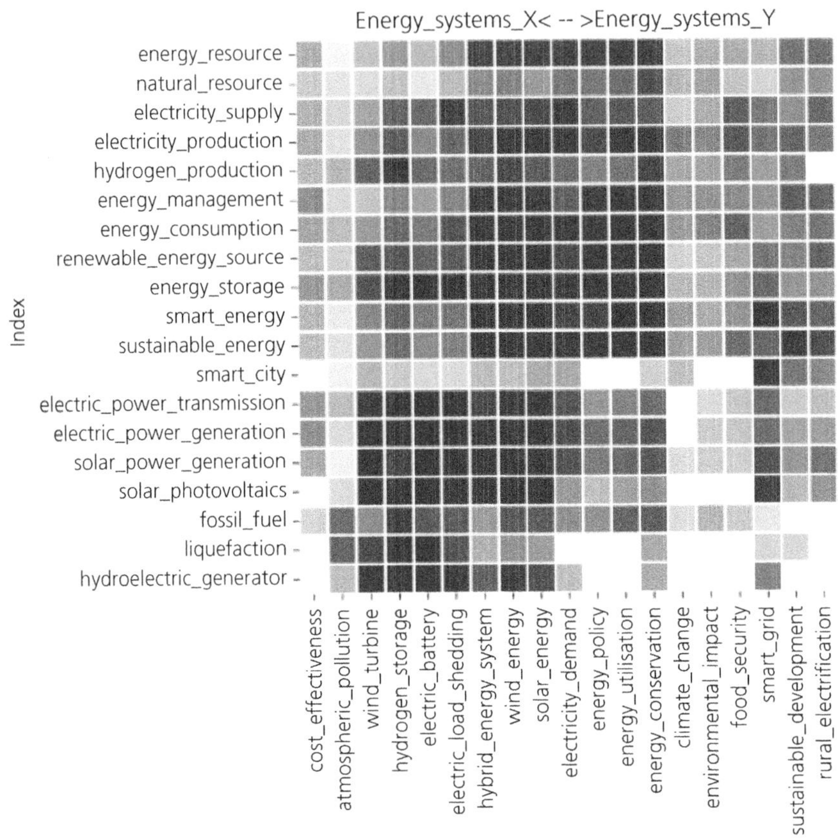

Figure 10.5 Healthcare systems

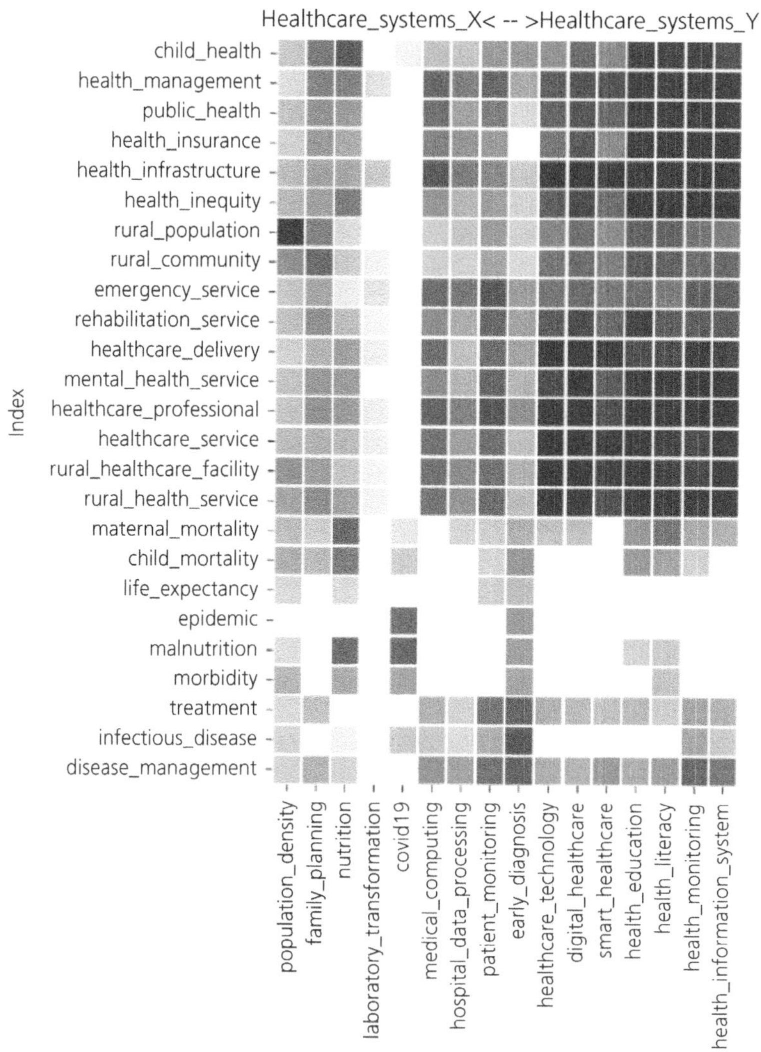

Figure 10.6 Science and technology systems

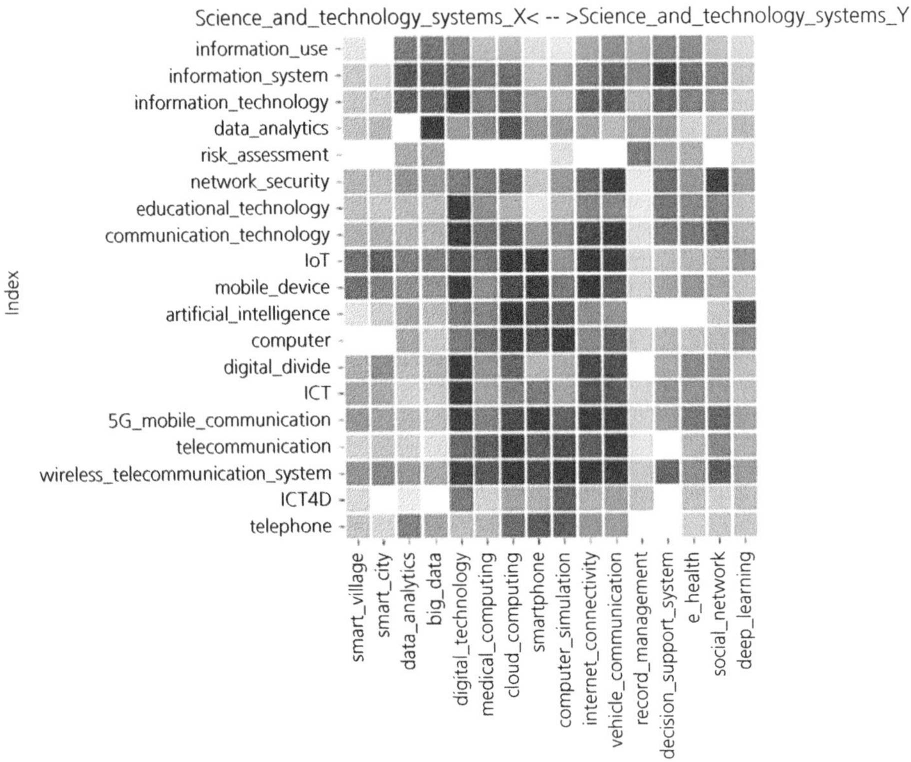

Figure 10.7 Transport systems

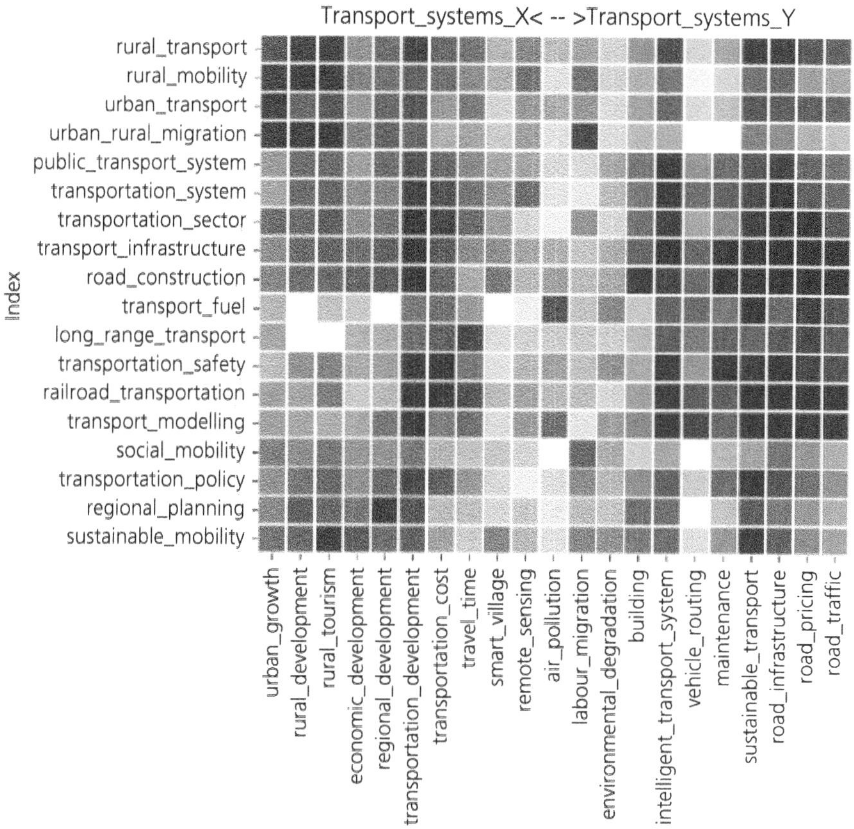

Figure 10.8 Water systems

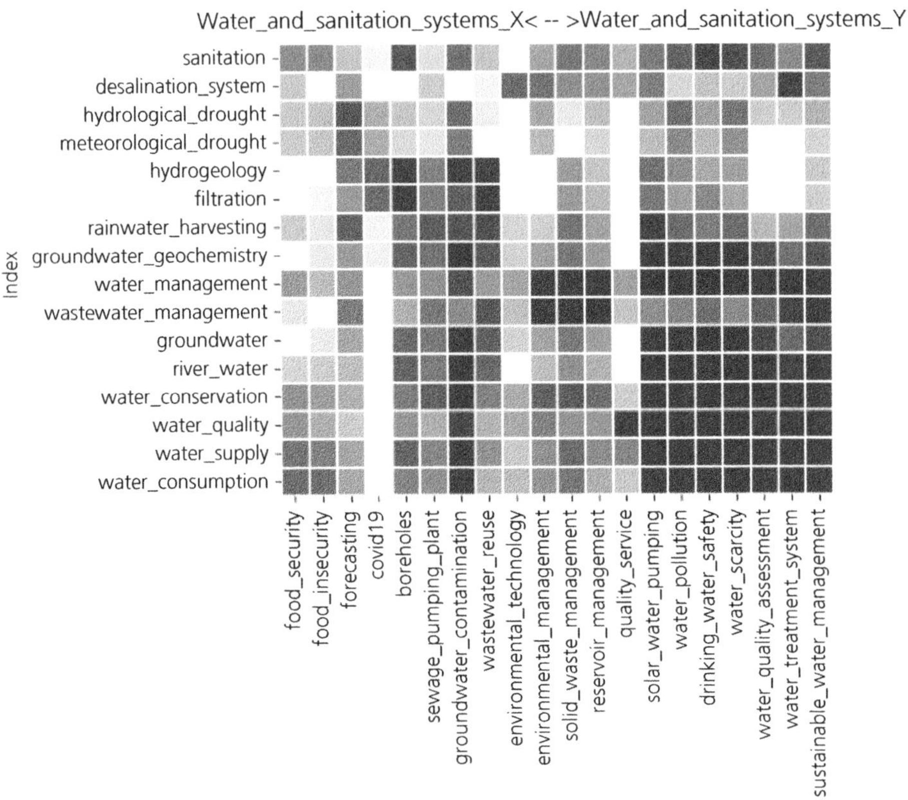

Table 10.1 Rural systems (continued on next page)

Rural development system	
Demographic systems Components: Living in rural areas, retirement, low-income, livestock, low-resource setting, policymaker roles, income, and inequity. Interdependence: Positive correlations suggest a potential for rural leadership. Strategies addressing multiple facets simultaneously are crucial for comprehensive rural development.	**Agriculture systems** Components: Agricultural technology, automation, food security, food production, economic growth, and employment growth. Interdependence: Strong positive correlations between agricultural technology and positive outcomes underscore the potential benefits of automation in critical areas. Environmental sustainability is interconnected with overall development goals, emphasising the need for effective water resource management.

Table 10.1 Continued

Rural development system	
Education systems Components: Child welfare, broader developmental initiatives, financial literacy, rural education, and professional teacher development. Interdependence: The interdependence of child welfare on broader developmental initiatives underscores the significant role of educational approaches. Financial literacy, rural education, and teacher professional development are crucial dimensions for rural settings.	**Electricity/energy systems** Components: Renewable and smart energy solutions, efficient energy management, sustainable practices in electricity generation, and integration of intelligent technologies. Interdependence: Embracing renewable and smart energy solutions and efficient energy management holds promise for comprehensive and sustainable rural development. The shift towards sustainable practices emphasises the integration of intelligent technologies.
Healthcare systems Components: Reliable healthcare infrastructure, integrated healthcare delivery. Interdependence: The importance of reliable healthcare infrastructure in rural areas is highlighted, emphasising the need for an integrated approach to healthcare delivery.	**Science and technology systems** Components: Interconnected technologies, collaborative and interconnected nature of technological ecosystems. Interdependence: The complex web of interdependencies emphasises technological ecosystems' collaborative and interconnected nature for effective rural development strategies.
Transport systems Components: Various modes of transportation, well-functioning, and comprehensive transport network. Interdependence: Positive correlations with various modes of transportation highlight the integral role of a well-functioning and comprehensive transport network in fostering positive outcomes in both rural and economic development.	**Water systems** Components: Addressing water quality, contamination, and environmental aspects. Interdependence: Positive associations emphasise that addressing water quality, contamination, and ecological aspects is crucial for fostering comprehensive rural development.

Enhancing sustainable rural development through collaborative innovation: a case study

Teaching notes

Synopsis

The most fertile area of innovation, creativity, and productivity in many rural regions of South Africa cannot take full root and grow owing to many challenges. It is because of a shortage of critical resources – that is, human capital and financial and technological infrastructure – that stand as a wall in front of the dreams and ambitions of the people dwelling in these calm corners of the world. However, it was during these challenges that came forth an awareness that the path to progress was not a journey that one man could explore, but was a shared responsibility. Acquiring that realisation of the urgency to evolve and taking confidence in the capacities of the rural people, the management of the CHIETA and the University of Johannesburg ventured into a daring and confident partnership that would result in inventive ways of triggering rural development. This collaboration aimed to harness the idea of interconnectivity as it pertains to the systems and subsystems within the rural communities with the intent to develop a system model. This future comprehensive system model will not only map the systems but also show their interrelationships and the systems hierarchy, ensuring that the right strategy can be applied in rural development and that resource allocation can be done more effectively. Thus, through the vision and strategic thinking of the UJ and the CHIETA management, a new chapter is being written in the history of rural development, which reflects the needs and interests of the people residing in rural communities. The goal of this study is not only the analysis of quantitative aspects but also the story of weaving together all the complexity of rural areas. The partnership result of this synergy would give birth to a system prioritisation model, which, among other features, would bring out where interventions are most needed. As a result, they would follow a much more strategic and practical approach to rural development and the allocation of resources. The deeper the story goes, so do the desired implications, acting as shining guides of hope.

Teaching objectives

This case study is written for use in a rural development and strategic planning module. The module could be part of a broader programme in areas such as community development, sustainable development, as well as social innovation and entrepreneurship. As such, the students should be able to do the following.

- Develop an understanding of the challenges hindering innovation, creativity, and productivity growth in rural communities.
- Apply strategic thinking to develop solutions for addressing challenges in rural communities.

- Identify challenges within rural communities and demonstrate the ability to prioritise them strategically in the context of rural development.
- Understand the significance of a system model in providing a strategic approach to addressing challenges and allocating resources effectively.
- Apply theoretical knowledge to real-world scenarios by discussing the practical implications of the developed system model for rural development.
- Demonstrate how strategic approaches can be translated into actionable plans for positive change in rural communities.

Research methods

The methodology adopted in this study represents a collaborative and multidisciplinary approach, showing a combined desire of rural communities to overcome problems hindering innovation, creativity, and productivity. The process follows a systematic series of steps to gather robust data on various facets of rural communities, encompassing their existing systems. The initial phase of the methodology involved an extensive literature review, wherein a thorough exploration of existing academic papers, research articles, and relevant documents was conducted. The literature review was used to come up with keywords; these keywords formed a search string for use on Scopus. These papers were later loaded into the natural language processing tool called word2vec to handle the quantitative analysis. One irreplaceable element of the research approach is quantitative analysis, which enabled capturing the multitude and intricacies of rural systems. By using the working mechanism of word2vec, the study gets the opportunity to pick up the numbers from qualitative data that is concealed within the books. This approach gives the opportunity for an efficient exploration and exposition of the interaction webs present at the level of rural ecosystems, as well as providing a total vision that is beyond the capabilities of traditional qualitative analyses.

The last procedure is building a full-scale system model. This model is not only useful for pinpointing established reinforcement systems but also for determining subsystems that are inside the infrastructure of rural communities. In addition, the model takes it a step further by showing the complex interaction of these subsystems at the same time. The model does not stop after the identification but also provides order to prioritise these subsystems. The critical factor here is that the model is discovering the cause and formulating a systematic plan where these subsystems are strategically prioritised, functioning as a blueprint for targeted interventions and resource allocation.

Keywords and subject code

Rural development, system model, rural communities, systems, innovation, digital innovation.

Assignment questions

1. What are the leading difficulties to the growth of innovation, creativity, and productivity in rural communities, mainly focused on human capital, financial support, and technological infrastructure?

2. How does the lack of crucial resources contribute to the shadow cast over the dreams and aspirations of rural communities, and what role does shared responsibility play in addressing these challenges?
3. What is the untapped potential within rural societies that can make a difference to development, and what would be the role of a joint project between an academic institution and an industrial partner in bringing this potential to the forefront?
4. In what ways do the identified challenges within rural communities relate to various systems and subsystems, and how can an exploration of these interrelations inform the development of a comprehensive system model?
5. What role can this system prioritisation model assume as a practical guide for interventions, and to what extent does it shed light on the issue of intervention areas that are particularly needed in rural development?
6. What is the prediction for the result of this joint venture, and how can the effects of the predicted results bring hope in rural areas?

Additional materials

The following readings are recommended for more information and insights.

Hariram NP, Mekha KB, Suganthan V and Sudhakar K (2023) Sustainalism: an integrated socio-economic-environmental model to address sustainable development and sustainability. *Sustainability (Switzerland)* **15(13)**: 10682, 10.3390/su151310682.

ILO (2022) Advancing Social Justice and Decent Work in Rural Economies. https://kebijakankesehatanindonesia.net/images/2019/policy_brief_wujudkan_keadilan_sosial_dalam_jkn.pdf (accessed 02/11/2023).

Kaiser N and Barstow CK (2022) Rural transportation infrastructure in low-and middle-income countries: a review of impacts, implications, and interventions. *Sustainability (Switzerland)* **14(4)**: 1–48, 10.3390/su14042149.

Mensah J (2019) Sustainable development: meaning, history, principles, pillars, and implications for human action: literature review. *Cogent Social Sciences* **5(1)**: 1–21, 10.1080/23311886.2019.1653531.

Shrestha M (2021) Access to basic services and its linkage with ending poverty. In *No Poverty, Encyclopedia of the UN Sustainable*. Springer Nature, Berlin, Germany, pp. 1–12, 10.1007/978-3-319-95714-2_1.

Stanef-Puică M, Badea L Şerban-Oprescu G, *et.al.* (2022) Green jobs – a literature review. *International Journal of Environmental Research and Public Health* **19**: 248–271, 10.3390/ijerph19137998.

Vorobyov S and Bugai Y (2019) Factors of socio-economic development of rural areas. In *IOP Conference Series: Earth and Environmental Science*, 10.1088/1755-1315/395/1/012109.

Wu Q, Guan X, Zhang J, and Xu Y (2019) The role of rural infrastructure in reducing production costs and promoting resource-conserving agriculture. *International Journal of Environmental Research and Public Health* **16(18)**: 3493, 10.3390/ijerph16183493.

Suggested core readings

Ngumbela X (2022) Challenges and tensions faced by rural-based municipalities in delivering basic services in South Africa. *Administratio Publica* **13(2)**: 136–159, 10.15651/JPAPR.2.

Ovchinnikova N and Vodolazskaya T (2021) Challenges and perspectives for rural development. *E3S Web of Conferences* **273:** 06007, 10.1051/e3sconf/202127306007.

Teaching plan and timing

The teaching plan aims to facilitate a comprehensive understanding of rural development challenges and strategies through the exploration of the provided case study. The primary focus is on fostering critical thinking, collaborative problem-solving, and practical application of theoretical knowledge in the context of rural development. This teaching plan is designed for four sessions divided into two days and can be adjusted based on the specific needs of the course and the depth of exploration desired.

Day 1

Session 1: Introduction to rural development challenges

Lecture (15 min)

- Introduce students to the context of rural development challenges.
- Familiarise them with the case study background and key concepts.
- Overview of rural development challenges.
- Discussion on the significance of addressing innovation, creativity, and productivity in rural communities.
- Explore the concept of shared responsibility.

Group discussion (5 min)

- Small groups discuss their thoughts on challenges faced by rural communities.
- Groups identify additional challenges not mentioned in the case study.
- Discuss how shared responsibility can address these challenges.

Class discussion (10 min)

- Share group insights and identify common themes.

Session 2: Potential within rural communities – systems and subsystems exploration

Lecture (20 min)

- Discussion on the benefits of collaborative ventures between academia and industry.
- Investigate the interconnected systems and subsystems within rural communities.
- Understand the need for a comprehensive system model.

Group exercise (5 min)

Groups to brainstorm and identify specific systems and subsystems within rural communities.

Day 2

Lecture (15 min)

Session 1: Practical application and actionable plans

- Analyse how the system prioritisation model translates into actionable plans.
- Discuss the expected outcomes outlined in the case study.
- Understand the importance of collaboration in rural development.

Group activity (5 min)

- Groups develop actionable plans for a real or hypothetical rural development project.
- Discuss how the collaborative effort between the University of Johannesburg and the CHIETA represents a shared purpose.

Session 2: Case study reflection and future considerations (10 min)

- Reflect on the key learnings from the case study.
- Consider future considerations for sustainable rural development.
- Students reflect on their key takeaways from the case study.

Analysis of the questions

The questions of this case study are open-ended; students can provide their own interpretations. Below are some guidelines on how they can be answered.

Challenges hindering innovation, creativity, and productivity

- **Human capital:** Limited access to quality education and skill development opportunities in rural areas, leading to a shortage of skilled workforce.
- **Financial support:** Lack of financial infrastructure, limited access to loans or grants for entrepreneurs, hindering the development of innovative projects.
- **Technological infrastructure:** Insufficient technological resources, limited internet connectivity, and inadequate access to modern tools and equipment.

Impact of resource lack on dreams and aspirations

- The absence of crucial resources creates barriers to realising dreams and aspirations, contributing to a sense of stagnation and unfulfilled potential.
- Shared responsibility encourages collaborative problem-solving, fostering a sense of community and empowering individuals to collectively address challenges.

Unlocking potential through collaboration

- Rural communities possess untapped skills, traditional knowledge, and cultural strengths that, if harnessed, could drive local development.
- Collaboration between an academic institution and an industry authority can bring educational expertise and practical industry insights, creating a synergy that enhances the development potential.

Interrelations of challenges and system model development

- Challenges related to education, finance, and technology may intersect with agricultural, healthcare, and economic systems within rural communities.
- The exploration of these interrelations informs the development of a comprehensive system model that identifies key areas requiring intervention.

Practical guide for interventions

- The system prioritisation model serves as a roadmap, indicating where interventions are most needed based on identified challenges.
- It sheds light on specific areas such as education, healthcare, and technology where targeted interventions can yield the most significant positive outcomes.
- This unity contributes to a new chapter in rural development by fostering holistic and sustainable approaches that consider the interconnectedness of challenges.

Expected outcomes and beacons of hope

- Anticipated outcomes include improved educational infrastructure, increased entrepreneurial activities, and enhanced technological access.
- These outcomes act as beacons of hope by signalling a positive trajectory for rural communities, inspiring confidence in the potential for lasting change and development.

Key learnings

The key lessons learned from this case study are a comprehensive cognitive view of the challenges obstructing innovation, creativity, and productivity in rural surroundings. The case vividly sheds light on the problems posed to creativity innovation and ecosystems in the rural community. This would begin with the acknowledgment that the issue is multifaceted and entails not only the challenges of human capital but also the areas of financial support and technological infrastructure. Also, examining the interdependencies of different systems and subsystems mainly operating in rural areas provides a means of understanding the intricacies of the forces that shape these areas. The study highlights the need for deep consideration of these relationships in designing better policies for comprehensive rural development. The creation of a perfect system model enables a big pot, which is not only responsible for the diagnosis of challenges but also provides a strategic frame for timely and effective priority setting. This strategy will act as a roadmap for a deeper mechanism of how resources can be put in villages where there is a need. Collaboration between the University of Johannesburg and the CHIETA revealed the importance of 'shared workshops' in overcoming challenges in rural areas. Such joint venture depicts a positive convergence between the knowledge base and industry ripeness, revealing the capacity to eliminate numerous problems through a collective effort. Finally, the study is seen to bring about profound revelations that may not be apparent to the untrained eye, and it supplies ways forward to empower the inhabitants of the community.

REFERENCES

Hariram NP, Mekha KB, Suganthan V and Sudhakar K (2023) Sustainalism: an integrated socio-economic-environmental model to address sustainable development and sustainability. *Sustainability (Switzerland)* **15(13)**: 10682, 10.3390/su151310682.

ILO (2022) Advancing Social Justice and Decent Work in Rural Economies. https://kebijakankesehatanindonesia.net/images/2019/policy_brief_wujudkan_keadilan_sosial_dalam_jkn.pdf (accessed 02/11/2023).

Mensah J (2019) Sustainable development: meaning, history, principles, pillars, and implications for human action: literature review. *Cogent Social Sciences* **5(1)**: 1–21, 10.1080/23311886.2019.1653531.

Shrestha M (2021) Access to basic services and its linkage with ending poverty. In *No Poverty, Encyclopedia of the UN Sustainable Development Goals*. Springer Nature, Berlin, Germany, pp. 1–12, 10.1007/978-3-319-95714-2_1.

Stanef-Puică M, Badea L, Şerban-Oprescu G *et.al.* (2022) Green jobs – a literature review. *International Journal of Environmental Research and Public Health* **19**: 248–271, 10.3390/ijerph19137998.

Vorobyov S and Bugai Y (2019) Factors of socio-economic development of rural areas. In *IOP Conference Series: Earth and Environmental Science*, 10.1088/1755-1315/395/1/012109.

Wu Q, Guan X, Zhang J, and Xu Y (2019) The role of rural infrastructure in reducing production costs and promoting resource-conserving agriculture. *International Journal of Environmental Research and Public Health* **16(18)**: 3493, 10.3390/ijerph16183493.

Prof. Arnesh Telukdarie
ISBN 978-1-83662-645-9
https://doi.org/10.1108/978-1-83662-642-820251011

Chapter 11
Sustainable waste management solutions for urban settlements: a case study of the City of Johannesburg, South Africa

Arnesh Telukdarie, Inderasan Munien and Noluthando Gamede

Introduction

The City of Johannesburg (CoJ) has faced significant challenges in waste management, a critical issue that highlights its broader sustainability goals. In July 2023, this challenge came into sharper focus for Vuyani Khumalo, a manager within the City of Johannesburg Metropolitan Municipality. Having assumed his role six months prior, Khumalo was keenly aware of the magnitude of the waste management crisis confronting the city. The municipality generates substantial volumes of waste daily, much of which remains improperly managed, contributing to environmental degradation and social inequalities. Despite extensive research and numerous studies over the years, the issue persists, with little progress in translating recommendations into actionable strategies. Upon assuming his position, Khumalo undertook a thorough review of the available studies and reports addressing the city's waste management issues. His findings revealed a striking consensus: CoJ faces a systemic waste management problem that demands immediate and comprehensive intervention. These studies detailed the inadequacies in the existing waste management systems, including inefficient collection processes, insufficient recycling initiatives, and limited engagement with informal waste reclaimers. The latter group, who play a critical role in recovering recyclable materials from waste streams, often operate without adequate support or recognition, highlighting gaps in the municipality's inclusive waste management policies.

Despite the clarity provided by these studies, the challenge lay not in identifying the problem but in determining the way forward. The abundance of recommendations – from enhancing recycling programmes to improving waste collection infrastructure – had not translated into decisive action. This inaction stemmed from various factors, including limited financial resources, bureaucratic hurdles, and competing municipal priorities. For Khumalo, the absence of a clear framework to evaluate and implement these recommendations further compounded the difficulty of addressing the issue. The complexity of the waste management ecosystem, with its myriad stakeholders and intersecting environmental, social, and economic dimensions, underscored the need for a systems-based approach to develop effective solutions.

The persistence of these challenges reflects broader structural issues within urban waste management systems. CoJ, like many other metropolitan municipalities in developing regions, struggles with balancing rapid urbanisation, population growth, and resource constraints. The informal sector, particularly waste reclaimers, fills critical gaps in the system but remains marginalised in policy discussions and interventions. Recognising this, Khumalo acknowledged that any meaningful solution would require integrating informal actors into the formal waste management framework, fostering collaboration among stakeholders, and leveraging innovative technologies. Furthermore, the city's challenge to act on previous recommendations points to a broader issue of implementation paralysis, often linked to political and administrative inertia. Addressing this requires not only technical solutions but also institutional reform to enhance accountability and streamline decision-making processes. For Khumalo, the task ahead involved navigating these complexities to devise a strategy that was not only practical and sustainable but also inclusive and actionable. This would entail prioritising evidence-based interventions, fostering partnerships with community and private-sector actors, and advocating for increased investment in waste management infrastructure.

The waste management crisis in Johannesburg represents a microcosm of the challenges faced by urban centres worldwide. It underscores the need for proactive leadership, integrated approaches, and a commitment to translating policy into action – an imperative Khumalo aimed to address during his tenure.

Background: geographical context and economic profile of the City of Johannesburg

The CoJ manages the local governance of South Africa's largest city, Johannesburg. Johannesburg, the capital of Gauteng Province, is home to nearly five million people and serves as the industrial and financial hub of South Africa. CoJ is responsible for the city's overall administration and the provision of services, including waste management, to its inhabitants. In terms of the Constitution of South Africa, CoJ is responsible for providing efficient and effective waste management services to the inhabitants of the city. To fulfil this responsibility, CoJ must develop a waste disposal strategy, generate waste disposal regulations, and establish, operate, and control waste transfer and disposal facilities. The management of CoJ's operations required the development and presentation of comprehensive solutions to the City Council. These solutions needed to include technical and regulatory proposals aligned with advancing CoJ's commitments to achieving Sustainable Development Goal (SDG) 6 (clean water and sanitation), SDG 11 (sustainable cities and communities), and SDG 15 (life on land).

CoJ's waste challenges

The solid waste crisis confronting CoJ was created primarily by the rapid urbanisation and development the city experienced since the advent of democracy. The City of Johannesburg generates approximately 2 150 000 tonnes of waste per year, of which only 3.2% is not landfilled. By December 2022, the city had effectively exhausted its waste landfill capacity, forcing the landfills to operate under contingency conditions. Early in 2023, CoJ commissioned a study to identify waste processing options that would divert waste from the five landfill sites that were currently operating. In August 2023, the waste management approached the University of Johannesburg (UJ) to review the findings and recommendations from the plethora of studies examining the CoJ waste management crisis. The UJ investigation aimed to place the city's current waste management practices in the context of current research findings, thereby pinpointing potential areas for enhancement. Another project objective was to assess the potential for implementing pay-as-you-throw (PAYT) strategies as a waste management solution for CoJ. Considering the size and complexity of Johannesburg's urban landscape and demographics, as well as the urgency of implementing alternative waste disposal solutions, it was quite a stretch to imagine that UJ would be able to identify a suitable solution for CoJ.

Landfilling

Landfilling is a waste disposal practice that significantly harms our environment, leading to loss of biodiversity, economic strain, health problems, and air, soil, and water pollution. Managing municipal solid waste (household, commercial, and industrial waste) is complex, requiring consideration of social, cultural, and ecological factors. Landfilling remains the most common disposal method owing to its low cost and simplicity. However, landfills are unsustainable owing to their limited space, unpleasant odours, harmful gas emissions, and toxic liquid leaks. Recognising these drawbacks, significant research is underway to develop sustainable waste management solutions that minimise environmental and societal harm. These solutions include waste-to-energy processes, lifecycle assessments, improved measurement and monitoring, advanced treatment technologies, and integration strategies. Global cities are major consumers, generating vast amounts of waste that pollutes the environment and fuels climate change. Waste ranks as the fourth-largest global emitter of greenhouse gases, accounting for over 3% of greenhouse gases and a significant 20% of methane emissions (Messina *et al.*, 2023). Landfills are a key culprit, as organic waste decomposes without oxygen and releases methane, a potent greenhouse gas. Beyond organic waste, construction, demolition, residential, industrial, and healthcare waste all contribute to the problem.

Sustainability

Effectively managing these waste streams is essential for a sustainable future. The European Union (EU) exemplifies this with its goal of achieving net zero greenhouse gas emissions by 2050. Their circular economy package, which is a core element of the European Green Deal, emphasises resource efficiency and recycling to reduce waste and emissions. For instance, despite generating 4.8 tonnes of waste per person annually, the EU only recycles 40%, with 33% still going to landfills (Ranjbari *et al.*, 2022). Developing countries like South Africa face hurdles in establishing sustainable waste management practices. National waste policies often come from the top down, failing to consider local needs. A greener future requires transforming municipal waste treatment by focusing on better sorting, composting, and recycling. The key to success lies in a holistic strategy that minimises waste emissions while also decoupling economic growth from resource consumption (Messina *et al.*, 2023). One promising approach is implementing PAYT systems in municipal solid waste management, which are billing programmes where residents are charged based on the amount of waste they throw away. This is different from traditional systems where waste collection fees are fixed and not connected to how much trash you generate. PAYT systems incentivise waste reduction and simplify waste stream separation for processing, ultimately reducing landfill usage and boosting resource recovery.

Pay-as-you-throw

A successful PAYT system requires three key elements: identifying the waste generators, measuring waste generation, and setting fair prices based on that amount (Elia *et al.*, 2015).

This entails determining the method of charging residents per unit of waste, determining the recipient of the charge, and devising a method for measuring the waste amount. On top of a good waste collection system, clear information for residents and a transparent pricing plan are crucial for a PAYT programme to work well. Research shows that when citizens understand the pricing clearly, they are more likely to participate. That is why it is important to have a well-structured, fair, and easy-to-understand pricing system (Ukkonen and Sahimaa, 2021). Finally, it is important to consider how people view PAYT compared with the old way of paying a fixed fee. Research suggests that the choice between these systems can affect how people behave (Dahlén and Lagerkvist, 2010).

It is important that switching to a PAYT system does not cost more than it saves. The new fees need to be high enough to cover the programme's costs. Weight-based PAYT systems rely on accurate data on how much waste people throw away, therefore record-keeping is essential. Just focusing on weight might not be enough; successful PAYT programmes also need well-designed municipal policies. Previous research has shown that waste reduction might only happen in the first year and could rise again without proper policy support (Campitelli *et al.*, 2023). Countries that have used PAYT for a long time have been able to refine their models and update them to comply with changing laws. Some countries, like Denmark, use a combination of a base fee and extra charges for special situations. These extra charges may be based on the type of waste you have (in some cases), or for things like carrying long distances, needing someone to unlock doors, or using a lift to collect the trash (Farahbaksh *et al.*, 2023). In other places, these additional fees depend on the level of service chosen. For instance, some residents might pay more for having their trash bin collected from inside their property rather than taking it to the curb themselves. A volume-based PAYT system bases the cost of waste collection on the amount of space the trash takes up, not its weight. This is different from systems that focus on the weight of waste. This can be achieved by selling waste collection bags whose price already includes a collection fee. Also, the size of the waste collection bin can be used to manage and measure the waste volumes. By connecting the amount of trash discarded with the cost of collecting, volume-based PAYT systems aim to incentivise residents to reduce waste and make more conscious choices about what they throw away.

Ukkonen and Sahimaa (2021) developed a pricing model for weight-based PAYT systems. This model calculates waste collection fees based on the weight of trash discarded and the level of service provided (e.g. curbside pick-up against in-home collection). It improves on volume-based PAYT systems, which are known to be less effective in encouraging recycling and waste reduction. The model incentivises residents to sort their waste and throw away less overall. It is also practical for waste management companies to implement. With this model, PAYT increases participation by rewarding residents who sort their waste properly, leading to fairer cost allocation. The model uses data from waste management companies, including estimated annual collection and transport costs for weight-based pricing. Additionally, it considers the waste management area's demographics (residents, housing types), service levels, and average waste generation per person. Finally, the model factors in the average composition of household waste. All costs and revenue go towards the waste management company, ultimately determining the price a resident pays for waste collection. In this system, residents become customers of their local waste management company. In a volume-based PAYT system, the amount paid for waste collection depends on how much space the trash occupies, not how much it weighs. This is different from systems that focus on the weight of waste. This can be achieved by selling waste collection bags whose price includes the cost of waste disposal, encouraging proper waste segregation and ensuring that the fees contribute to sustainable waste management practices.

PAYT in CoJ

The UJ found that implementing a PAYT system in Johannesburg faced significant challenges, making it a poor fit for solving the city's waste crisis in the medium term. Here is why.

- **Costly infrastructure:** PAYT requires extensive and expensive waste management and billing systems to accurately measure individual waste volumes and set fair fees.
- **Public resistance:** According to a 2017 CSIR report (Kadyamadare, 2017), residents currently oppose PAYT. Gaining public acceptance would likely require a significant upfront investment in education and outreach campaigns.
- **Balancing affordability and efficiency:** Setting transparent pricing that is fair to residents while still covering the programme's costs is a complex task.
- **Policy and partnerships:** Designing and implementing effective PAYT policies requires a lot of effort, potentially involving partnerships with private companies, adding further complexity.

Given the urgency of Johannesburg's waste management issues, UJ determined that concentrating on swift, implementable solutions might be a more effective strategy. To identify different options, UJ decided to identify cities around the globe with similar waste compositions to CoJ and review their waste management practices. CoJ categorises its waste into seven types: organic materials, paper, plastics, glass, metals, textiles, and a miscellaneous category. To compare Johannesburg's waste composition with other countries, researchers looked for countries that use the same classification system when reporting waste data. UJ conducted a literature review, identifying 16 countries that met these criteria for the comparison. To make the comparison accurate, researchers needed a representative sample of Johannesburg's municipal solid waste (MSW). CoJ itself had previously commissioned studies to obtain this data. Four prior reports generated for CoJ in 2001, 2014, 2015, and 2022 analysed the waste composition (as shown in Figure 11.1). Importantly, these studies revealed a consistent waste composition over the past two decades. This consistency reduced the risk of choosing inappropriate waste management technologies owing to changes in waste types over time.

To ensure a fair comparison, UJ only considered countries that categorise their waste into the same seven categories used by CoJ. The researchers combined the waste composition of all 16 countries, along with Johannesburg's, into the table shown in Figure 11.2 after finding those whose waste analysis data matched this format. To identify countries with waste most similar to Johannesburg's, the researchers then highlighted the two closest matches for each waste category shown in Figure 11.2.

After comparing waste composition across 16 countries, which revealed significant differences in waste composition across the globe, UJ identified nine countries that shared similar waste compositions per category (the two nearest compositions per category) with Johannesburg. These nine countries came from a diverse range of regions, including Africa, Europe,

North America, Australia, Asia, and South America. None of the countries shared more than two waste category compositions with CoJ, highlighting the global variation in waste generation. However, this categorisation approach proved valuable in identifying potential waste management strategies for Johannesburg. This analysis showed how the CoJ's counterparts with similar compositions managed waste. Burning waste (incineration), biological treatment, and recycling were their main methods, along with landfills. However, all these processes required sorted waste streams to function properly.

A phased approach

Based on their research, UJ recommended an approach involving a phased transition towards a circular economy in CoJ. This approach entailed the following.

- **Sorting waste streams:** This is the first step, separating waste into different categories for proper treatment.
- **Reducing landfill use:** By diverting waste from landfills through recycling and other methods, reliance on landfills can be reduced.
- **Partnering with private recycling:** Utilising existing recycling infrastructure in the private sector can be beneficial.
- **Waste processing:** Sorted waste can then be recycled, biologically treated, or incinerated to recover energy.

This approach leveraged South Africa's strong private sector recycling industry and included technologies proven effective in countries with waste similar to CoJ. However, all these methods required sorted waste.

Waste sorting

Since the existing resident participation rates in sorting programmes were low, and sorted waste was shown to be crucial to divert trash from landfills, UJ found the best option for transitioning towards a circular economy was to use material recovery facilities (MRFs). These facilities sort waste into consistent types, providing the necessary input for recycling, incineration, and biological treatment. Therefore, UJ proposed that CoJ build an MRF to sort collected waste. Recyclable materials would then be sent to recyclers, and the remaining waste would be processed further. This approach even allows waste pickers to participate as sorters or suppliers of sorted materials. The UJ solution was supported by various previous investigations that CoJ commissioned. The estimated cost in 2023 (considering inflation) could be managed through partnerships with the private sector. Once sorted, the waste would be directed to different processing facilities (Table 11.1). Several factors needed to be considered when designing an effective waste management system, with multiple processing options available for some categories like paper, organic waste, and plastic. Analysing various scenarios is crucial to identifying the optimal processing configuration. An optimisation study based on a scenario analysis would help achieve minimal landfill usage, maximise the utilisation of existing waste processing infrastructure, and extract the maximum value from the resources present in the waste stream.

Scenario

To comprehensively evaluate the implementation of different technologies, simplified scenarios were developed using only two processing methods for each waste category. For instance, in the case of organic waste, simulations were run with combinations like digester and composter, digester and incinerator, and digester and landfill. Additionally, a combined scenario including all available technologies for organic waste processing was also modelled (Table 11.2). This approach was similarly applied to paper and plastics. Apart from direct landfilling, paper can be treated through recycling or incineration. For organic waste, the proposed methods are incineration, composting, and digestion. Plastic waste is either recycled or incinerated. Glass and rubble are crushed, while metals are recycled. These processes are depicted in Figure 11.3.

For each scenario, the researchers varied the proportions of waste directed to each treatment option. Essentially, they looked at different combinations of processing methods for each waste type. Finding the most efficient solution involves complex calculations, considering these various processing paths and waste proportions. Table 11.3 displays the simulation outcomes, indicating the total amount of waste landfilled. The findings indicate that the minimum amount of waste sent to landfill is 34%. Table 11.3 demonstrates that waste processing significantly influences diversion rates. The closest landfilled quantity to the baseline of 84% is 64%, representing the highest landfill rate in a processing scenario. This worst-case scenario results from a high reliance on direct landfilling. Scenario 5 of Case 2 achieved the lowest overall amount of landfilled waste. This included both the waste directly landfilled and any ash or inorganic materials left over from incineration. The specific processing details for this scenario are provided in Table 11.4. Nearly half of the waste was recycled in this scenario, amounting to 812 424 tonnes of material diverted from landfills. How the waste was processed is shown in Table 11.5. This scenario significantly reduced landfilling. Out of the total waste collected, only 466 876 tonnes (or 30%) ended up in landfills. Determining the best processing route is essential for maximising the amount of waste treated and diverted from landfills. Scenario 5 achieves the highest waste diversion with a comparable investment to other, less effective approaches. Furthermore, it involves the lowest level of incineration among all scenarios that include incineration.

Conclusion

The analysis of the CoJ report highlighted that the capital costs for establishing waste processing facilities remained consistent across most scenarios, except for the incineration-heavy option, which emerged as the most expensive at R5.43 billion. This finding underscored the importance of selecting the most cost-effective and environmentally sustainable waste treatment methods. Among the evaluated scenarios, Scenario 5 demonstrated the best outcomes in minimising landfilled waste while maintaining a cost (approximately R2.68 billion) comparable to less effective alternatives. Furthermore, it required minimal reliance on incineration, thereby reducing overall emissions.

The UJ proposal built on extensive research into CoJ's waste management challenges, integrating insights from several key perspectives.

- **Global best practices:** examining successful waste treatment strategies from around the world.
- **Comparable countries:** analysing waste management approaches in nations with similar waste compositions.
- **Local context:** tailoring solutions to Johannesburg's unique waste generation patterns and urban dynamics.

While previous experts had recommended many of the technologies included in the UJ proposal, the proposed approach stood out for its innovative integration of these technologies into a comprehensive system. This holistic strategy represented a departure from earlier piecemeal recommendations, providing a cohesive framework for tackling the city's waste management challenges. Given the complexity of factors involved, a detailed computational analysis was conducted to determine the optimal waste processing pathways. The results of this analysis demonstrated significant progress toward a sustainable waste management model for CoJ. The findings point to a clear path forward: implementing an integrated waste management system based on Scenario 5 and the recommendations of the UJ proposal. The next crucial step is the systematic execution of this solution, requiring collaboration across stakeholders, investment in infrastructure, and alignment with CoJ's sustainability goals. This integrated approach has the potential to transform waste management in Johannesburg, paving the way for a more sustainable urban environment.

Exhibits

Figure 11.1 Solid waste composition of CoJ

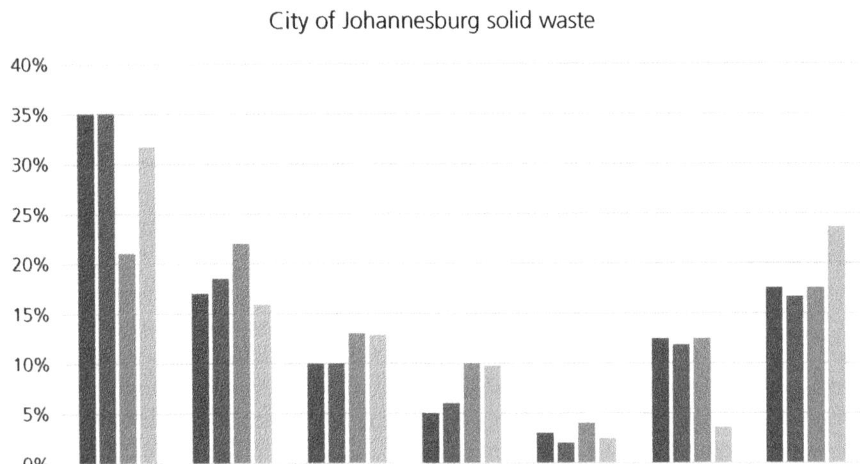

	Organic	Paper	Plastic	Glass	Metals	Textiles	Other
2001	35%	17%	10%	5%	3%	12%	18%
2014	35%	19%	10%	6%	2%	12%	17%
2015	21%	22%	13%	10%	4%	12%	18%
2022	32%	16%	13%	10%	2%	4%	24%

Figure 11.2 Waste composition across 16 countries alongside CoJ

Country	Organic	Paper	Plastic	Glass	Metals	Textiles	Other
CoJ	31.5	19.0	10.8	6.8	2.8	12.2	17.1
Singapore	28.9	18.8	29.9	2.14	0.34	5.4	14.5
China	38.6	22.2	21.5	3.6	2.3	3	8.8
Germany	30	24	13.0	10	1.4	4	17.7
USA	14.9	26.6	12.9	4.4	9	9.5	3.2
Australia	48.4	17.3	7.61	3.81	19.38	0	3.46
Japan	36	33	11.0	5	4	1	6
Denmark	12.78	11.57	1.61	5.8	6.29	0	23.86
Turkey	64	6	3	2	1	0	24
Malaysia	51.9	8.45	13.2	3.32	2.75	3.1	17.31
Kazakhstan	28.5	12.15	16.7	3.99	2.93	-	35.7
Argentina	40.94	14.24	20.2	5.2	2.18	3.07	14.15
Colombia	74.19	2.58	1.42	0.97	2.13	0.35	18.36
Indonesia	53.75	14.92	14.02	2.45	1.82	1.11	11.93
Morocco	60	7.5	7.5	2.5	2.5	0	7
Algeria	54.4	9.75	16.88	1.16	2.84	12.62	2.35
Bangladesh	77.7	4.48	7.35	0	0.44	2.53	7.5

Figure 11.3 Processing paths of waste

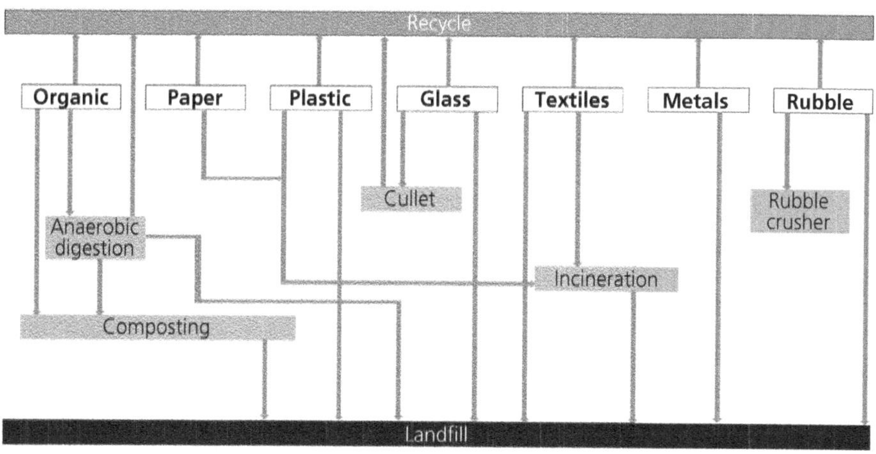

Figure 11.4 Lifecycle of waste collection

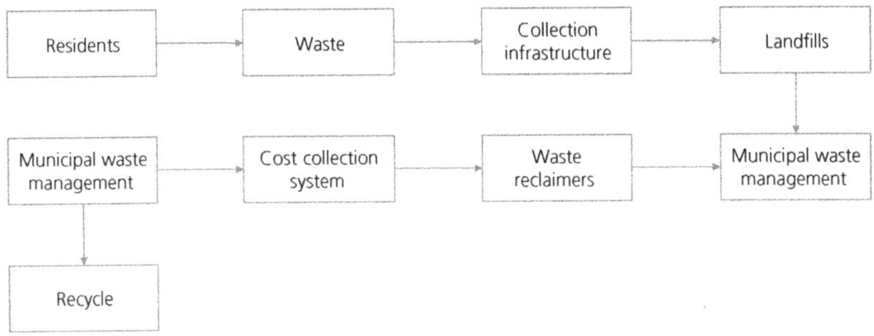

Figure 11.5 Co-digestion process

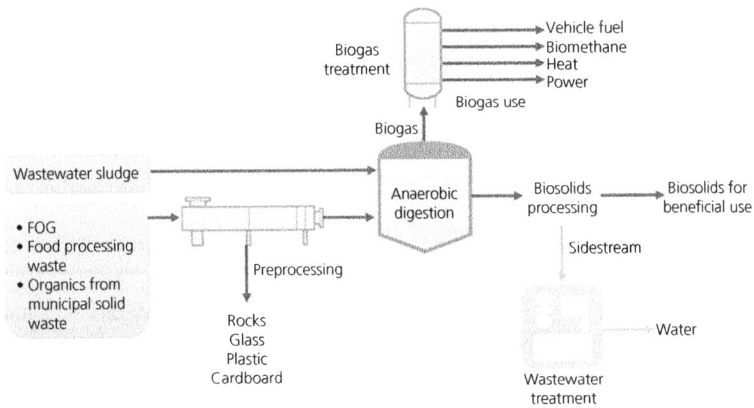

Figure 11.6 Turbine generator

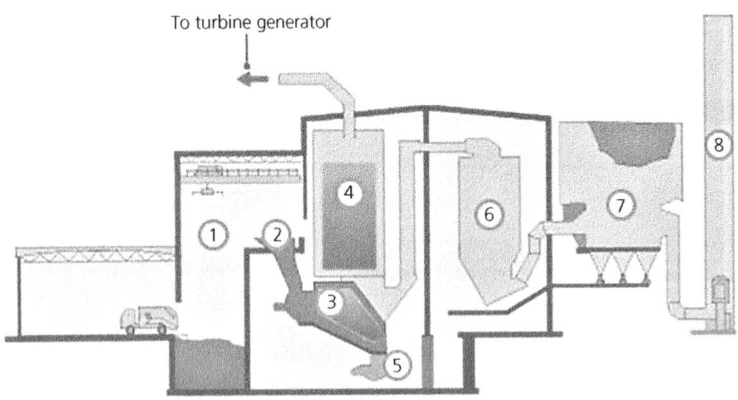

Figure 11.7 Compost generation from the anaerobic treatment process

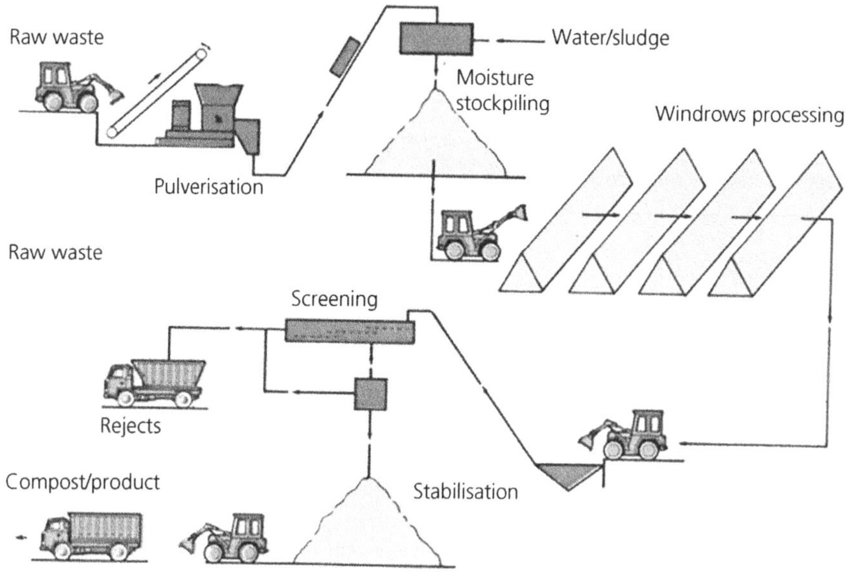

Figure 11.8 Crushing plant process

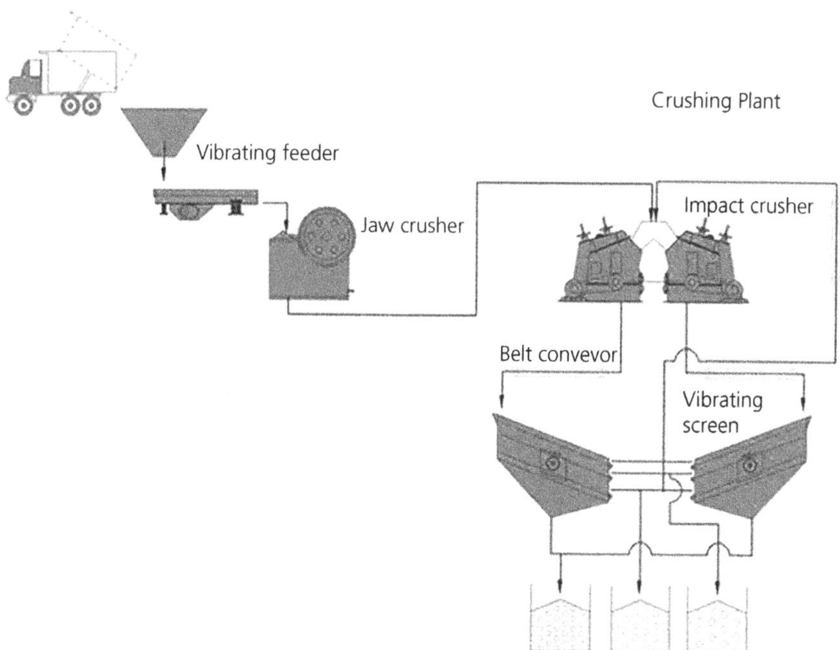

Table 11.1 Result of waste collected

Organic waste	Anaerobic digesters or composters
Mixed waste (organic, paper, plastic, textiles)	Incineration (with energy recovery)
Glass	Crushed into cullet for reuse
Building and construction rubble	Crushed for reuse
Leftover ash and any unprocessable materials	Landfill

Table 11.2 Processing pathway per category

Organic waste	Scenarios explored using only digesters and composters, digesters and incinerators, and digesters and landfills. A scenario combining all organic waste treatment options was also examined
Paper	Recycling and incineration were evaluated as alternatives to direct landfilling
Plastics	Recycling and incineration were considered
Glass and rubble	Crushing was assumed for processing
Metals	Recycling was assumed

Table 11.3 Amount of landfilled waste

	Scenario 1	Scenario 2	Scenario 3	Scenario 4	Scenario 5
Total % landfilled case 1	55	48	42	37	34
Total % landfilled case 2	45	41	52	32	30
Total % landfilled case 3	84	64	55	52	45
Total % landfilled case 4	84	50	44	38	32

Table 11.4 Amount of landfilled waste by the type of process

Process	Volume: tonnes	Proportion of total: %
Recycled	812 424	49
Digested	259 284	16
Composted	259 284	16
Incinerated	131 700	8

Table 11.5 Waste processing proportions per waste category

Paper	60% recycled, 40% incinerated
Organics	50% digested, 50% composted
Plastics	70% recycled, 30% landfilled
Glass	45% recycled into cullet (crushed glass for reuse), 55% landfilled
Textiles	40% incinerated, 60% landfilled
Metals	40% recycled, 60% landfilled
Rubble	40% recycled, 60% landfilled

Sustainable waste management solutions for urban settlements: a case study of the City of Johannesburg, South Africa

Teaching notes

Synopsis

This case study delves into the response of the City of Johannesburg (CoJ) to its escalating waste management crisis. It explores the strategies devised and implemented by CoJ, drawing comparisons with other global cities with similar waste generation profiles. Additionally, the study examines the perspectives of stakeholders involved in the CoJ waste value chain. Both the separate-at-source and pay-as-you-throw strategies were analysed in the global context and specifically in relation to CoJ. The study employs scenario analysis to identify optimal solutions and showcases the potential role of the fourth industrial revolution in addressing the impending waste management challenges. The significance of combining multiple technologies and aligning diverse stakeholders to redirect waste from landfills towards beneficial utilisation is highlighted in the study. Students of environmental management, urban planning, analytical techniques, and waste processing technologies may find this case study relevant.

Teaching objectives

This study is suitable for students pursuing undergraduate studies up to master's level of postgraduate studies and is applicable to students engaged in studies involved in environmental management, urban planning, civil engineering, and project management. Reviewing the City of Johannesburg waste management case study should inspire students to do the following.

- Understand the complexities associated with waste management by identifying the components and interactions defining the waste management system.

- Define the route to a circular economy within the context of the constraints imposed on CoJ.
- Analyse the methodology used to identify the optimal processing path for CoJ's municipal waste.
- Assess and critically review the proposed solution within CoJ's existing context.

Research methods

The case study was compiled using data provided by the City of Johannesburg, reviewing internal and external technical and strategic reports commissioned by the City of Johannesburg, and conducting an extensive desktop study of global waste management practices. The study is based on historical and contemporary analyses and data. The simulation results presented in this case study are based on models built to conduct scenario analysis and optimise the waste disposal strategy.

Keywords and subject code

Comparative review, scenario analysis, treatment solutions, waste management, waste separation.

Assignment questions

1. Identify and diagram the key components of Johannesburg's waste management system and their relationships.
2. Evaluate UJ's technology selection to minimise landfill disposal and justify any alternative choices.
3. Perform a SWOT analysis based on the case study to develop an action plan for achieving a circular economy in the City of Johannesburg.

Additional materials

Highlighting the reality of the problems and challenges mentioned in the case study is important to increase the engagement of students. In this regard, YouTube is a wonderful resource. The following YouTube videos are very informative in terms of contextualising the waste crisis facing the City of Johannesburg.

Alliance to End Plastic Waste (2023) Closing the gap with reclaimers/Empowering waste workers in Johannesburg. https://youtu.be/9jiFwq0kEN0 (accessed 09/06/2024).

eNCA (2024) Joburg's waste woes: City fast running out of landfill space. https://www.youtube.com/watch?v=5SBJCMaRcg0 (accessed 16/06/2024).

Suggested core readings

Dunne L, Convery FJ and Gallagher L (2008) An investigation into waste charges in Ireland, with emphasis on public acceptability. *Waste Management* **28(12)**: 2826–2834, 10.1016/j.wasman.2008.07.007.

Jiang L, Wang H, Tong AA, Song J and Chen X (2020) The measurement of green finance development index and its poverty reduction effect: dynamic panel analysis based on improved entropy method. *Discrete Dynamics in Nature and Society* **2020(1)**, 10.1155/2020/8851684.

Laurijssen, Marsidi M, Westenbroek A, Worrell E and Faaij A (2010) Paper and biomass for energy? The impact of paper recycling on energy and CO2 emissions. *Resources, Conservation and Recycling* **54(12)**: 1208–1218, 10.1016/j.resconrec.2010.03.016.

Teaching plan and timing

- *20 min*: The first 20 minutes should be spent on the introduction and background. The discussion should be supplemented by an internet search for newspaper articles and relevant informative websites, and the YouTube video mentioned in 'additional materials' should be watched. Students should gain an understanding of the scope and discuss how it affects the city's residents.
- *25 min*: The teacher should solicit feedback from students on the components of CoJ's waste management system and expand the model answer with learnings from students.
- *30 min*: Students should work in small groups and critique the various technologies they have found, and then groups should present different technologies and their functioning to the class.
- *15 min*: The teacher should facilitate a SWOT session and generate an action plan for the City of Johannesburg.

Analysis of assignment questions

Question 1. The City of Johannesburg's waste management system is characterised by a straightforward, linear flow of waste from generation to disposal, with minimal interaction between stakeholders. The process begins with residents, which include both households and commercial establishments, as the primary source of waste. These residents generate waste and deliver it into a single waste stream without separation at the source. The municipality is responsible for collecting this waste using city-managed infrastructure. Once collected, the waste is transported to various landfills across the city, which serve as the primary destination for disposal. This approach results in most waste being deposited in landfills, reflecting the limitations of the existing waste management framework. The relationships between various stakeholders are extremely limited and lack sophistication. A beneficial topic for discussion is whether enhancing relationships between various stakeholders will improve the waste management situation for the city (Figure 11.4).

A notable feature of Johannesburg's system is the role of informal waste reclaimers, who operate outside of the municipality's formal structure. These reclaimers extract valuable recyclable materials from the mixed waste stream deposited in bins or at landfills. They manually sort through the waste to recover items such as plastics, metals, and paper, which are then diverted to recyclers. Recyclers process these materials for reuse, contributing to the circular economy. However, this critical role is performed under informal conditions, with waste reclaimers receiving little to no institutional recognition or support from the municipality. This gap highlights the absence of integration and formal partnerships within the system.

The relationships between stakeholders in Johannesburg's waste management system are limited and unsophisticated. Residents interact with the municipality solely by disposing of waste and paying for waste collection services. Similarly, waste reclaimers and recyclers operate independently of the municipality, with no formal mechanisms to coordinate their activities or integrate them into a broader waste management strategy. This lack of collaboration reflects the system's broader challenges, including inefficiencies in collection, low recycling rates, and over-reliance on landfills. As a result, the potential benefits of stakeholder interaction – such as improved waste diversion, enhanced resource recovery, and reduced environmental impacts – remain unrealised.

A critical area for improvement in Johannesburg's waste management system is the enhancement of relationships between various stakeholders. Establishing formal partnerships between the municipality, residents, waste reclaimers, and recyclers could lead to a more efficient and sustainable waste management model. For instance, introducing source separation programmes, where residents sort recyclables from general waste, could significantly reduce the burden on reclaimers and increase recycling rates. Similarly, providing reclaimers with logistical and financial support could improve their working conditions and efficiency, benefitting the entire system. In addition, engaging recyclers in the design of municipal waste policies could ensure that recycling infrastructure is optimised to handle the materials recovered.

The current system's reliance on a linear model, coupled with a lack of stakeholder integration, limits its effectiveness and sustainability. By shifting to a more collaborative, circular model, the City of Johannesburg could address its waste management challenges while advancing environmental sustainability and social equity. Strengthening these relationships and implementing coordinated strategies could transform the city's waste management landscape, reducing its dependency on landfills and creating a more resilient system.

Question 2. A review of the various technologies proposed for the city of Johannesburg is shown below.

Anaerobic digestion

This process, known as co-digestion, uses existing wastewater treatment digestors to break down the organic fraction of municipal solid waste. Microbes under anaerobic conditions use this process to break down organic waste, generating biogas for electricity or heat. We refer to the solid fraction as digestate, which undergoes further processing to produce compost for use as fertiliser.

The City of Johannesburg is able to use its existing wastewater infrastructure or invest in new anaerobic treatment facilities dedicated exclusively to municipal solid waste. In addition to diverting waste from landfills, this technology will enable the city to acquire stable sources of gas for electricity and heating. This solution allows for more sustainable uses of municipal solid waste, facilitating the displacement of nonrenewable sources of electricity. The process shown in Figure 11.5 is a co-digestion process where existing wastewater treatment digestors are used to digest the organic fraction of municipal solid waste. This process involves using microbes under anaerobic conditions to break down the organic waste to generate biogas which can be used for electricity or heat. The solid fraction is called digestate and can be processed further to generate compost to be used as fertiliser.

Incinerating the organic portions of the solid waste, such as paper and plastics, that anaerobic digestion cannot treat, is an option. A furnace combusts the paper and plastics, producing ash, fluoride gas, and heat. The turbine generator uses this heat to generate electricity (Figure 11.6).

Composting

Instead of being disposed into landfills, the digestate from the anaerobic treatment process can be further treated to generate compost that can be used as fertiliser (Figure 11.7).

Rubble and glass crushing

The rubble and glass constituents of the municipal solid waste can be separated and independently processed in crushing plants. The coloured glass is crushed into powder and used as an augmentation stream in existing manufacturing plants. The crushed rubble is used for brick making. This material is thus recycled into useful applications and diverted from the landfills (Figure 11.8).

The solution proposed for the City of Johannesburg accounts for the majority of the landfilled waste. Despite the need to dispose of components from some processes in landfills, this solution will not eliminate landfilling, nor will it contribute to a circular economy. However, it is the first step to achieving a circular economy.

Question 3. The City of Johannesburg can shift toward a sustainable, circular waste management system that maximises resource efficiency, reduces environmental impact, and fosters economic growth.

SWOT analysis

Strengths

- **Functioning waste collection infrastructure:** The City of Johannesburg (CoJ) has an operational waste collection system, ensuring that waste generated by residents and businesses is collected and managed centrally.

- **Functioning payment collection system:** The municipality has an established mechanism to collect payments for waste management services, ensuring a steady revenue stream to support operations.
- **Legislative command over waste management:** CoJ has regulatory control over waste management, enabling it to implement new policies and strategies to address current challenges effectively.

Weaknesses

- **Depleted landfill capacity:** The city's landfills are nearing their maximum capacity, which poses an urgent challenge for waste disposal.
- **Lack of integration of recycling:** Recycling activities are not formally incorporated into the city's waste management system, limiting resource recovery and contributing to inefficiencies.
- **Absence of a culture of waste separation:** Residents and businesses do not separate recyclables from general waste at the source, which complicates the recycling process and increases landfill dependency.

Opportunities

- **Reclaiming value from waste streams:** By recovering and reusing materials from waste, the city can reduce landfill use, conserve resources, and create value.
- **Augmenting electricity and gas supply:** Technologies such as anaerobic digestion and incineration can convert organic waste into energy, providing additional electricity and gas to meet demand.
- **Transitioning to a circular economy:** Moving from a linear waste management model to a circular economy can minimise waste, maximise resource efficiency, and promote sustainability.
- **Creating business and employment opportunities:** Developing waste processing facilities and formalising recycling activities can stimulate economic growth and generate jobs.

Threats

- **Waste disposal crisis:** If landfill capacity is fully depleted without alternative solutions in place, the city could face a severe waste disposal crisis.
- **Unavailability of finance:** Limited access to funding may delay or prevent the implementation of critical technologies and infrastructure upgrades.

Action plan: transitioning to a circular economy

- Secure capital for investment in technology and infrastructure
  - Collaborate with public and private stakeholders to raise funds for waste processing technologies such as anaerobic digesters and advanced recycling facilities.
  - Apply for grants and subsidies aligned with national and international sustainability goals.

- Review and upgrade infrastructure requirements
  - Assess the current waste management infrastructure and identify gaps that hinder the transition to a circular economy.
  - Develop plans to upgrade collection, processing, and recycling systems to meet future demands.
- Bridge infrastructure gaps
  - Identify critical gaps in waste processing and recycling infrastructure and prioritise investments to address these.
  - Ensure the establishment of transfer stations, sorting facilities, and waste-to-energy plants.
- Legislative reforms
  - Review existing legislation and propose amendments to facilitate source separation, waste reclaiming, and energy recovery.
  - Introduce incentives for residents and businesses to adopt sustainable waste practices, such as tax breaks for recycling participation.
- Foster partnerships
  - Create formal partnerships with waste reclaimers and recyclers to integrate them into the waste management system.
  - Provide training and financial support to reclaimers, enhancing their role in the recycling value chain.
- Forecast financial and socio-economic impacts
  - Develop comprehensive cost forecasts for future waste management services, including operational and capital expenditure.
  - Quantify the economic value of waste streams and potential revenue from energy recovery technologies.
  - Assess the impact of the circular economy transition on job creation and local businesses.
- Promote resource recovery and energy generation
  - Quantify the potential for recovering materials from waste streams and their market value.
  - Estimate the electricity, gas, or heat generation potential from technologies like anaerobic digestion and incineration.
- Build public awareness and encourage waste separation
  - Launch public awareness campaigns to promote waste separation at the source.
  - Implement pilot programmes in selected neighbourhoods to demonstrate the benefits of separation and recycling.

Key learnings

- The waste management system is complex, and decomposing the components of complex systems is central to resolving complex problems.
- A key input to strategic analysis is data that are applied to methodology.
- Optimisation studies are valuable in determining solutions to real-world problems.

Postscript

The City of Johannesburg is busy reviewing the recommendations presented by the University of Johannesburg. As yet, no determination has been made as to the future course the city will undertake to resolve its waste management crisis.

REFERENCES

Campitelli A, Aryoug O, Ouazzani N, Bockreis A and Schebek L (2023) Assessing the performance of a waste management system towards a circular economy in the Global South: the case of Marrakech (Morocco). *Waste Management* **166(1)**: 259–269, 10.1016/j.wasman.2023.05.003.

Dahlén L and Lagerkvist A (2010) Pay as you throw. Strengths and weaknesses of weight-based billing in household waste collection systems in Sweden. *Waste Management* **30(1)**: 23–31, 10.1016/j.wasman.2009.09.022.

Elia V, Gnoni MG and Tornese F (2015) Designing Pay-As-You-Throw schemes in municipal waste management services: a holistic approach. *Waste Management* **44**: 188–195, 10.1016/j.wasman.2015.07.040.

Farahbaksh S, Snellinx S, Mertens A *et al.* (2023) What's stopping the waste-treatment industry from adopting emerging circular technologies? An agent-based model revealing drivers and barriers. *Resources, Conservation & Recycling* **190(1)**.

Jiang L, Wang H, Tong A *et al.* (2020) The measurement of green finance development index and its poverty reduction effect: dynamic panel analysis based on improved entropy method. *Discrete Dynamics in Nature and Society* **2020(1)**, 10.1155/2020/8851684.

Kadyamadare GA. (2017). Assessment of waste separation at source by residential households as a tool for sustainable waste practices: a case study of the City of Johannesburg. University of the Witwatersrand, Johannesburg, South Africa.

Messina G, Tomasi A, Ivaldi G and Vidoli F (2023) 'Pay as you own' or 'pay as you throw'? A counterfactual evaluation of alternative financing schemes for waste services. *Journal of Cleaner Production* **412**: 137363, 10.1016/j.jclepro.2023.137363.

Ranjbari M, Shams Esfandabadi Z, Shevchenko T *et al.* (2022) Mapping healthcare waste management research: past evolution, current challenges, and future perspectives towards a circular economy transition. *Journal of Hazardous Materials* **422(6)**, 10.1016/j.jhazmat.2021.126724.

Ukkonen A and Sahimaa O (2021) Weight-based pay-as-you-throw pricing model: encouraging sorting in households through waste fees. *Waste Management* **135(1)**: 372–380, 10.1016/j.wasman.2021.09.011.

THEME 4

SUSTAINABLE DIGITAL TRANSFORMATION IN EDUCATION AND SKILLS DEVELOPMENT

Prof. Arnesh Telukdarie
ISBN 978-1-83662-645-9
https://doi.org/10.1108/978-1-83662-642-820251013

Chapter 12
Green learning management systems: the integration of digitalisation into technical and vocational education

Arnesh Telukdarie and Mangaliso Maduna

Introduction

The Chemical Industries Education and Training Authority (CHIETA) is a prominent Sector Education and Training Authority (SETA) in South Africa, dedicated to promoting skills development within the chemical industry. Like many graduate students across the nation, individuals often encounter difficulties in finding employment that matches their qualifications, mainly owing to the scarcity of job openings and frequent requirement for prior experience. To address these challenges, the CHIETA partnered with the University of Johannesburg and a team of skilled software engineers in June 2023 to create a viable solution: the skill supply and demand database (SSDD). This initiative, which falls under the CHIETA's management, aims to enhance job prospects for students. Established in accordance with the Skill Development Act of 1998, the CHIETA's mission is to bridge the skills gap in the chemical and manufacturing sectors through targeted training and development initiatives. Therefore, the SSDD platform was designed to connect job seekers with potential employers by posting relevant job opportunities, facilitating a smoother transition into careers that match their education. The SSDD is expected to play a significant role in matching qualified graduates with suitable job opportunities, facilitating a smoother transition into careers that match their education. Thus, the case study explores the development and implementation processes of the SSDD project and its potential for success. It also highlights how digital tools like the SSDD contribute to Sustainable Development Goal (SDG) 8 (decent work and economic growth) by helping qualified individuals secure meaningful employment.

Industry background

The chemical industry is a vital component of South Africa's industrial sector, contributing significantly to areas such as petrochemicals, pharmaceuticals, and agrochemicals. Despite its importance, the industry faces numerous challenges that hinder its growth and efficiency. Key among these challenges are resource constraints, skill shortages, talent mismatches, and operational inefficiencies. These issues not only impede the sector's productivity but also limit its ability to capitalise on growth opportunities and technological advancements. One of the most pressing issues is the shortage of skilled workers, which often surpasses the available talent pool. This imbalance creates significant operational challenges for chemical companies and exacerbates broader socio-economic issues, including unemployment and underemployment. Moreover, there is a frequent disconnect between the skills possessed by job seekers and those required by employers, highlighting a gap between academic qualifications and industry needs. Traditional methods of skills development and recruitment have proven inadequate in addressing these challenges. Therefore, innovative solutions, such as the SSDD platform, are essential to bridge the gap between education and employment. By promoting full and productive employment and ensuring decent work for all, the SSDD platform contributes directly to SDG 8. It facilitates connections between job seekers and employers, helping to mitigate the issue of unemployment in South Africa and ensuring that industries have access to the talent they need to thrive.

Description of the skills supply and demand database

The SSDD addresses the challenges of talent acquisition and development in the chemical sector. The CHIETA collaborated with the University of Johannesburg to develop the SSDD web application, marking a significant advancement in identifying, matching, and cultivating skills to meet the industry's evolving needs. The SSDD platform serves as a centralised repository for skills, qualifications, and job opportunities within the chemical sector, seamlessly integrating with the CHIETA's management information system (MIS). This platform enables employers, learners, and training providers to interact meaningfully through its user-friendly interface, fostering a collaborative ecosystem that promotes talent optimisation and strategic workforce planning. For a detailed overview of the SSDD's benefits and key features, refer to Figure 12.1.

Implementation and methodology

The journey of implementing the SSDD was meticulously documented and structured, a process that is clearly illustrated in Figure 12.2. It required careful planning, coordination, and teamwork to turn the idea into reality. One of the first steps was to organise training sessions and workshops to teach people how to use the SSDD effectively. These sessions aimed to familiarise users with the platform's features and functionalities, ensuring that they could navigate it with ease. Ensuring that the SSDD complied with legal regulations and safeguarded user privacy was another critical aspect of the implementation process. This involved putting in place robust data protection measures in line with laws like the Protection of Personal Information (POPI) Act. Encryption protocols, access controls, and regular audits were implemented to maintain the confidentiality and integrity of user information, instilling trust and confidence in the platform. Despite the challenges encountered along the way, including technical complexities and logistical hurdles, the implementation of the SSDD was driven by a shared vision and commitment to improving the workforce landscape in the chemical sector. By fostering open communication and collaboration among stakeholders, barriers were overcome, and progress was made toward achieving common goals. The SSDD, now operational, serves as a vital tool for connecting job seekers with employment opportunities and training programmes in the chemical sector. Through its user-friendly interface and comprehensive database, individuals can register their qualifications, search for suitable jobs, and access relevant training resources with ease. Employers, in turn, benefit from a streamlined recruitment process, enabling them to identify and engage with qualified candidates more efficiently.

Process flow

When individuals register on the platform, they will be required to provide their login credentials, such as their ID or passport number, email address, contact number, and password. In addition, they will need to agree to the terms and conditions of usage, which will inform them about the POPI Act compliance with the system. Once they have successfully logged in, they can create their profiles, and the system will automatically generate a resume based on the information they have provided. Figure 12.3 illustrates a resume generated by the system.

- When reviewing a candidate's resume, the CHIETA SSDD administrator verifies that the profile and resume/CV meet professional and stakeholder standards, while also checking qualifications with employers.
- The SSDD platform allows learners to upload copies of their certificates and resumes to their profile. When employers download the resumes of learners from the SSDD, their resumes contain a hyperlink leading to these certificates and professional registrations; then CHIETA levy-paying companies can shortlist candidates they are interested in and schedule interviews for them.
- Companies must accept the SSDD Terms and Conditions of Usage; one such term stipulates that employers notify the SSDD administrator about placements to facilitate tracking and reporting purposes.
- Learners have the responsibility of updating their profiles, notifying current placement status, and organising placement. With both learner and employer reporting, the CHIETA can cross-reference the information for accuracy before updating its database by notifying those learners who may have forgotten to update their details about placement opportunities.

Impact and outcomes

The SSDD is a powerful tool that has a significant positive influence on employment outcomes, workplace development, and industry competitiveness. One of the most impactful aspects of the SSDD is the way it makes it easier for companies to find and identify suitable candidates for the right job. By providing a centralised platform where job seekers can showcase their skills and qualifications, companies can quickly identify candidates who match their requirements. This has led to a much faster and more efficient recruitment process, saving time and resources for both the company and job seekers. Companies are not the only ones reaping the benefits of the SSDD. Job seekers, especially graduates, also benefit greatly from the platform. One of the major advantages for graduates is having access to a wide range of job opportunities within the chemical sector, all in one place. This eliminates the need for graduates to spend time searching for jobs across various platforms, enabling them to effortlessly apply for positions that match their qualifications and interests. The platform also allows graduates to showcase their skills and qualifications, making it easier for them to stand out to potential employers. This is particularly important for graduates who may not have a lot of work experience yet. They can showcase their potential to employers and boost their chances of employment by emphasising their skills and qualifications. But the SSDD is not just about connecting people with jobs. It also plays a vital role in addressing the skills gap within the industry. By analysing the data collected from the platform, the CHIETA can identify areas where there is a shortage of skills and take appropriate actions to address them.

This ensures that the industry remains competitive and sustainable, and graduates can find jobs that are aligned with their qualifications and interests. Overall, the SSDD is a valuable tool for both job seekers and companies, making the recruitment process faster, more efficient, and more effective. It is a prime example of how technology can be used to create a more sustainable and prosperous future for all, especially for graduates who are just starting their careers.

Conclusion

The SSDD is making a significant impact in the chemical sector by aligning with SDG 8: decent work and economic growth. By effectively connecting job seekers and graduates with suitable employment opportunities and linking employers with qualified candidates, the SSDD addresses critical skills gaps and enhances the efficiency of the job market. Looking ahead, the SSDD has the potential to expand its influence beyond the chemical sector by collaborating with other industries. Such expansion could revolutionise the job search process and the acquisition of necessary training across various sectors, fostering a more dynamic and responsive labour market. With continued support and collaboration, the SSDD can play a pivotal role in building a more inclusive and robust workforce in South Africa, contributing significantly to sustained economic growth and the promotion of decent work for all. By leveraging advanced technology and data analytics, the SSDD not only improves recruitment outcomes but also supports continuous learning and upskilling, ensuring that the workforce remains competitive and capable of meeting industry demands. This innovative approach underscores the importance of strategic workforce planning and skills development in achieving broader socio-economic goals. Through initiatives like the SSDD, South Africa can move closer to realising the aspirations of SDG 8, ensuring that economic growth is inclusive, sustainable, and beneficial for all members of society.

Exhibits

Figure 12.1 The SSDD benefits and key features

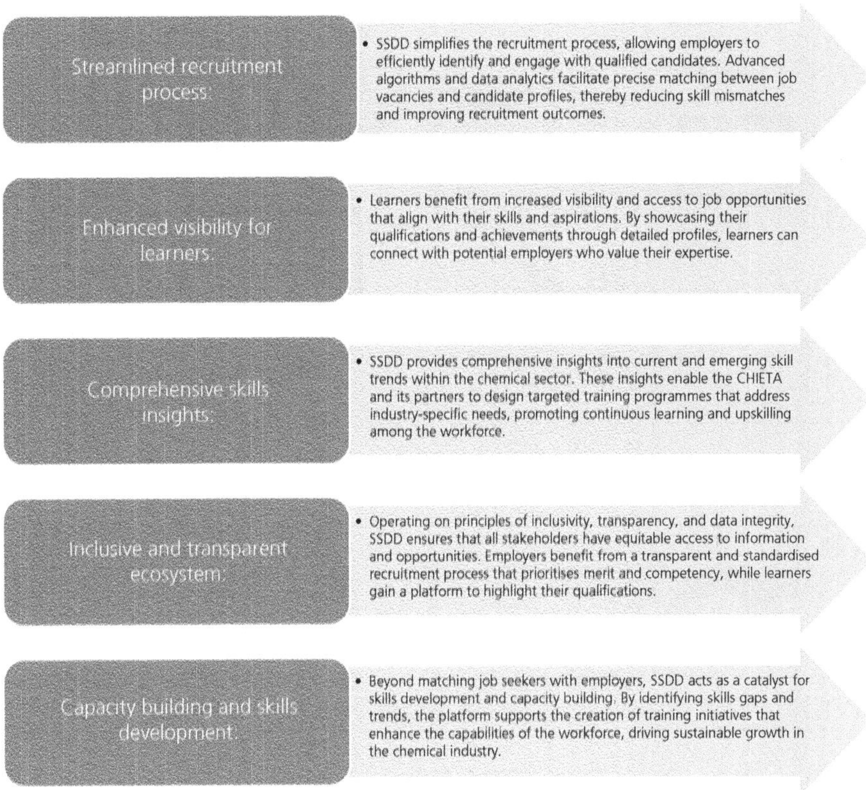

Figure 12.2 Process flow of the SSDD application

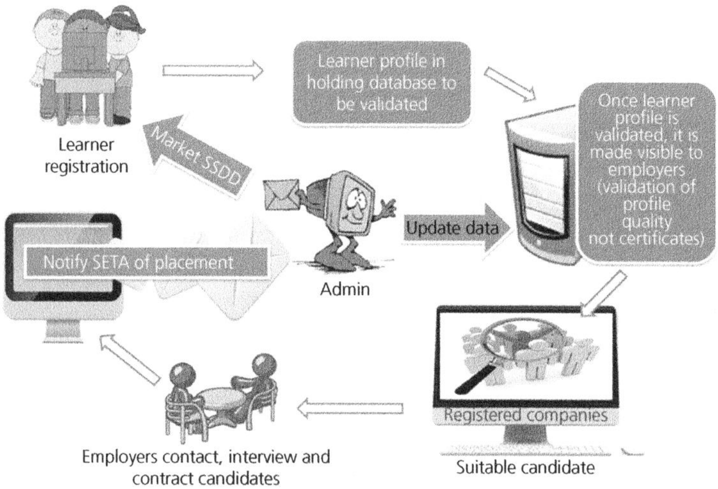

Figure 12.3 Snapshot of a fictional resume generated by the application

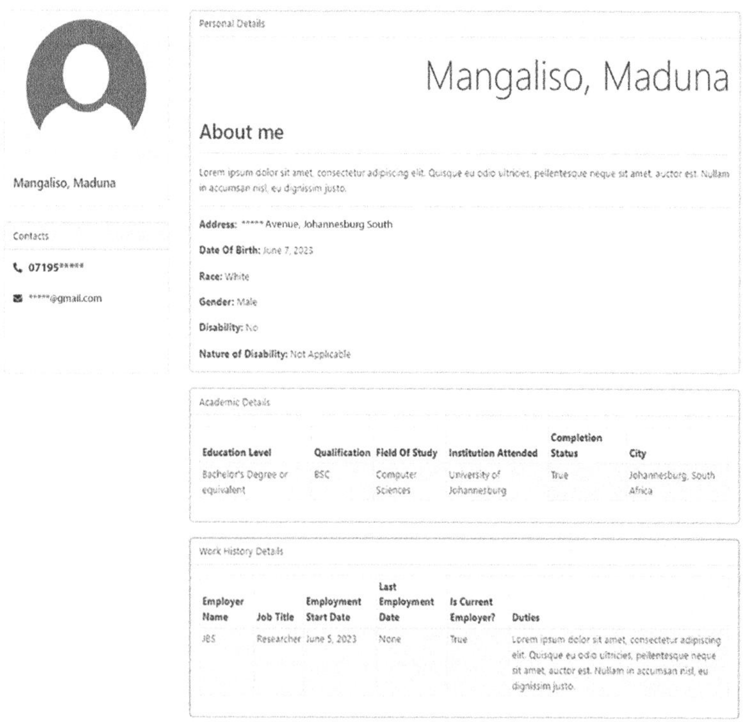

Green learning management systems: the integration of digitalisation into technical and vocational education

Teaching notes

Synopsis

This case study explores the challenges of structural unemployment, a significant issue aggravated by a surplus of graduates who lack sector-specific qualifications. The skills supply and demand database (SSDD) is introduced as a transformative solution to this problem, serving as a centralised platform for individuals who are qualified in various sectors. This web-based platform facilitates seamless interactions among employers, learners, and the Chemical Industries Education and Training Authority (CHIETA). It offers numerous benefits, such as improved absorption rates, increased employment contracts, and better employability outcomes. The platform ensures compliance with the Protection of Personal Information (POPI) Act, which safeguards data integrity, while its logging features enhance transparency in the recruitment process. Beyond its application in the chemical sector, this case study also examines the broader implications of the SSDD, contributing to a significant shift in labour demand and supply systems. By highlighting its unique features and methodologies, the study provides valuable insights into how the SSDD can foster a more efficient and effective employment landscape in South Africa.

Teaching objectives

The SSDD case study is designed for students and professionals in the fields of economics, labour market analysis, workforce development, and technology. The study aims to explain the challenges faced by South Africa's chemical sector and how the SSDD solves these problems through its various functions. For a better understanding of the subject matter, readers should have basic knowledge of labour market dynamics, structural unemployment, and data management concepts.

- To understand the challenges of structural unemployment in South Africa's chemical sector.
- To analyse the role of the SSDD in addressing these challenges.
- To evaluate the impact of the SSDD on key metrics such as absorption rates, employment contracts, and employability outcomes.

Keywords and subject code

Structural unemployment, absorption rates, labour market dynamics, transferability of solutions, ethical considerations in technology, critical thinking in problem solving.

Assignment questions

1. What are the key challenges contributing to structural unemployment in South Africa's chemical sector?
2. How does the SSDD facilitate seamless interactions among employers, learners, and the CHIETA?
3. What metrics are appropriate for evaluating the impact of the SSDD, and why are they important?
4. How does the SSDD comply with the POPI Act, and why is this compliance crucial?
5. Discuss the potential broader implications of the SSDD beyond the chemical sector.

Additional materials

Adams C and Yu D (2022) *Labour Market Trends in South Africa in 2009–2019*. Stellenbosch Economic Working Paper: WP03/2022. Department of Economics, University of Stellenbosch, and the Bureau for Economic Research, Stellenbosch, South Africa.

Agarwal P (2022) Structural unemployment. *Intelligent Economist*. https://www.intelligenteconomist.com/structural-unemployment/ (accessed October 2024).

Bhorat H (2004) Labour market challenges in the post-apartheid South Africa. *South African Journal of Economics* **72(5)**: 940–977.

Forsyth E (2023) *Structural Unemployment: Definition, Policies & Examples*. Study.com. https://study.com/academy/lesson/structural-unemployment-definition-causes-examples.html (accessed October 2022).

Hendrickx F (n.d.) *Protection of Workers' Personal Data: General Principles*. International Labour Organization (Working papers). https://webapps.ilo.org/static/english/intserv/working-papers/wp062/index.html (accessed October 2022).

HR Future (n.d.) *What Are Challenges Still Facing the South African Labour Market?* HR Future. https://www.hrfuture.net/talent-management/hiring/what-are-challenges-still-facing-the-south-african-labour-market/ (accessed October 2022).

Jayakrishnan A (n.d.) *Exploring the Evolution of Digital Workplace Platform*. Infince Digital Workspace. https://www.infince.com/blog/exploring-the-evolution-of-digital-workplace-platform#:~:text=A%20digital%20workplace%20platform%20is,and%20customer%20relationship%20management%20software (accessed October 2022).

Kenton W (2023) *Structural Unemployment: Definition, Causes, and Examples*. Investopedia. https://www.investopedia.com/terms/s/structuralunemployment.asp (accessed October 2022).

Mester LJ (n.d.) *Successful Workforce Development Programs: Four Lessons from Four Decades of Federal Reserve Research and Outreach*. Federal Reserve Bank of Cleveland. https://www.clevelandfed.org/collections/speeches/2022/sp-20220414-successful-workforce-development-programs (accessed October 2022).

Millennium Infotech (2023) *Impact of Technology on Employment*. Linkedin. https://www.linkedin.com/pulse/impact-technology-employment-millennium-info-tech-inc/ (accessed October 2022).

POPI Act Compliance (n.d.) *POPIA General Information.* POPI Act Compliance. https://www.popiact-compliance.co.za/popia-information/2-popia-general-information#:~:text=%E2%80%9CPOPIA%E2%80%9D%20THE%20PROTECTION%20OF%20PERSONAL,storage%20process%20of%20Personal%20Information (accessed October 2022).

Protection of Personal Information Act (POPI Act) (n.d.) POPIA South Africa. https://popia.co.za/ (accessed October 2022).

StudySmarter (n.d.) Structural Unemployment. https://www.studysmarter.co.uk/explanations/macroeconomics/economic-performance/structural-unemployment/ (accessed October 2022).

Works R (2017) The impact of technology on labor markets. *Monthly Labor Review (MLR)*: U.S. Bureau of Labor Statistics. https://www.bls.gov/opub/mlr/2017/beyond-bls/the-impact-of-technology-on-labor-markets.htm (accessed October 2022).

Teaching plan and timing

90-min teaching plan for the SSDD case study.

Note: The timings provided are approximate and may vary based on class dynamics and participation.

Understanding challenges and introducing the SSDD (20 min)

Objectives

- Introduce the case study and its relevance.
- Present the challenges faced by the chemical sector.
- Briefly introduce the SSDD platform.

Activities

- Brief lecture on the context and challenges.
- Quick overview of the SSDD functionalities.
- Q&A to address initial questions.

Analysing the role of the SSDD (30 min)

Objectives

- Break down the specific functionalities of the SSDD.
- Explore how the SSDD addresses the identified challenges.

Activities

- In-depth exploration of the SSDD functionalities.
- Case study analysis in small groups.
- Rapid-fire class discussion on the advantages and limitations.

Evaluating impact and broader implications (20 min)

Objectives

- Discuss the impact of the SSDD on key metrics.
- Explore the broader implications of the SSDD.

Activities

- Brief lecture on the impact of the SSDD metrics.
- Group discussion on the potential transferability of the SSDD.

Critical thinking and technology's role (15 min)

Objectives

- Encourage critical thinking about the SSDD's limitations and potential improvements.
- Emphasise the transformative role of technology.

Activities

- Group activity proposing improvements to the SSDD.
- Brief discussion on ethical considerations.

Student presentations and class discussion (5 min)

Objectives

- Provide a brief opportunity for student groups to share key insights.

Activities

- Quick round of student group presentations.
- Open-floor discussion on immediate takeaways.

Analysis of assignment questions

Question 1. What are the key challenges contributing to structural unemployment in South Africa's chemical sector?

Structural unemployment in South Africa's chemical sector is a complex issue rooted in various systemic challenges. Structural unemployment refers to a type of unemployment that occurs owing to underlying structural changes in the economy, rather than temporary fluctuations in demand (Amani, 2017). The first key challenge contributing to structural unemployment in South Africa is the mismatch between the skills possessed by job seekers and the skills required by available job opportunities (Betcherman, 2007). Tanzanian researchers found that a job/skills mismatch is one of the barriers to employment for graduates (Amani, 2017). This implies that graduates may possess skills that are not in demand or relevant to the current job market, leading to their unemployment. Furthermore, ineffective career guidance and a lack of access to relevant labour market information exacerbate the skills mismatch problem.

Another key challenge contributing to structural unemployment in South Africa is the lack of investment in industries that have the potential to create employment opportunities. This can be attributed to factors such as insufficient infrastructure, limited access to capital, and a lack of government support for entrepreneurship and innovation. Additionally, globalisation and technological advancements have also played a significant role in contributing to structural unemployment in South Africa. Globalisation, particularly technological innovations such as artificial intelligence and robotics, has disrupted economic trajectories and business models in both developed and developing countries (Amani, 2017). As a result, the knowledge and skills requirements for entering the labour market have changed, leading to a mismatch between the skills possessed by job seekers and those required by available job opportunities. Moreover, other factors contributing to structural unemployment in South Africa include a lack of confidence and poor communication skills among job seekers. These factors can hinder their ability to effectively market themselves and secure employment (Adams *et al.*, 2000). Furthermore, the characteristics of employers and their recruitment practices can also contribute to structural unemployment. Employers may have unrealistic expectations or preferences that are not aligned with the skills and attributes of job seekers, leading to a 'frictional mismatch' (Adams *et al.*, 2000). This mismatch between employer and job-seeker expectations, combined with an asymmetry of information, can result in difficulties in matching job openings with qualified candidates. Apart from the misalignment between workforce skills and employer demands, other factors exacerbate the problem. For instance, South Africa's historical legacy of apartheid has left deep-seated inequalities in access to education and training opportunities. This historical context disproportionately affects marginalised communities, perpetuating a cycle of unemployment and poverty. In addition, the education system in South Africa plays a role in contributing to structural unemployment. Many students in South Africa face barriers to employability owing to the choice of their field of study (Adams *et al.*, 2000). These barriers can include pursuing degrees or qualifications that are not in high demand in the job market, leading to a mismatch between their skills and the available job opportunities. Moreover, the rapid evolution of technology in the chemical sector requires continuous upskilling and adaptation. However, the education system often fails to keep pace with these advancements, leaving graduates ill-prepared for the modern workforce. Additionally, the lack of investment in research and development within the sector hampers innovation and limits job creation opportunities.

Explore further: To delve deeper into the multifaceted nature of structural unemployment in South Africa's chemical sector, organise a class discussion or seminar. Encourage students to explore historical and contemporary factors contributing to unemployment, such as apartheid's legacy, technological advancements, and educational disparities. Assign readings from academic journals and policy reports focusing on South Africa's labour market dynamics. Encourage students to critically analyse these sources and identify potential solutions to address structural unemployment in the chemical sector.

Question 2. How does the SSDD facilitate seamless interactions among employers, learners, and the CHIETA?

The SSDD not only acts as a catalyst for seamless interactions among employers, learners, and the CHIETA, but it also embodies a paradigm shift in how stakeholders engage in the labour market. The platform serves as a centralised hub where employers can access a comprehensive database of sector-qualified individuals, thereby streamlining the recruitment process and reducing search costs for both employers and job seekers. Learners, equipped with specific industry skills, can easily showcase their qualifications and connect with potential employers, transcending traditional barriers to entry into the workforce. This accessibility empowers learners to navigate the labour market more effectively and pursue career opportunities aligned with their skills and interests. Furthermore, the CHIETA, as a regulatory body, benefits from improved visibility into the skills landscape of the chemical sector facilitated by the SSDD. This enhanced data-driven decision-making enables the CHIETA to identify emerging skills gaps, prioritise training initiatives, and foster greater collaboration between educational institutions and industry stakeholders. By leveraging the SSDD as a strategic resource, the CHIETA can effectively align training programmes with industry needs, thereby enhancing the relevance and responsiveness of the education and training system to labour market demands. Moreover, the SSDD's role in facilitating seamless interactions extends beyond the chemical sector, serving as a blueprint for similar initiatives in other industries and regions. The platform's adaptability and scalability make it a valuable tool for addressing structural unemployment challenges and promoting skills development across diverse sectors. By harnessing technology to bridge the gap between education and employment, the SSDD exemplifies a holistic approach to workforce development that prioritises collaboration, innovation, and inclusivity. Figure 12.2 shows the process flow for the SSDD.

Explore further: Using the SSDD as a case study, conduct a workshop or seminar on the role of technology platforms in enhancing labour market efficiency and promoting skills development. Invite guest speakers from industry, academia, and government agencies to share their perspectives on the benefits and challenges of implementing such platforms. Divide students into small groups and assign each group a specific aspect of the SSDD to research, such as its impact on employer recruitment practices, learner engagement, or regulatory oversight. Encourage students to critically analyse the strengths and limitations of the SSDD and propose recommendations for improving its effectiveness and scalability.

Question 3. What metrics are appropriate for evaluating the impact of the SSDD, and why are they important?

To assess the impact of the SSDD, several key metrics should be considered. Absorption rates, indicating the successful integration of qualified individuals into the workforce, are crucial. This metric reflects the platform's effectiveness in connecting learners with employment opportunities. Employment contracts and employability outcomes are equally important, providing insights into the sustainability and quality of placements facilitated by the SSDD. Monitoring these metrics is essential to gauge the platform's contribution to reducing structural unemployment in the chemical sector. Additionally, tracking the number of successful matches and assessing the duration of employment contracts ensures a comprehensive evaluation of the SSDD's impact on the labour market. Another metric that can be used is the geographical distribution of job placements facilitated by the platform. Analysing the regional distribution of employment opportunities can highlight disparities in access to jobs and identify areas with persistent unemployment challenges, thereby informing targeted interventions and resource allocation strategies (Levin and Rouse, 2012). Furthermore, tracking the diversity and inclusivity of job placements can assess the platform's effectiveness in promoting equitable access to employment opportunities for historically marginalised groups, such as women, people with disabilities, and individuals from disadvantaged socio-economic backgrounds. Monitoring demographic indicators – such as gender, age, and ethnicity – can help identify potential barriers to employment and inform strategies to promote diversity and inclusion within the workforce. Moreover, assessing the alignment between learners' skills profiles and industry demand trends can provide valuable insights into emerging skills gaps and future workforce needs. By analysing data on the types of skills in high demand and those in short supply, policymakers and training providers can tailor education and training programmes to address industry-specific needs effectively (ILO, 2017). Additionally, tracking the scalability and sustainability of the SSDD over time can evaluate the platform's long-term viability and its ability to adapt to evolving labour market dynamics.

Explore further: Organise a workshop or training session on data analytics and performance metrics in workforce development, focusing specifically on the metrics for assessing the impact of the SSDD. Invite guest speakers from academia, industry, and government agencies to share their expertise on data-driven decision-making and programme evaluation. Divide students into small groups, and assign each group a specific metric to research and analyse in the context of the SSDD. Encourage students to explore real-world data sets and case studies to understand how these metrics can inform policy and practice in addressing structural unemployment.

Question 4. How does the SSDD comply with the POPI Act, and why is this compliance crucial?

The SSDD's compliance with the POPI Act entails several key components that are essential for safeguarding the privacy and rights of individuals. One such component is data minimisation, which involves collecting only the minimum amount of personal information necessary for the intended purpose. By limiting the scope of data collection to essential qualifications and employment history, the SSDD reduces the risk of unauthorised access to or misuse of personal data. Furthermore, the SSDD ensures data accuracy and integrity by regularly updating and verifying the information stored on the platform. Employers, learners, and regulatory bodies can rely on the accuracy of the data provided by the SSDD to make informed decisions regarding recruitment, skills development, and regulatory compliance. Additionally, the SSDD facilitates data transparency and accountability by providing users with clear information on how their personal data are collected, processed, and shared, as well as their rights regarding data access, rectification, and deletion. Moreover, the SSDD employs encryption and other security measures to protect personal data from unauthorised access, disclosure, or alteration. By encrypting sensitive information both during transmission and storage, the SSDD ensures that personal data remain confidential and secure, thereby enhancing user trust and confidence in the platform. Additionally, the SSDD conducts regular audits and risk assessments to identify and mitigate potential security vulnerabilities, ensuring ongoing compliance with the POPI Act and other relevant data protection regulations.

Explore further: Organise a workshop or training session on data protection and privacy regulations, focusing specifically on the POPI Act and its implications for technology platforms like the SSDD. Invite guest speakers from legal firms, regulatory agencies, or data protection authorities to provide insights into the regulatory framework and compliance requirements. Divide students into small groups, and assign each group a scenario or case study related to data protection compliance in the context of the SSDD. Encourage students to analyse the scenario, identify potential compliance issues, and develop strategies for addressing them in accordance with the POPI Act.

Question 5. Discuss the potential broader implications of the SSDD beyond the chemical sector.

The implementation of the SSDD in the chemical sector holds significant potential for broader implications in workforce development and labour market efficiency beyond its initial scope. One notable implication is the platform's role as a catalyst for innovation and collaboration across various industries facing similar structural unemployment challenges. By showcasing the success of the SSDD in connecting skilled workers with employment opportunities, other sectors can emulate the SSDD's model to effectively address their own labour market needs. This cross-sectoral collaboration can lead to the development of interconnected networks of skills databases, fostering a more dynamic and responsive labour market ecosystem. Furthermore, the SSDD has the potential to enhance the effectiveness of government-led skills development initiatives and vocational training programmes. By providing policymakers with real-time data on industry demand trends, skills shortages, and employment opportunities, the SSDD enables targeted interventions and resource allocation strategies to address emerging workforce challenges proactively. Additionally, the SSDD can facilitate partnerships between educational institutions and industry stakeholders to co-design curriculum and training programmes that align with evolving industry needs, thereby enhancing the relevance and quality of education and training provision. Moreover, the technology-driven approach of the SSDD can serve as a model for global workforce development strategies, particularly in countries grappling with high levels of unemployment and skills mismatches. By leveraging digital platforms and data analytics tools, countries can create their own versions of the SSDD tailored to their unique labour market dynamics and regulatory frameworks. This global proliferation of SSDD-like platforms can contribute to a paradigm shift in how countries address workforce challenges, moving towards more agile, data-driven, and inclusive approaches to skills development and employment facilitation.

Explore further: Organise a seminar or debate on the potential broader implications of the SSDD beyond the chemical sector, inviting guest speakers from diverse backgrounds, including academia, industry, government, and international organisations. Assign students to research and present specific case studies or examples of SSDD-like initiatives in different countries or industries worldwide. Facilitate discussions on the opportunities and challenges associated with scaling up and replicating SSDD models in diverse socio-economic contexts. Encourage students to critically evaluate the transferability and adaptability of SSDD-like platforms and propose recommendations for maximising their impact on global workforce development strategies.

Key learnings

- Understanding structural unemployment dynamics: Students will get a grasp of the root causes of structural unemployment in South Africa, including skills mismatches, historical inequalities from apartheid, underinvestment in critical sectors, and rapid technological changes. This understanding highlights systemic barriers to employment and the need for targeted interventions.
- Role of digital platforms in labour market efficiency: The SSDD exemplifies how centralised digital plaforms can bridge gaps between employers, job seekers, and training bodies. Learners will analyse how the SSDD streamlines recruitment, reduces search costs, and aligns education with industry needs, putting an emphasis on technology's transformative role in workforce development.
- Metrics for assessing employment impact: Students will identify key performance indicators such as job placement rates, skills alignment, and employment contract sustainability. These measures underscore the importance of data-driven decision-making to assess labour market interventions and ensure accountability.
- Scalability and cross-sectoral implications: The SSDD's success offers a model for addressing structural unemployment globally. Students will explore how similar platforms can be adapted across industries and regions, fostering collaboration between education, industry, and policymakers to create agile, inclusive labour markets.

REFERENCES

Adams J, Greig M and McQuaid RW (2000) Mismatch unemployment and local labour-market efficiency: the role of employer and vacancy characteristics. *Environment and Planning A: Economy and Space* **32(10)**: 1841–1856.

Amani J (2017) Prevalence of, and factors associated with, unemployment among graduates: evidence from Tanzania. *Africa Education Review* **14(3–4)**: 230–244.

Betcherman GG (2007) *A Review of Interventions to Support Young Workers: Findings of the Youth Employment Inventory*. World Bank Social Protection Discussion Paper. World Bank, Washington, DC, USA.

International Labour Organization (ILO) (2017) *Skills Needs Anticipation: Systems and Approaches*. ILO, Geneva, Switzerland.

Levin H and Rouse C (2012) *The Geography of Opportunity: A Framework for New Directions in Policy and Research*. Brookings Institution Press, Washington, DC, USA.

Prof. Arnesh Telukdarie
ISBN 978-1-83662-645-9
https://doi.org/10.1108/978-1-83662-642-820251014

Chapter 13
Modernising vocational welding education: the CHIETA's success and challenges in blended learning and virtual reality integration

Arnesh Telukdarie, Andile Nokwethaba Mtshali and Nonhlanhla G Sithole

Introduction

In the wake of the COVID-19 pandemic, industries globally faced unprecedented challenges. The pandemic impacted daily activities and education, including the vocational education sector. The welding industry within South Africa's chemical sector faced a unique set of challenges. It was against this backdrop that a pioneering initiative unfolded, spearheaded by the Chemical Industries Education and Training Authority (CHIETA), which plays a pivotal role in supporting technical and vocational education and training (TVET) institutions in South Africa. In April 2021, the CHIETA presented a pilot project on blended learning. Six colleges from different provinces were selected through a grant application process, and training began at different times. Of the six TVETs, three started their training in November 2021 and three in mid-January 2022. The CHIETA centrally set final assessments for October 2022, officially closing the pilot project and giving it a unique status.

With the transformative effects of the fourth industrial revolution (4IR) heightened by the COVID-19 pandemic, the pilot project unfolded in contemporary times, highlighting the urgent need for innovative solutions. The integration of virtual reality (VR) into educational systems and industrial training can significantly advance the objectives of the United Nations' (UN) sustainable development goals (SDGs) 4 and 9 by providing immersive, interactive, and inclusive learning environments. VR enhances quality education (SDG 4) by creating simulated classrooms and virtual field trips, which make learning engaging and accessible, especially for remote or disadvantaged students. It also facilitates practical skills development through realistic simulations of vocational tasks and soft skills scenarios. Moreover, in industry and infrastructure (SDG 9), VR offers innovative solutions for workforce training, including safety protocols and industrial operations, as well as urban and sustainable infrastructure planning. These advancements not only promote inclusive education and lifelong learning but also drive sustainable industrialisation and foster innovation, supporting a resilient infrastructure framework. At the centre of this narrative stands the CHIETA organisation, with its commitment to advancing vocational education and skills development within the chemical industry. The project takes place within the diverse landscape of South Africa, where the chemical industry is a vital economic sector and TVET institutions play a crucial role in developing skilled professionals. The central challenge faced by the CHIETA is modernising vocational welding education to meet the industry's evolving demands. As a technological solution to advance traditional teaching methods, there is a pressing need for innovative approaches to skills development. Through collaboration, innovation, and adaptability, the organisation embarked on a transformative journey to not only recover from the pandemic but also pave the way for a technologically advanced future of vocational education. This chapter delves into the intricacies of this groundbreaking initiative and uncovers the lessons learned in revolutionising vocational welding education in South Africa.

Transforming chemical industry skills and welding education

The TVET sector plays a crucial role in addressing skills gaps and unemployment rates. TVET institutions offer vocational and occupational training programmes that equip students with practical skills and knowledge relevant to various industries (Mutebi, 2023). Historically, TVET education in South Africa has faced challenges, including outdated curricula, inadequate infrastructure, and a lack of alignment between training programmes and industry needs. However, the sector has undergone significant reforms in recent years to enhance its relevance and effectiveness in meeting the demands of the labour market (Adams, 2019). Despite these challenges, TVET institutions play a critical role in providing accessible and affordable education and training opportunities to a diverse range of learners, contributing to the country's economic development and social mobility (Hodson and Kaufman, 2017).

In supporting TVET training institutions, the CHIETA focuses on developing skills within the chemical industry while ensuring that training programmes meet high industry standards by providing accreditation and maintaining a quality assurance framework. Another significant aspect of the CHIETA's role is to encourage collaboration between TVET institutions and industry partners. This ensures that the training provided aligns with the specific needs of the chemical industry, making graduates more attractive to employers. The organisation stays informed about industry trends and skills requirements through ongoing research activities, guiding the development of relevant training programmes. Through collaboration with government bodies, industry associations, and educational institutions, the organisation creates a well-coordinated effort to advance vocational education and skills development within the chemical industry.

This differentiation is important because it emphasises the applied nature of welding training, which is more appropriate for the facilities and methods of TVET colleges. Universities mainly emphasise the accumulation of knowledge and research-based learning, which may not be compatible with the vocational training of the welding profession. In placing welding programmes in colleges, the CHIETA ensures that students are equipped with practical experience and training that will enable them to meet the needs of the chemical industry. This also helps to foster better relations between colleges and industries, making the training offered relevant and timely. Concentrating welding education within colleges enables the curriculum to be more fluid and adaptable, which is important given the fast pace of technological change that defines modern industry practices.

Moreover, the welding trade faces multiple challenges within the chemical industry, including the need for skilled professionals and the transformative effects of 4IR. The challenges brought about by the COVID-19 pandemic prompted training service providers to seize opportunities for expediting digital-based skills development strategies. These strategies were designed to re-skill, cross-skill, and up-skill a substantial workforce, enabling them to capitalise on opportunities within existing technological solutions. The CHIETA organisation embarked on a pioneering initiative to modernise vocational welding education. The organisation implemented a dynamic blended learning programme as a possible solution to recover from the pandemic, paving the way for technologically advanced ways of teaching and training.

The blended learning approach through a coded welding skills programme

In alignment with the imperative to fully embrace 4IR and prioritise the digitalisation of skills development, the CHIETA launched a pilot programme. This initiative sought to not only benefit the participating TVET institutions but also contribute to the broader skills development sector within the chemical operations field. Executed in 2022, the pilot programme adopted a blended learning approach and involved 115 students from six TVET colleges situated across four provinces. These colleges were West Coast College in Vredenburg (Western Cape), Boland College in Worcester (Western Cape), College of Cape Town in Thornton (Western Cape), Umfolozi College in Richards Bay (KwaZulu-Natal), Maluti College in Kwetlisong (Free State), and Gert Sibande College in Standerton (Mpumalanga).

Participation in the pilot project was initiated through applications from the colleges, responding to the 2021 discretionary grant funding window. The training, originally scheduled to start in November 2021 with a one-year duration, experienced delays in recruitment and procurement procedures. Consequently, training began at different times, with the last college starting in mid-January 2022 and the programme officially concluding on 27 February 2023. Following the selection and signing of the programme contract, the participating colleges attended workshops covering a variety of topics, including curriculum presentation, assessment processes, and welding codes and standards. Support sessions, including one-on-one interactions to monitor programme progress, were provided. Virtual sessions were conducted, complemented by at least two on-site visits to each college. In instances where on-site visits were not feasible, relevant reports were submitted to ensure the achievement of programme deliverables.

The programme's standard requirements were formulated through collaboration between the CHIETA and stakeholders registered with the Quality Council for Trades and Occupations, with backing from the Department of Higher Education and Training. Essential qualifications for programme staff encompassed a welder trade certificate, diploma, and International Institute of Welding (IIW) programme qualifications, accompanied by a minimum of three years of experience in coded welding. Facilitators brought diverse industry experiences, spanning TVET colleges, mining, piping, steel, engineering, tubing companies, power stations, and participation in World Skills Competitions. Some facilitators held additional qualifications, such as higher diplomas and IIW programmes. While five colleges employed internal staff that met the qualifications, one enlisted an external individual. The CHIETA provided additional training for uniformity through sector skills trainers.

The learner recruitment process considered varied entry requirements, ranging from Grades 9 and 10 (NQF Level 1) with a background in mathematics and science to experienced welders with full trade or short welding course completion. Colleges could also conduct entry-level assessments through recognition of prior learning (RPL) processes. Support mechanisms for learners, such as human resources, student liaison officers, placement officers, or support and guidance business units, were reported by all colleges. Recruitment platforms included word of mouth, social media channels, internal bulletins, bulk SMS messages, and college websites, with durations ranging from two weeks to one month.

The skills programme, categorised at NQF Level 04, comprised 96 credits (960 hours or 120 days of training), resulting in a statement of results (SOR). To ensure quality, the programme curriculum adhered to Quality Council for Trades and Occupations (QCTO) standards, featuring eight knowledge modules (18 credits, approximately 180 hours, or 22.5 days to complete) and ten application modules (78 credits, 780 hours, or 97.5 days of training). These modules focused on welding trade introduction, career opportunities, coded welder certification requirements, occupational profiles of welders (60%), specialised positions (10%), artisan recognition of prior learning (ARPL) trade test procedures (5%), and the work environment of a welder (25%), detailed in Table 13.1. The training programme modules were developed with different themes for facilitator support and self-learning activities. Learners need to grasp these themes for skills development. Colleges received orientation and capacity-building before signing agreements, covering curriculum presentation, assessment processes, welding codes, and standards. Moreover, the main goal was to prepare learners to produce welded test coupons by following specific guidelines quickly. The programme aimed to ensure the produced welds met the visual and evaluation criteria defined by welding codes (Table 13.1).

The programme's restrictions posed challenges, particularly for learners without prior welding, maths, and science experience. One college considered this a disadvantage. Repetitive introductory welding modules caused some learners to lose interest, especially those with prior welding expertise who dropped out after finding job opportunities. The limited coverage of welding methods, focusing on SMAW/MMA and GTAW/TIG, hindered learners seeking work beyond these specifications. Challenges varied across colleges, including difficulties accessing machinery, load-shedding disruptions, and the need for additional facilitator training. Discrepancies in equipment selection and virtual welding machine software comprehension led to a low success rate in some colleges. Issues with licensing for virtual computers incurred significant expenses for one college, affecting overall progress.

Blended learning primarily occurred through Moodle, supplemented by textbooks and hardcopy notes. However, limitations in technology tools and access to platforms like Avi-edu and Moodle posed challenges, especially for students without Wi-Fi connectivity. Some colleges faced problems with welding equipment calibration, hindering effective demonstrations and outcomes. Unexpected obstacles throughout the programme disrupted the evaluation and training of virtual welding equipment, resulting in unsatisfactory outcomes. College workshops required repairs and maintenance, and some learners lacked proper personal protective equipment, posing safety risks. Procurement challenges further complicated the programme's execution.

Collaborative programme assessment with the University of Johannesburg

The UN's adoption of the SDGs emphasises the importance of fostering partnerships based on shared visions and goals for sustainable development solutions. The CHIETA pilot programme achieved notable success in providing students with a comprehensive and dynamic learning experience. Blended learning allowed for flexibility, accommodating diverse learning styles and preferences. The integration of welding simulation enabled students to enhance their skills in a controlled environment before transitioning to hands-on applications. This approach not only increased proficiency but also reduced the risk of errors in real-world scenarios. However, despite its success, the programme encountered challenges primarily stemming from outdated equipment and limited online capacity within the TVET institutions. These obstacles hindered the full realisation of the programme's potential, highlighting the need for continuous adaptation to technological advancements. In a proactive response to the identified challenges, the CHIETA initiated a collaborative effort with the University of Johannesburg (UJ)'s research team. The collaboration aimed to assess the programme's impact, identify challenges, and explore potential solutions. Leveraging the university's research expertise provided a holistic understanding of the programme's effectiveness and the hurdles faced by TVET institutions.

The UJ research team reviewed the blended learning programme documents, looking into the programme learner guide, key learning and development areas, and finding possible solutions. Through this analysis, the team outlined all the programme challenges and conducted a focus group interview with four of the six colleges that were part of the programme. The findings from these interviews highlighted some of the challenges already discovered by the documental analysis, among others.

The welding programme's findings across multiple colleges provided important insights into the initiative's additional and college-specific challenges and successes. Initially enrolling 20 students at each college, the occurrence of dropouts at West Coast College and Maluti College raises questions about potential difficulties faced by students during the programme (Table 13.2). This signals the need for a closer examination of the programme's elements that might have contributed to these dropouts. The variation in equipment availability among participating colleges is a noteworthy aspect. While West Coast College demonstrated a strong commitment by purchasing the augmented reality (AR) welding machine, Boland and Maluti Colleges enjoyed unhindered access to the Soldamatic machine, contributing to a smoother learning experience. However, the College of Cape Town's reliance on borrowed equipment suggests resource challenges, potentially influencing the extent of practical training and the overall educational experience (Table 13.2).

Although all the trainers received curriculum training before the start of the programme for the uniform learning system, diverse approaches to curriculum and assessment methods also emerged among colleges. Some adhered to the CHIETA's guidelines, aiming for standardised evaluation, while others, such as Boland and Maluti Colleges, embraced a more flexible, technology-driven educational model, emphasising online learning. Understanding the impact of these varied approaches on student engagement, performance, and overall programme outcomes is crucial for refining future implementations. The challenges encountered by colleges throughout the programme shed light on systemic issues. Accommodating students without prior welding experience emerged as a common hurdle, suggesting a need for tailored support mechanisms.

Technical difficulties, including those related to load-shedding, impacted practical sessions and online resource access, emphasising the importance of addressing infrastructure challenges. Resource constraints, particularly the costs associated with procuring welding equipment, emerged as a significant obstacle for some colleges. This financial challenge has broader implications for the overall quality of technical education, raising questions about the sustainability of such programmes without adequate financial backing. However, there was a notable outcome: eight students at West Coast College and Boland College secured employment. This underlined the programme's effectiveness in preparing students for future welding careers. This positive result serves as a testament to the programme's impact and suggests areas of strength that can be leveraged for further improvement.

This analysis provided a comprehensive understanding of the challenges and successes of the welding programme and was essential for refining its structure, addressing identified issues, and enhancing its overall efficacy. This logical discussion forms the basis for informed decision-making and future improvements in vocational education programmes. The UJ research team used these findings, along with literature research, to propose an economically and socially sustainable solution that leverages technological advancements.

The chemical industry could experience a significant transformation with the use of technological advancements, particularly in education and skills development. Immersive and interactive learning experiences can be provided through VR training programmes and classroom facilities. This type of educational approach provides students and workers with opportunities to learn practical skills without risking their safety, as it simulates real-life scenarios for welding techniques, chemical experiments, or safety protocols without the need for physical equipment or materials, ultimately reducing costs while minimising potential hazards associated with hazardous workplace environments (Radianti *et al.*, 2020). Additionally, technology provides remote access to education and training programmes that were previously restricted by location limitations.

To revolutionise chemical industry education, it is crucial to implement sustainable technological solutions that facilitate skill-building and uphold environmental sustainability. One approach could be utilising renewable energy resources for powering VR training facilities while also incorporating environmentally friendly principles throughout the curriculum to decrease the overall carbon footprint of educational programmes. Another potential solution involves integrating virtual laboratories into practical experiments as a means of reducing reliance on hazardous materials or chemicals, thereby minimising any negative ecological effects while promoting safety measures for both instructors and students.

The TVET sector in South Africa encounters various obstacles, which include substandard curricula, inadequate amenities, and restricted resources. Nevertheless, there are promising prospects for growth and progress. The partnership between UJ and the CHIETA endeavours to tackle these issues by modernising curriculum plans, improving infrastructure facilities, and investing in instructive training programmes (Akoojee, 2016). With the application of Industry 4.0 technologies, such as VR and AR television, institutions can revolutionise vocational education systems while advancing skills development measures. Interactive learning experiences offer deeper immersion into practical contexts, bridging theory gaps with encountered practices (Bacca *et al.*, 2015). By embracing digital transformations within their frameworks, these establishments could significantly advance educational standards, resulting in favourable student employment outcomes.

The solution from this partnership integrates cutting-edge technologies to revolutionise competency training across South Africa's vocational domain. These initiatives not only tackle current challenges within TVET but also promote sustainable development objectives related to education, employment opportunities, and overall economic prosperity. UJ proposed integrating the blended learning training programme into a VR environment, making training completely safe, easily accessible, and technologically advanced for students. To achieve this proposed development, the UJ team used a mixed-methods approach to provide a comprehensive analysis of the CHIETA welding programme. The study began with a literature review to identify global best practices in vocational education and training and the integration of VR technology. This was followed by a thorough examination of the CHIETA report on the welding programme, which included qualitative and quantitative data on the programme's implementation and outcomes. Focused group sessions were conducted with TVET participants to gather firsthand insights into their experiences and challenges. Additionally, the development and integration of VR-based welding training content were assessed through pilot implementations and feedback from both instructors and students. This multifaceted approach allowed for a detailed understanding of the programme's effectiveness, challenges, and opportunities for future improvements.

The UN's SDGs 4 and 9 have influenced the approach taken towards advancing vocational welding education by promoting inclusive, equitable, and quality learning opportunities. Under the CHIETA's dynamic blended learning programme, traditional classroom instruction is fused with digital technologies, creating immersive and interactive educational experiences that improve overall teaching standards. United Nations Educational, Scientific and Cultural Organization (UNESCO) reports from both 2017 and 2019 (Ganter de Otero, 2019; Ossiannilsson, 2017) state that the solution aligns with facilitating lifelong learning, which enhances educational excellence, by cultivating essential digital literacy skills in students relevant to modern workplaces. Furthermore, with resilient educational and skills development infrastructure, inclusive and sustainable industrialisation, and innovation by utilising technological advancements and partnerships within the industry, sustainable progress is guaranteed. Welding training now uses VR technology to modernise educational infrastructure and promote the cost-effectiveness, safety, and scalability of advanced training methods. Industry stakeholders' partnerships keep the programmes relevant to meet current needs leading to sustainable economic growth as well as achieving goals mentioned under SDGs 4 and 9 by facilitating a holistic approach that strengthens education practices and lays the foundations for development-industry advancement.

Integration of VR for welding training

The collaboration's pivotal focus was on integrating technological solutions to elevate training methodologies substantially. The objective was to overcome challenges associated with outdated equipment and broaden the scope of immersive learning experiences. The introduction of VR aimed to replicate real-world welding scenarios, providing a secure and controlled environment conducive to skills development. The virtual classroom is focused on the application module, as indicated in Table 13.1. The knowledge modules were reviewed and presented in PowerPoint, allowing students to access and study the content at their own pace. A thorough procedure for creating the VR environment is built simultaneously for the application modules. This workflow entails careful planning, design, and integration of interactive features. This methodical approach guarantees thorough coverage of both academic information and practical skills, offering a comprehensive and long-lasting educational framework. The UJ research team played a crucial role in identifying the requisite skills, tools, and materials necessary for developing this VR environment.

Following the identification of these resources, the blended learning programme curriculum underwent thorough revisions to seamlessly integrate each module into the VR system. This process extended over several months, with each module subjected to a minimum of two review cycles before implementation and final approval. The evolution of welding education, particularly the introduction of the VR welding facility, represents a significant milestone in response to the challenges faced by CHIETA and TVET institutions during their collaborative programme (Table 13.3). The research team showcased a commitment to innovation; the VR welding facility emerges as a groundbreaking advancement, offering a secure and immersive virtual environment for welding education. Its purpose expands beyond traditional training limitations, providing an environment where learning is not only effective but also engaging.

The research team's tailored approach, which ensures the VR facility specifically addresses identified needs and challenges within the collaborative programme, is noteworthy. The research team strategically designed this tailored solution to optimise skill acquisition, enhance safety measures, and eliminate logistical barriers.

The team adopted the Unity game engine, which serves as the foundation for orchestrating the virtual experience by handling physics and integrating three-dimensional (3D) models. Blender contributed its 3D modelling capabilities, seamlessly integrating detailed welding artifacts into Unity. The collaboration of Steam and SteamVR ensures a seamless VR experience, accommodating various hardware devices and fostering a smooth connection between the virtual environment and the user's hardware. The Streaming Assistant quietly plays a crucial role in streamlining technical aspects. The PICO VR headset was also procured and utilised as a pivotal element, providing an immersive experience with head tracking and hand controllers, creating a lifelike sensation during welding activities. In essence, the VR welding facility transcends being a mere technological project; it signifies a bold paradigm shift in welding education, intending to revolutionise traditional training methodologies and embrace a future where welding education is not only imperative but also culturally significant. This solution goes beyond problem-solving; it represents a redefinition of how we approach learning the art of welding in a world poised for transformative advancements.

The digital classroom offers students crafted and detailed replications of real-life welding scenarios. This allows them to delve into an interactive learning experience that mimics the intricacies of actual welding tasks with great accuracy (Figure 13.1). The simulations offered provide opportunities for individuals to enhance their skills in fundamental techniques used in welding, including torch manipulation, joint preparation, and bead placement, all within a safe environment specifically designed for experimentation without physical constraints or safety hazards present. The platform also includes comprehensive safety training modules that provide students with essential knowledge of critical safety procedures and protocols in welding environments. By emphasising the importance of personal protective equipment (PPE) usage and instilling a solid foundation in proper safety practices, this digital environment adequately prepares learners to competently navigate real-life welding situations. Another crucial benefit is flexibility, which allows learners to redo exercises as they require without fear of wasting resources or jeopardising their safety. This method allows students to gradually perfect their abilities and techniques by repeated practice over time. Additionally, the virtual medium offers a safe environment for apprentices, eliminating risks associated with real-life welding practice, such as exposure to harmful substances and extreme temperatures. Furthermore, the platform provides prompt feedback on students' welding skills, assessing weld quality, precision, and adherence to safety protocols. This instant evaluation enables them to evaluate their progress promptly and recognise areas for improvement in real time for a more effective approach to learning. Employing the potential of digital technology can create an interactive learning process that hones skills development while fostering continuous growth, leading to excellence within the welding education culture.

Conclusion

The integration of VR technology in welding education, particularly with the development of the VR welding facility, stands as a transformative and pioneering initiative. This collaborative effort, driven by the CHIETA and TVET institutions in partnership with the UJ's research team, exemplifies a commitment to innovation in overcoming challenges. The VR welding facility, designed to simulate real-world welding scenarios, offers a secure and controlled environment for skills development, effectively addressing limitations associated with outdated equipment. The tailored approach taken by the research team ensures that this VR solution is finely attuned to the specific needs and challenges identified in the collaborative programme, additionally reaching sustainable educational development inclusive of SDGs 4 and 9. Technological components, including the Unity game engine, Blender, Steam, SteamVR, Streaming Assistant, and PICO VR headset, synergistically contribute to creating an immersive and effective learning experience. This initiative goes beyond rectifying challenges; it signifies a paradigm shift in welding education, aiming to redefine traditional training methodologies and pave the way for a future where learning the art of welding is not only essential but also culturally significant. As this innovative journey unfolds, the VR welding facility is poised to optimise skills acquisition, enhance safety protocols, and eliminate logistical barriers. It represents a leap forward in welding education, demonstrating the readiness of the industry to embrace transformative advancements. The commitment to continuous improvement, coupled with a forward-thinking approach, ensures that the welding education landscape evolves to meet the demands of the dynamic and ever-changing technological era.

Exhibits

Figure 13.1 Training and engagement snapshots from the VR welding environment

Table 13.1 Module outline from the blended learning programme

Theory modules	Practical modules
KM-01: Introduction to the welding trade, NQF Level 2, Credits 1	AM-01: Safely use, care and store engineering hand tools, NQF Level 2, Credits 4
KM-02: General health, safety, and environmental protection concepts, NQF Level 2, Credits 1	AM-02: Safely use, care and store engineering power tools, NQF Level 2, Credits 2
KM-03: Welding codes, standards, and parameters, NQF Level 4, Credits 2	AM-03: Safely use, care and store engineering measuring equipment, NQF Level 2, Credits 2
KM-04: SMAW/MMA process, NQF Level 3, Credits 2	AM-04: Produce fillet welds on carbon steel material using the shielded metal arc welding process, NQF Level 3, Credits 8
KM-05: GTAW/TIG process, NQF Level 3, Credits 3	AM-05: Produce fillet welds on carbon steel material using the gas tungsten arc welding process, NQF Level 3, Credits 4
KM-06: Safely use, care and store engineering hand tools, NQF Level 2, Credits 4	AM-06: Produce plate butt welds on carbon steel material using the shielded metal arc welding process, NQF Level 4, Credits 6
KM-07: Safely use, care and store engineering power tools, NQF Level 2, Credits 2	AM-07: Produce plate butt welds on carbon steel material using the gas tungsten arc welding process, NQF Level 4, Credits 8
KM-08: Safely use, care and store engineering measuring equipment, NQF Level 2, Credits 3	AM-08: Produce plate butt welds on carbon steel material using the gas tungsten arc welding and shielded metal arc welding process combination, NQF Level 4, Credits 4
	AM-09: Produce pipe welds on schedule 40, ½-inch Ø, 2-inch Ø and 6-inch Ø carbon steel material using the gas tungsten arc welding process, with carbon steel and carbon steel welding consumables in the 2G, 5G and 6G positions, NQF Level 4, Credits 20
	AM-10: Produce pipe welds on schedule 40, 6-inch Ø carbon steel material using the gas tungsten arc welding (GTAW/TIG) and shielded metal arc welding (SMAW/MMA) processes combination, with carbon steel and carbon steel welding consumables, NQF Level 4, Credits 20

Table 13.2 Feedback from the programme colleges

College	Students enrolled	Dropouts	Welding equipment access	Challenges	Students employed
West Coast College	20	7	Hired and later purchased	The programme did not account for people without welding experience	4
Boland College	20	–	Soldamatic	Lack of technological experience from the facilitators and load-shedding	4
Maluti College	20	5	Soldamatic	Cost of procuring the machine, lack of time on the borrowed machine	5
College of Cape Town	20	–	Borrowed from Boland	COVID-related challenges, lack of software experience, remote access to classes	–

Table 13.3 Participating TVET colleges, geographical location, accreditation scope, and number of learners

TVET college	Geographical location	Accreditation scope	Number of participating learners
Gert Sibande College	Standerton (Mpumalanga)	OC: Welder NQF Level 4	15
West Coast College	Vredenburg (Western Cape)	OC: Welder NQF Level 4	20
Boland College	Worcester (Western Cape)	OC: Welder NQF Level 4	20
Umfolozi College	Sundumbili and Alton (KwaZulu-Natal)	OC: Welder NQF Level 4	20
Collage of Cape Town	Thornton (Western Cape)	OC: Welder NQF Level 4	20
Maluti College	Kwetlisong (Free State)	Carbon Welding SP, NQF Level 2	20

Total learners enrolled were 115, while 83 remained with training and were assessed, and 33 dropped out for various reasons, mainly finding job opportunities.

Modernising vocational welding education: the CHIETA's success and challenges in blended learning and virtual reality integration

Teaching notes

Synopsis

In response to the increasing need for affordable and advanced technical skills training, the CHIETA introduced a dynamic blended learning programme for coded welding skills. Aligned with the challenges posed by the COVID-19 pandemic and the 4IR, this initiative integrated face-to-face sessions, online learning, welding simulation, and practical training to enhance skills development, job creation and economic prospects. The six-month programme ran successfully across six TVET colleges in different provinces, enrolling 115 students during a 2022 pilot. Despite its success, TVET institutions encountered challenges such as outdated equipment and limited online capacity. Seeking solutions, the CHIETA collaborated with the University of Johannesburg (UJ)'s research team to assess the programme's impact, challenges, and potential solutions. This collaboration extended to integrating virtual reality (VR) technology, aiming to further elevate training methodologies while aligning SDGs 4 and 9, developing sustainable education solutions and leveraging Industry 4.0 advancements. The case study sheds light on the transformative role of immersive technologies in vocational education, offering insights for educators and institutions navigating the evolving landscape of technical skills training. It provides a detailed case study of blended learning as the new advanced solution to the traditional classroom teaching and learning era. It highlights some of the challenges, successes, and collaborative efforts, and the transformative impact of immersive technologies in vocational education. The focus of this study is to inform students of education and training advancement through technological solutions.

Teaching objectives

This case study is written for Strategic Management and Development in Emerging Markets modules at the undergraduate and postgraduate level. It provides a comprehensive understanding of the CHIETA blended learning case, encompassing the challenges, successes, collaborative efforts, and the transformative impact of immersive technologies in vocational education. The purpose of this study is to inform the students on education and training advancement through technological solutions. From this case, students will be able to do the following.

- Delve into the concept and complexities associated with blended learning in vocational education.
- Explore the dynamics of collaborative and partnership initiatives for problem solving.
- Assess the transformative impact of integrating VR technology into vocational welding education.
- Examine the process of tailoring technological solutions to address specific challenges within vocational education.

From this case study, students will gain an understanding of the multifaceted nature of blended learning within vocational education, including its intricacies and challenges. They will explore the dynamics of collaborative initiatives and partnerships aimed at resolving complex issues. Moreover, they will tap into the intricacies of tailoring technological solutions to address specific challenges encountered within vocational education settings, thus broadening their perspective on educational advancement through technology.

Research methods

This case study encompasses a comprehensive approach including secondary and primary data collection. The secondary data were collected by conducting a desktop analysis and review of literature in vocational education, welding education, technological solutions to teaching and learning, and by conducting an in-depth review of the CHIETA blended learning programme, including a detailed assessment of the coded welding training materials. The primary data were collected from engaging with four of the six TVET colleges who were part of the CHIETA pilot study to comprehend challenges, benefits, and opportunities for enhancement. Following this, the development phase involves creating learning and training packages. The knowledge modules are formatted in PowerPoint for asynchronous learning, while application modules undergo review, and a workflow is established for crafting the VR environment. Considerations during VR environment development encompass prerequisites, PPE, equipment, consumable requirements, and technical welding aspects. Subsequently, an evaluation of the new learning packages involves sharing the VR training package with the CHIETA and selected TVETs, including those from the blended learning programme and others with potential benefits. The feedback garnered is then used to update the VR training package, emphasising a proactive, user-friendly approach rooted in data-driven principles. The methodology is designed to replicate and evaluate existing training methods, set quality benchmarks, establish assessment frameworks, and collect pertinent data.

Keywords and subject code

Blended learning, welding education, virtual reality, collaborative skills programme, sustainability, education.

Assignment questions

1. How did the CHIETA initiate collaboration between TVET institutions and the University of Johannesburg, and what was the specific focus of this collaboration on enhancing vocational education?
2. What were the key challenges encountered during the implementation of the CHIETA blended learning programme, and how did the collaboration with the University of Johannesburg contribute to overcoming these challenges?
3. How did the collaboration with the University of Johannesburg contribute to aligning the VR welding facility with the specific needs and challenges identified in the CHIETA collaborative programme?
4. What outcomes and successes resulted from the CHIETA blended learning programme and the integration of VR, particularly in terms of student employment and skills development?
5. How does the integration of VR in vocational welding education signify a paradigm shift, and what broader implications does it have for the future of skills development in the context of the fourth industrial revolution (4IR)?
6. How does the integration of VR in vocational welding education signify a paradigm shift, and what broader implications does it have for the future of skills development in the context of SDGs 4 and 9?

Additional materials

For additional background on the CHIETA blended learning programme and VR welding environment, learners are encouraged to look at the following.

Engineering News (2023) CHIETA launches blended learning programme for chemical industry welding artisans. Engineering News. https://www.engineeringnews.co.za/article/chieta-launches-blended-learning-program-for-chemical-industry-welding-artisans-2023-03-02 (accessed 14/12/2023).

Suggested core readings

Additional reading from papers on modernising education and blended learning.

Bacca J, Baldiris S, Fabregat R, Graf S and Kinshuk (2014) Augmented reality trends in education: a systematic review of research and applications. *Educational Technology & Society* **17(4)**: 133–149.

Bacca J, Baldiris S, Fabregat R and Graf S (2015) Mobile augmented reality in vocational education and training. *Procedia Computer Science* **75**: 49–58, doi: 10.1016/j.procs.2015.12.203.

Bisami NA, Tasir Z and Kew SN (2023) Innovative pedagogical principles and technological tools capabilities for immersive blended learning: a systematic literature review. *Education and Information Technologies* **28(2)**: 1373–1425.

Boughzala I *et al*. (2017) Blended learning: the new normal and emerging technologies. *International Journal of Information and Communication Technology Education* **13(1)**: 43–53.

Chemical Industries Education and Training Authority (2020) Innovation to Be at the Epicenter of How We Facilitate Skills Development and Training in the Future. https://www.chieta.org.za/2020/11/18/innovation-to-be-at-the-epicenter-of-how-we-facilitate-skills-development-and-training-in-the-future/

Ganter de Otero JP (2019) Innovation in TVET: UNESCO-UNEVOC Trend Mapping. UNESCO-UNEVOC International Centre for Technical and Vocational Education and Training, Bonn Germany.

Harmon SA (2019) Virtual reality in education: a tool for learning in the experience age. *Tech Trends* **63(5)**: 13–21.

McGonigal J (2011) *Reality is Broken: Why Games Make Us Better and How They Can Change the World*. Penguin, London, UK.

McGrath S and Russon JA (2023) TVET SI: towards sustainable vocational education and training: thinking beyond the formal. *Southern African Journal of Environmental Education* **39**.

Merchant Z, Goetz ET, Cifuentes L, Keeney-Kennicutt W and Davis TJ (2014) Effectiveness of virtual reality-based instruction on students' learning outcomes in K-12 and higher education: a meta-analysis. *Computers & Education* **70**: 29–40.

Ossiannilsson E (2017) Blended learning. State of the nation. The International Council for Open and Distance Education, Oslo, Norway.

Sithole MD, Wissink H and Chiwawa N (2022) Enhancing the management systems and structures of technical vocational education and training colleges in South Africa. *Administration Publica*, **30(3)**: 86–105.

Stefan H, Mortimer M and Horan B (2023) Evaluating the effectiveness of virtual reality for safety-relevant training: a systematic review. *Virtual Reality* **27(4)**: 2839–2869.

Teaching plan and timing

The total teaching time for this case is 1 h, 55 min. This can be done in one long session or in two fairly timed sessions divided into different discussions. The times can also be adjusted according to the advancement of the class.

Introduction to the CHIETA (10 min)

- Overview of the CHIETA's role in skills development within the chemical industry.
- Importance of collaboration between the CHIETA and industry partners.
- Challenges faced by the welding industry, especially in the context of the 4IR and the COVID-19 pandemic.

The CHIETA blended learning approach (15 min)

- Introduction to the dynamic blended learning programme for coded welding skills.
- Structure of the programme, including face-to-face sessions, online learning, welding simulation, and practical training.
- Challenges encountered during the pilot programme across six TVET colleges.
- Successes and positive outcomes, such as employment opportunities for students.

Collaborative assessment with the UJ (10 min)

- Overview of the CHIETA's collaboration with the UJ's research team.
- Assessment of the programme's impact, challenges, and potential solutions.
- Discussion of findings from the focus group interviews with participating colleges.
- Finding sustainable solutions that align with UN's SDGs.

Introduction to VR integration (10 min)

- The need for technological solutions in welding education.
- Introduction to the integration of VR technology.
- Key objectives and goals behind adopting VR in welding training.

Development, evaluation, and technological components and impacts of VR training packages (40 min)

- Review of the CHIETA blended learning programme as a foundation for VR development.
- Discussion on the creation of knowledge modules, theoretical components, and practical VR environments.
- Considerations for prerequisites, PPE, equipment, and technical welding aspects in VR development.
- Sharing the VR training package with the CHIETA and selected TVETs for review.
- Discussion of feedback received and its incorporation into the VR training package.
- Importance of ongoing evaluation and adaptation in response to stakeholder input.
- Deep dive into the technological components, including Unity game engine, Blender, Steam, SteamVR, Streaming Assistant, and PICO VR headset.
- Discussion on the impact of VR on skill acquisition and the overall learning experience.

Innovation and future readiness (15 min)

- Discussion on how the CHIETA approaches innovation in response to challenges, emphasising future readiness and adaptability.
- Alignment of educational initiatives with evolving industry demands and technological advancements.

Synthesis and discussion (15 min)

- Synthesis of key learnings from the case study.
- Open discussion on the broader implications of integrating VR in welding education.
- Reflection on the transformative potential of innovative educational approaches.

Analysis of assignment questions

Question 1. How did the CHIETA initiate collaboration between TVET institutions and the University of Johannesburg, and what was the specific focus of this collaboration on enhancing vocational education?

The CHIETA's collaboration initiative involved a strategic approach to addressing the evolving needs of vocational education. The organisation fostered partnerships with TVET institutions and the UJ to create a holistic framework for skills development within the chemical industry. The specific focus of this collaboration was to enhance vocational education in response to the multifaceted challenges faced by the welding industry. The CHIETA acknowledged the pivotal role of technology and innovation in reshaping vocational education, particularly in the context of the 4IR. By engaging with the UJ, the CHIETA sought to leverage the expertise of a research institution to integrate cutting-edge solutions into its training programmes. The collaboration aimed to implement a blended learning approach with a strong emphasis on VR technology.

This forward-looking strategy was designed not only to recover from the challenges posed by the COVID-19 pandemic but also to establish a technologically advanced mode of teaching and training. The UJ's involvement provided a research-driven perspective, ensuring that data-driven principles underpinned the collaborative efforts. The focus on VR technology was strategic, offering a dynamic and immersive learning experience that aligns with the industry's demand for skilled professionals who can navigate the complexities of modern welding practices.

The CHIETA initiated collaboration by recognising the need for innovation in vocational education, with a focus on integrating VR technology to modernise welding education. The UJ played a key role in providing research-backed insights to guide the implementation of this transformative initiative, ensuring that vocational education meets the high standards demanded by the chemical industry.

Teaching tip: In exploring the CHIETA's collaboration initiative, encourage students to delve into the strategic reasoning behind forging partnerships. Discuss the significance of aligning vocational education with industry needs and the role of partnerships in achieving this alignment. Use this question to prompt critical thinking about the practical applications of collaboration in educational settings and the value of research institutions in driving innovation in vocational education.

Question 2. What were the key challenges encountered during the implementation of the CHIETA blended learning programme, and how did the collaboration with the University of Johannesburg contribute to overcoming these challenges?

The implementation of the CHIETA blended learning programme encountered several key challenges that necessitated a collaborative approach for effective resolution. One significant challenge was the discrepancy in equipment availability among participating TVET colleges. This challenge had implications for the overall consistency and quality of the training experience. Additionally, diverse approaches to curriculum and assessment methods emerged among colleges, introducing variations that required attention for standardised evaluation. The collaboration with the UJ played a pivotal role in addressing these challenges. Through a thorough analysis of the blended learning programme documents and collaborative discussions with the participating colleges, the UJ's research team identified the root causes of the challenges and proposed viable solutions. The focus on learner dropout rates, equipment access, and curriculum approaches provided a comprehensive understanding of the issues. The findings from the collaborative assessment informed tailored interventions. For instance, the research team proposed strategies to enhance technological literacy among facilitators, ensuring a more uniform and technology-driven educational model. Recommendations for improving equipment access and addressing discrepancies were also part of the collaborative solutions. The collaborative effort resulted in actionable insights that were crucial for refining the programme structure, enhancing curriculum presentation, and aligning assessment processes across colleges.

In essence, the collaboration with the UJ contributed to overcoming key challenges by providing a research-driven approach to problem-solving. The partnership facilitated a holistic assessment of the programme, leading to practical solutions that addressed specific issues faced by TVET colleges during the implementation of the blended learning programme.

Teaching tip: When addressing challenges and collaboration, guide students in critically analysing the interconnection of challenges and collaborative solutions. Discuss how the UJ's research-driven insights provided a holistic understanding of the challenges and practical implications of collaborative problem-solving in real-world vocational education. Emphasise the role of research in identifying root causes and tailoring solutions, fostering a deeper appreciation for collaborative approaches to addressing challenges.

Question 3. How did the collaboration with the University of Johannesburg contribute to aligning the VR welding facility with the specific needs and challenges identified in the CHIETA collaborative programme?

The integration of a VR welding facility was a transformative initiative within the CHIETA collaborative programme. The collaboration with the UJ was instrumental in ensuring that the VR welding facility was finely tuned to address the specific needs and challenges identified in the implementation of the blended learning programme. The collaboration began with a meticulous analysis of the challenges faced by TVET colleges, as evidenced by the feedback from the programme colleges. The UJ's research team delved into aspects such as student dropout rates, welding equipment access, and specific challenges faced by each participating college. This detailed understanding formed the basis for aligning the VR welding facility with the identified needs. The tailored approach adopted by the research team ensured that the technological components, including the Unity game engine, Blender, Steam, SteamVR, Streaming Assistant, and PICO VR headset, were integrated seamlessly. Each of these components played a specific role in creating an immersive and effective learning experience within the VR environment. For instance, the use of the PICO VR headset provided an immersive experience with head tracking and hand controllers, creating a lifelike sensation during welding activities. Moreover, the collaborative effort focused on addressing systemic issues identified during the programme, such as accommodating students without prior welding experience and overcoming technical difficulties related to load-shedding. The VR welding facility was designed not only to rectify these challenges but also to optimise skills acquisition, enhance safety measures, and eliminate logistical barriers.

The collaboration with the UJ ensured that the VR welding facility was not a generic technological solution but a finely tuned response to the specific needs and challenges identified in the CHIETA collaborative programme. This alignment was crucial for the success and effectiveness of the VR welding facility in providing a secure and controlled environment conducive to skills development.

Teaching tip: Focus on the intricacies of tailoring solutions to specific needs. Encourage students to explore the process of aligning technology with identified challenges in vocational education. Discuss the role of detailed analysis and understanding in the VR welding facility's development. Encourage discussions about how to tailor technology to address systemic problems and individual college obstacles. Use this question to highlight the importance of a collaborative and research-driven approach to developing innovative solutions that are finely tuned to the unique requirements of vocational education programmes.

The CHIETA's collaborative initiation involved a strategic approach to addressing the evolving needs of vocational education. The organisation fostered partnerships with TVET institutions and the UJ to create a holistic framework for skills development within the chemical industry. The specific focus of this collaboration was to enhance vocational education in response to the multifaceted challenges faced by the welding industry. The CHIETA acknowledged the pivotal role of technology and innovation in reshaping vocational education, particularly in the context of the 4IR. By engaging with the UJ, the CHIETA sought to leverage the expertise of a research institution to integrate cutting-edge solutions into its training programmes. The collaboration aimed to implement a blended learning approach with a strong emphasis on VR technology. This forward-looking strategy was designed not only to recover from the challenges posed by the COVID-19 pandemic but also to establish a technologically advanced mode of teaching and training. The UJ's involvement offered a research-driven perspective, ensuring that the collaborative efforts were underpinned by data-driven principles. The focus on VR technology was strategic, offering a dynamic and immersive learning experience that aligns with the industry's demand for skilled professionals who can navigate the complexities of modern welding practices.

The CHIETA initiated collaboration by recognising the need for innovation in vocational education, with a focus on integrating VR technology to modernise welding education. The UJ played a key role in providing research-backed insights to guide the implementation of this transformative initiative, ensuring that vocational education meets the high standards demanded by the chemical industry.

Question 4. What outcomes and successes resulted from the CHIETA blended learning programme and the integration of VR, particularly in terms of student employment and skills development?

The outcomes and successes resulting from the CHIETA blended learning programme, coupled with the integration of VR, are multifaceted and reflect the transformative impact of the collaborative efforts. One significant success is evident in the positive employment outcomes for students, as exemplified by the feedback from West Coast College and Boland College. The blended learning programme, with its innovative approach, created a conducive environment for skills development, aligning with industry demands. The integration of VR technology, particularly the VR welding facility, played a pivotal role in optimising skills acquisition. The immersive and dynamic nature of the VR environment allowed students to practice and perfect various welding techniques in a controlled setting. This not only increased proficiency but also reduced the risk of errors in real-world scenarios. The success stories of eight students securing employment at West Coast College and Boland College underscore the programme's effectiveness in preparing students for welding careers.

This positive outcome demonstrates the programme's effectiveness and identifies areas of strength for future enhancement. The discussion on outcomes should delve into the specifics of how the programme, augmented by VR technology, contributed to increased employability and skills development among participants. The UJ's collaborative assessment was crucial in evaluating these outcomes. By analysing feedback from programme colleges and assessing factors such as dropout rates, equipment access, and employment statistics, the collaborative research provided valuable insights into the tangible successes achieved.

In essence, the outcomes and successes of the collaborative programme and VR integration extend beyond theoretical advancements. They manifest as tangible employment opportunities and enhanced skills development, validating the effectiveness of the innovative approaches adopted by the CHIETA and its collaborators.

Teaching tip: Encourage students to think beyond the immediate context and reflect on the broader implications of technological integration in vocational education. Discuss the concept of a paradigm shift, and guide students in exploring how VR technology can redefine traditional training methodologies. Use this question to spark conversations about the transformative potential of technology in the 4IR. Foster discussions about how innovative approaches in vocational education prepare students for a future where technology is integral to skills development and industry relevance.

Question 5. How does the integration of virtual reality in vocational welding education signify a paradigm shift, and what broader implications does it have for the future of skills development in the context of the fourth industrial revolution (4IR)?

The integration of VR in vocational welding education signifies a profound paradigm shift, marking a departure from traditional training methodologies and heralding a transformative era in skills development. Unlike conventional approaches, VR introduces an immersive and dynamic learning environment, fundamentally altering how students acquire and apply welding skills. Firstly, VR enables a departure from the constraints of physical resources and safety concerns. Traditional welding education relies heavily on live practices with real metals and welding equipment, posing inherent risks. VR provides a secure and controlled setting for learners to practice various welding techniques without exposure to hazardous materials. This not only ensures student safety but also allows for repetitive practice, fostering mastery of skills over time.

Moreover, the VR environment facilitates a level of interactivity and engagement unattainable in traditional settings. Students can explore different welding processes, such as arc welding, and receive instantaneous feedback on their performance. This immediate feedback loop is invaluable for skills refinement, as it allows learners to quickly adjust and improve, contributing to a more efficient and effective learning process.

The paradigm shift extends beyond the immediate benefits of vocational welding education. In the context of 4IR, where technological advancements redefine industries, the integration of VR in skills development becomes a strategic response to the evolving needs of the workforce. It aligns with the demand for digital literacy and adaptability – crucial components for success in the 4IR landscape. The broader implications for the future of skills development are significant. VR-equipped vocational education not only prepares students for specific jobs but also cultivates a mindset of continuous learning and adaptation to technological changes. This aligns with the 4IR's emphasis on skills that transcend specific job roles, focusing on agility and versatility.

The integration of VR in vocational welding education represents a paradigm shift by offering a safe, interactive, and technologically advanced learning environment. Its broader implications lie in preparing learners for the demands of the 4IR, where adaptability and digital literacy are paramount for success in the ever-changing landscape of skills development.

Teaching tip: When introducing the concept of the paradigm shift in vocational welding education through VR integration, start by fostering an understanding of the traditional methods and challenges faced in live welding practice. Illustrate the inherent risks, resource limitations, and constraints associated with conventional approaches. This sets the stage for highlighting the transformative potential of VR. Next, engage students by explaining the fundamental principles of VR technology and how it addresses the identified challenges. Use relatable examples to help students visualise the immersive learning experience that VR offers. Emphasise the safety, interactivity, and immediate feedback aspects of VR that distinguish it from traditional methods.

Encourage critical thinking by initiating discussions on the broader implications of VR in the context of the 4IR. Discuss how technological advancements are reshaping industries, as well as the skills needed to thrive in this dynamic landscape. Facilitate conversations about the role of vocational education in preparing students for not just specific jobs but also for a future where adaptability and digital literacy are key.

Integrate hands-on experiences where possible, allowing students to explore basic VR welding simulations or demonstrations. This practical engagement enhances comprehension and reinforces the theoretical concepts discussed. Conclude by encouraging students to reflect on the significance of VR integration in vocational education and its role in shaping the future of skills development in the 4IR era.

Question 6. How does the integration of VR in vocational welding education signify a paradigm shift, and what broader implications does it have for the future of skills development in the context of SDGs 4 and 9?

The incorporation of VR technology signifies a noteworthy progression in vocational education and correlates with SDGs 4 and 9. VR facilitates immersive, interactive, and inclusive learning settings that satisfy the crucial aims of both SDGs. VR technology offers simulated classrooms and practical skills development scenarios, which significantly enhance the quality of education delivery by directly contributing to SDG 4 (quality education). Through immersive virtual environments, VR facilitates experiential learning opportunities that surpass traditional classroom restrictions. As a result, students can practically engage in hands-on exercises with prompt feedback that stimulates deep understanding and aids memory retention. Additionally, VR mitigates socio-economic and geographic hurdles when accessing educational resources, so all learners receive equal access to high-quality training experiences regardless of their backgrounds.

VR provides modern answers for employee training, safety measures, and infrastructure blueprinting. This actively supports the objectives of SDG 9 (industry, innovation, and infrastructure). Within welding industries found in chemical sectors specifically, VR simulations offer staff a safe place to develop skills needed for intricate tasks, which makes accidents less likely while increasing overall workplace security. Additionally, by providing urban development tools like realistic simulations and data visualisation aids to transportation systems or energy infrastructure, VR can help plan sustainable projects better.

Teaching tip: Encourage students to engage in reflective discussions that delve into the broader implications of integrating VR technology into vocational education while considering SDGs 4 and 9. Encourage them to examine how this integration can enhance educational quality, accessibility, safety, and efficiency in workplaces, foster innovation, and contribute towards sustainable development. Furthermore, encourage dialogue on ways stakeholders can collaborate to maximise the potential of VR to address global challenges and advance SDG objectives. Also prompt learners to consider ethical issues such as data privacy concerns or equity access disparities regarding utilising this technology. Additionally, ask participants about possible ramifications relating to job losses arising from introducing VR technology. By fostering dialogue around these critical topics and encouraging collaboration among various parties involved, learners will develop a holistic understanding of responsible usage surrounding the incorporation of VR tools within their field-specific training programmes, curricula, and support measures.

Key learnings

The case study provides valuable insights into the dynamic landscape of technical skills training, showcasing the adaptability required to meet industry needs, the refined nature of blended learning challenges, and the transformative potential of collaborative efforts and technological integration in vocational education. Aligned with the objectives of

SDGs 4 and 9, this initiative not only addresses immediate challenges but also lays the groundwork for enduring resilience and innovation in vocational education. These learnings are pivotal for educators, policymakers, and industry stakeholders seeking to navigate the evolving terrain of technical skills development. This is achieved by highlighting the importance of adaptive strategies in technical training programmes. The CHIETA blended learning initiative showcases a responsive approach to external challenges, particularly in the face of the 4IR and the constraints imposed by the COVID-19 pandemic.

The integration of face-to-face sessions, online learning, and welding simulation reflects a commitment to evolving with industry demands and leveraging technology to enhance skills development. It also touches on the challenges and opportunities encountered during the blended learning programme, highlighting the nuanced nature of implementing such initiatives. Issues like outdated equipment, limited online capacity, and divergent approaches among participating colleges shed light on the intricacies of programme execution. However, the positive outcome of student employability showcases the potential impact of innovative training methods. This duality emphasises the need for a comprehensive understanding of the challenges and successes of navigating the complexities of blended learning in vocational education. Finally, it examines future training solutions through collaborative partnerships, considering the potential advancements that can be achieved by utilising technological solutions in education. The collaboration with the UJ to integrate VR into the training programme unveils a strategic response to identified challenges. By harnessing VR technology, the initiative aims to revolutionise traditional training methodologies. The case study brings to the forefront the significance of collaborative efforts between industry players and research institutions in driving innovation in education. This integration not only addresses current challenges but also signals a paradigm shift in how welding education is approached, emphasising the cultural significance of such advancements.

Postscript

The case study unveils the dynamic trajectory of the CHIETA blended learning programme, illustrating a strategic response to external challenges posed by industry evolution and the COVID-19 pandemic. The initiative adeptly combines face-to-face sessions, online learning, and practical training, showcasing a multifaceted approach to skills development. As the programme unfolded, it encountered challenges typical of educational initiatives, such as outdated equipment and limited online capacity within TVET institutions. The collaborative effort with the UJ brings a pivotal shift with the integration of VR technology. This move transcends mere problem-solving, signifying a paradigm shift in welding education. The tailored, sustainable educational and skills development approach to addressing specific needs and challenges within the collaborative programme stands out, emphasising the importance of bespoke solutions in educational innovation.

Challenges encountered during the programme implementation, including equipment discrepancies among colleges and diverse approaches to curriculum and assessment methods, offer valuable lessons. These challenges became catalysts for refining the programme structure and enhancing its overall efficacy. The case study unfolds as a narrative of resilience, adaptability, and a commitment to continuous improvement in vocational education. The integration of VR technology became the focal point of this transformative journey, with technologies like Unity game engine, Blender, Steam, SteamVR, Streaming Assistant, and PICO VR headsets playing instrumental roles. This technological integration is not merely about staying current; it is about redefining the entire approach to learning welding, signalling a cultural shift in the perception of vocational training. The case study calls for an ongoing dialogue on the challenges and successes of vocational education. It prompts stakeholders to navigate the evolving landscape with a keen eye on adaptability, technological integration, and a commitment to tailor solutions that meet the unique demands of vocational training in the twenty-first century.

REFERENCES

Adams WC (2019) *Factors that Influence the Employability of Technical Vocational Education and Training (TVET) Graduates. A Comparative Study of two TVET colleges in the Gauteng Province*. Doctoral dissertation, University of the Witwatersrand, Johannesburg, South Africa.

Akoojee S (2016) Private TVET in Africa: understanding the context and managing alternative forms creatively! *Journal of Technical Education and Training* **8(2)**: 38–51.

Bacca J, Baldiris S, Fabregat R and Graf S (2015) Mobile augmented reality in vocational education and training. *Procedia Computer Science* **75**: 49–58.

Mutebi R (2023) *Efficacy of an Online Pedagogy on TVET Practical Skills Training in Uganda*. Doctoral dissertation, University of Eldoret, Kenya.

Radianti J, Majchrzak TA, Fromm J and Wohlgenannt I (2020) A systematic review of immersive virtual reality applications for higher education: design elements, lessons learned, and research agenda. *Computers & Education* **147**: 103778.

Prof. Arnesh Telukdarie
ISBN 978-1-83662-645-9
https://doi.org/10.1108/978-1-83662-642-820251015

Chapter 14
Sustainable business intelligence in Master of Business Administration programmes: Johannesburg Business School's breakthrough with the Shiny web application

Arnesh Telukdarie, Mpho Godfrey Nkadimeng and Thabile Dube

Initiating digital transformation in MBA administration at Johannesburg Business School

In November 2023, Johannesburg Business School (JBS) encountered challenges in its Master of Business Administration (MBA) administrative practices. The challenges faced by the JBS MBA administration included the complexities of cohort management, examination monitoring, and regulatory compliance, which necessitated a robust solution capable of streamlining processes and ensuring accuracy. The school realised it relied on manual spreadsheet-based systems, which resulted in inefficiencies, data inconsistencies, and compliance challenges. It was at this moment that the school thought about embarking on a journey towards a comprehensive digital transformation drive. This journey was through a strategic implementation of modernised technology and software engineering mechanisms that could help with smoother programme operations. JBS aimed to develop a web-based application utilising Shiny to centralise administrative functions, minimise errors, and provide stakeholders with real-time access to high-quality data. The strategic objectives included rationalising paper-based processes, achieving regulatory compliance, and saving time through the implementation of a single digital ecosystem. Additionally, JBS aimed to incorporate sustainability by reducing paper usage and minimising the environmental footprint associated with traditional administrative methods. By leveraging software engineering best practices, including intentional modularisation, rigorous testing, and seamless integration with MS SQL databases, JBS sought to ensure scalability, performance, and data security. While success is never guaranteed, JBS stands a strong chance of succeeding in their transformative journey towards a more efficient and effective MBA administration system.

The Johannesburg Business School

Established as a department within the University of Johannesburg (UJ), JBS has become a player in the field of business education, shaping the landscape of professional growth in South Africa. The MBA programme, one of its offerings, receives recognition for its excellence and relevance to industry standards. Known for its dedication to providing education tailored to meet business needs, JBS has gained widespread praise for its innovative approach to executive learning. The journey of JBS began with a vision to bridge the gap between knowledge and practical application, preparing business leaders with the skills and understanding required to navigate today's complex global market. With a faculty consisting of scholars, experienced industry professionals, and forward-thinking leaders, JBS has fostered a learning environment marked by intellectual rigour, hands-on learning experiences, and collaborative interactions. Since its inception, JBS has remained committed to empowering individuals to achieve their potential, driving advancement and promoting positive societal change. Through collaborations with companies, government bodies, and international organisations, JBS has broadened its impact and influence in fostering regional innovation and entrepreneurship. The demand for business professionals is on the rise, and the future lies within JBS's high walls.

Streamlining administrative processes: the digital transformation journey

JBS relied a lot on manual processes, which caused a lot of chaos for the MBA programme administrators. Usually, the appointment process utilised by the JBS programme administrators required the completion of a paper-based form for the nomination of assessors. This form encompassed various sections, including details of the student, supervisor, co-supervisor (if applicable), and two assessors. Each section required the input of personal and professional information, such as names, qualifications, affiliations, contact details, and academic background. Additionally, the form requested information regarding the assessor's experience with postgraduate supervision, publications, and international assessment. But it was not too long until the sun beamed. The JBS devised a solution that would forever change MBA administration processes (Figure 14.1). They developed a Shiny enterprise application tailored specifically for appointment management. The digitalised process aimed to enhance efficiency, accuracy, and accessibility while maintaining compliance with university policies and data handling procedures.

The digital transformation involved the key steps shown in Figure 14.1.

Leveraging agile development and stakeholder collaboration for digital transformation

The methodology employed in this case study embodies the principles of the KISS (keep it simple, stupid) principle and agile methodology to address the complex challenges encountered by JBS in managing MBA administration processes. By adhering to the KISS principle, JBS prioritised simplicity and efficiency in the design and implementation of its digital transformation initiative. This approach ensured that the solution remained intuitive and user-friendly, minimising complexity and maximising usability for stakeholders. Furthermore, the agile methodology served as the guiding framework for the development and deployment of the Shiny enterprise application. Through iterative cycles of development, testing, and feedback, JBS embraced flexibility and adaptability, allowing for continuous refinement and improvement of the digital solution. Agile practices such as sprint planning, daily stand-ups, and regular retrospectives fostered collaboration among cross-functional teams and enabled rapid response to changing requirements. Having identified the necessity for digital transformation, JBS embarked on a methodical journey that commenced with a comprehensive needs assessment.

Needs assessment

The methodology began with a comprehensive needs assessment to identify pain points and inefficiencies in the existing MBA administration processes at JBS. This involved stakeholder consultations, process mapping, and analysis of current workflows. To ensure a holistic understanding of the challenges faced, extensive stakeholder consultations were carried out. This included soliciting feedback and insights from administrative staff, ICT personnel, and other relevant personnel involved in the MBA administration processes. By actively involving key stakeholders, valuable perspectives and insights were gained into the existing pain points and areas for improvement. In addition to stakeholder consultations, process mapping techniques were utilised to visualise and understand the current state of MBA administration workflows at JBS. Process maps provided a detailed overview of the various steps, tasks, and interactions involved in each administrative process. This visual representation helped identify bottlenecks, redundancies, and areas of inefficiency within the workflows. Furthermore, a thorough workflow analysis was conducted to delve deeper into the intricacies of MBA administration processes. This involved examining each step in the workflow, analysing data inputs and outputs, and identifying potential areas for optimisation. Through meticulous analysis, hidden inefficiencies and pain points that may have been overlooked during initial assessments were uncovered.

Technology evaluation

A thorough evaluation of available technologies and platforms was conducted after identifying the need for digital transformation. Considerations included scalability, suitability for the specific requirements of MBA administration, compliance with university regulations, and compatibility with existing systems. In the technology evaluation phase, the methodology extended beyond the mere identification of available technologies and platforms. A rigorous assessment process encompassing multiple dimensions was undertaken to ensure the selection of the most suitable solution for the unique requirements of MBA administration at JBS. Key considerations included scalability, regulatory compliance, compatibility with existing systems, and overall total cost of ownership (TCO).

Selection of a Shiny framework

Based on the evaluation, the decision was made to leverage the Shiny framework for the development of the enterprise application. Shiny was chosen for its versatility, ease of deployment, and compatibility with the R programming language, which was already being utilised by JBS for data analysis. However, the decision to adopt Shiny went beyond its compatibility with R. Given the nature of the MBA administration processes, which are inherently data-driven, Shiny's capabilities for connecting to multiple databases as well as its language-agnostic nature were particularly appealing. This allowed for flexibility in data manipulation and analysis, enabling JBS to leverage its existing data infrastructure effectively. Moreover, Shiny's support for dynamic data visualisation and manipulation made it well-suited for the interactive and user-centric nature of the enterprise application. With Shiny, JBS could create interactive dashboards, visualisations, and reports that provided stakeholders with real-time insights and analytics, empowering informed decision-making. Additionally, Shiny's ability to seamlessly integrate with other programming languages, such as Python and JavaScript, further expanded its utility and versatility. This meant that JBS had the flexibility to incorporate additional functionalities and modules as needed, without being constrained by the limitations of a single programming language.

Agile development

The development process at JBS adhered to agile methodologies, embracing iterative cycles of development, testing, and feedback, as illustrated in Figure 14.2. This agile approach fostered flexibility, adaptability, and continuous improvement throughout the development lifecycle, ensuring that the final solution effectively met the needs of stakeholders. The figure demonstrates how the agile process moves from planning (with a backlog of items) through collaborative iterations and daily reviews, to delivery, incorporating regular feedback loops to enhance the deliverables.

Backlogs and sprint planning

At the outset of each development cycle, the team collaborated to prioritise tasks and define the scope of work for the upcoming sprint. This involved creating and refining a backlog of user stories and tasks, organised based on their importance and complexity. Through collaborative discussions and input from stakeholders, the team established clear objectives and deliverables for the sprint.

Daily reviews and stand-ups

Throughout the development process, daily stand-up meetings were conducted to facilitate communication, alignment, and transparency among team members. During these brief sessions, team members provided updates on their progress, highlighted any challenges or impediments, and identified opportunities for collaboration. These daily reviews served as an opportunity to address issues promptly, adjust priorities as needed, and ensure that development efforts remained focused and on track.

Feedback and iterative improvement

Feedback from stakeholders played a crucial role in shaping the direction of development efforts. Regular demonstrations and showcases were held to solicit feedback on interim deliverables and gather insights into user preferences and requirements. This iterative feedback loop allowed the team to incorporate stakeholder input, refine features, and prioritise enhancements based on real-world usage and feedback.

Collaborative planning and retrospectives

Collaborative planning sessions were conducted at the end of each sprint to reflect on progress, review outcomes, and plan for the next iteration. These planning sessions provided an opportunity for the team to celebrate successes, identify areas for improvement, and adapt strategies based on lessons learned. Additionally, retrospective meetings were held periodically to assess team performance, discuss process improvements, and identify opportunities for optimisation.

Training and adoption

Training sessions were conducted to familiarise users with the new digital platform and ensure smooth adoption. User feedback was solicited during training sessions and incorporated into the final application to optimise usability and user experience.

Testing and quality assurance

Rigorous testing and quality assurance measures were implemented throughout the development process to identify and address any bugs or issues. This included unit testing, integration testing, and user acceptance testing to ensure the reliability and functionality of the application.

Deployment and rollout

Once testing was complete and the application was deemed ready for production, it was deployed to the live environment. A phased rollout approach was adopted to minimise disruption and ensure a smooth transition from manual processes to the digital platform.

Monitoring and evaluation

Post-deployment, the Shiny application was closely monitored to track usage, performance, and user feedback. Regular evaluations were conducted to assess the impact of the digital transformation on MBA administration processes and identify areas for further improvement.

Documentation and knowledge transfer

Comprehensive documentation was developed to guide users on how to use the Shiny application effectively. Knowledge transfer sessions were also conducted to ensure that stakeholders were equipped with the necessary skills and understanding to utilise the digital platform optimally.

Core features

Nominations

The nomination form serves as a pivotal component in streamlining the administrative process at JBS. As shown in Figure 14.3, the form is meticulously crafted with a focus on user-centric design to ensure ease of use and efficiency for administrators submitting nominations. The interface offers customisable fields that empower administrators to tailor the form to specific nomination requirements, fostering adaptability and flexibility within the system. Robust validation and error handling mechanisms are implemented to maintain data accuracy and completeness, minimising errors and ensuring data integrity. Data security is paramount, with stringent measures such as encryption and access controls safeguarding sensitive nomination data. Additionally, a notification system is integrated to provide timely alerts to administrators and stakeholders regarding new nominations and updates, enhancing communication and transparency throughout the process.

Cohort Analytics KPI dashboard

The Cohort Analytics KPI dashboard serves as a comprehensive tool for monitoring key performance indicators (KPIs) relevant to MBA programme administration at JBS. As depicted in Figure 14.4, the interface presents critical metrics such as total students, total supervisors, ethics submissions, and the number of students awaiting Human Development Committee (HDC) review. Interactive charts, graphs, and visualisations within the dashboard enable stakeholders to gain actionable insights from the data. Customisable dashboards empower users to tailor their views based on their roles and preferences, facilitating personalised data analysis and decision-making.

Student management for online classes

The student management module streamlines the allocation of supervisors to students within the online learning environment at JBS. Administrators can efficiently assign supervisors based on predefined criteria, facilitating the supervision process and ensuring effective supervision (Figure 14.5).

Examiner management and student–examiner database

The examiner management module simplifies the assignment of examiners to students for assessments and examinations. A centralised database maintains student–examiner relationships, allowing administrators to retrieve relevant records quickly and efficiently. Robust search and query functionalities enable administrators to retrieve specific records based on various criteria, enhancing usability and accessibility within the system. Workflow automation streamlines examiner appointment processes, reducing manual effort and enhancing efficiency. Data integrity and security measures ensure the confidentiality and integrity of student–examiner data, fostering trust and compliance with regulatory standards.

Conclusion

This chapter has examined JBS's strategic initiative to implement a Shiny web application in MBA administration. This endeavour underscores a significant shift from traditional, manual processes to a modern, integrated digital approach, highlighting a broader movement within educational administration towards embracing technology to enhance efficiency and decision-making.

Summary of insights

The transformation at JBS involved several critical steps.

- Identification of inefficiencies in existing manual processes.
- Strategic planning and stakeholder engagement to ensure the solution's alignment with administrative needs.
- Adoption of agile methodologies and the KISS principle to maintain simplicity and user-centricity in the application's development.

This case study contributes to the educational administration literature by illustrating the practical application of theory in a real-world setting. It demonstrates how theories of change management and organisational behaviour can be effectively applied to lead technological adoption in an educational institution. Furthermore, the discussion around the selection and implementation of the Shiny framework provides a concrete example of decision-making processes in technology adoption, offering valuable lessons on evaluating and choosing software solutions that align with organisational goals and capabilities.

Reflection on broader educational trends

The digital transformation at JBS is reflective of a wider trend in educational institutions towards leveraging technology to solve administrative and operational challenges. This shift is driven by the need to improve service delivery, enhance data-driven decision-making, and meet the growing expectations of digital-savvy student populations. By integrating advanced technologies into administrative processes, institutions like JBS are setting new standards for efficiency, transparency, and responsiveness in educational administration.

Concluding thoughts

In closing, the journey of JBS from manual to digital processes not only enhances the understanding of the complexities involved in digital transformation but also serves as an inspirational blueprint for other institutions navigating similar paths. This case study reaffirms the critical role of strategic planning, stakeholder engagement, and thoughtful technology selection in achieving successful transformation outcomes in educational settings.

Exhibits

Figure 14.1 Digital transformation steps

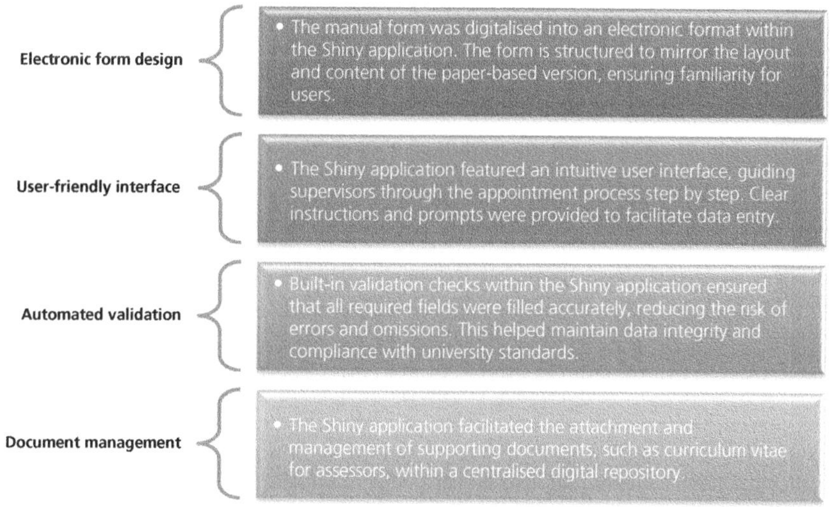

Figure 14.2 Agile development methods with iterative cycles

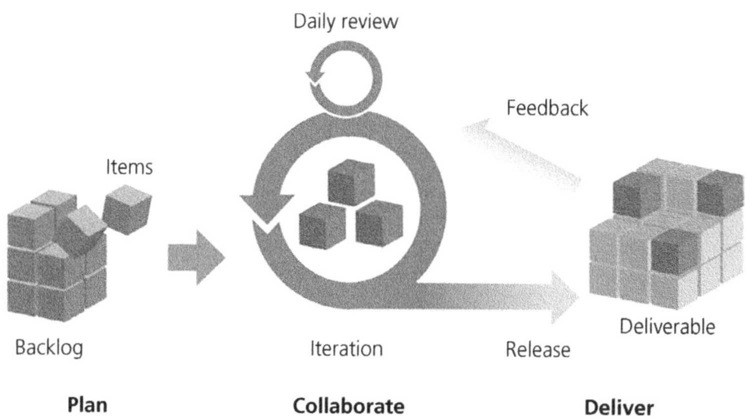

Figure 14.3 Assessor nomination

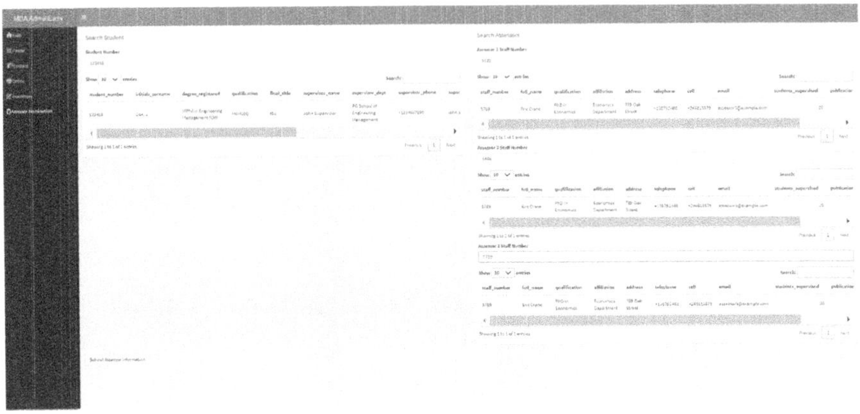

Figure 14.4 Snapshot of the analytics KPI dashboard

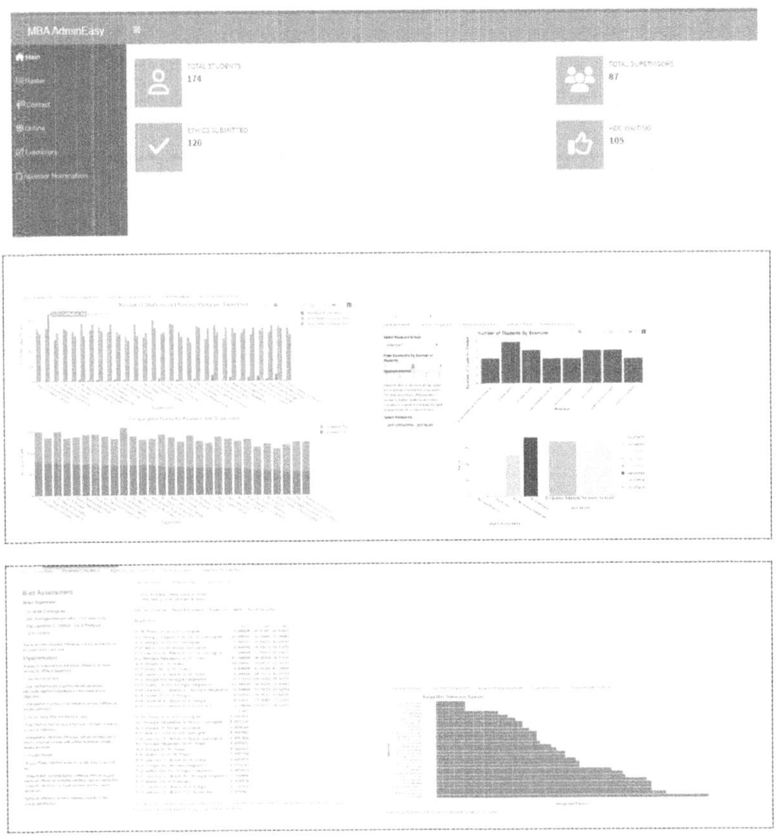

Figure 14.5 Online management dashboard

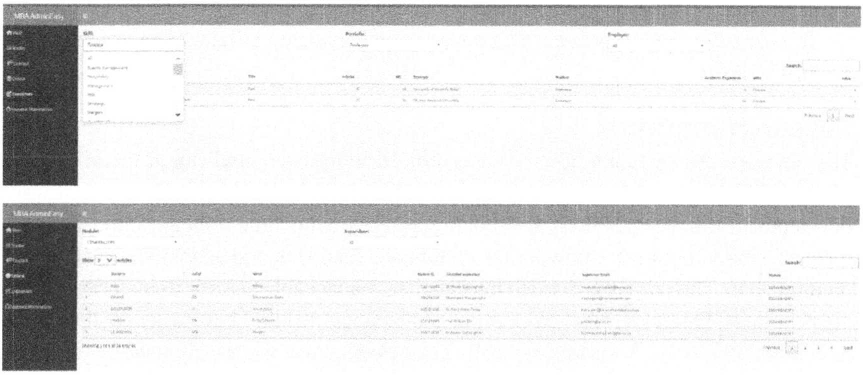

Sustainable business intelligence in Master of Business Administration programmes: Johannesburg Business School's breakthrough with the Shiny web application

Teaching notes

Synopsis

The case highlights the Johannesburg Business School (JBS)'s proactive approach to addressing challenges in MBA programme administration, particularly in handling student, examiner, and supervisor information. Despite its proven expertise in various fields, the institution faced difficulties with the complex structure of MBA administration. In response, the MBA lead team introduced a groundbreaking solution – the Shiny web application – in 2023. The application will make MBA administration easier to manage. Offering a user-friendly interface and a suite of functionalities tailored to simplify tasks and streamline processes, the Shiny web tool will revolutionise MBA administration processes. Furthermore, the application incorporates sustainable practices by reducing paper usage and improving energy efficiency through optimised digital workflows. With a firm commitment to delivering excellence in education, JBS embraced cutting-edge technology as a catalyst for continuous improvement, underscoring its dedication to providing the best possible experience with the school.

Teaching objectives

The case study's relevance depends on the specific focus and learning objectives of each module. It offers interdisciplinary applications, making it adaptable to various courses related to information systems, management, administration, and technology.

- Understanding the challenges in MBA administration.
- Appreciating the role of technology in addressing administrative complexities.
- Recognising the importance of innovation in enhancing efficiency in academic institutions.

Research methods

The development methodology for the advanced Shiny web application for MBA programme administration incorporates principles from both agile and traditional software development lifecycle (SDLC) methodologies. Following agile principles, the development process is iterative, allowing for continuous feedback and adaptation to changing requirements. The project is divided into smaller, manageable tasks or user stories, which are prioritised based on stakeholder input and business value. These tasks are then tackled in short iterations, or sprints, typically lasting one to two weeks, allowing for frequent releases and rapid progress. Daily stand-up meetings are held to ensure transparency, communication, and alignment among team members, while regular sprint reviews and retrospectives facilitate continuous improvement and adaptation. Simultaneously, the development process adheres to key phases of the SDLC, including requirements gathering, design, development, testing, deployment, and maintenance. The project begins with a comprehensive needs assessment, identifying pain points and requirements through stakeholder consultations, process mapping, and analysis. Based on these requirements, a detailed design phase follows, outlining the architecture, user interface, and functionality of the application. Development then proceeds iteratively, with each sprint focusing on implementing specific features or enhancements. Rigorous testing and quality assurance measures are integrated throughout the development process to ensure the reliability, functionality, and security of the application. Deployment follows a phased approach, with regular releases to production environments accompanied by thorough documentation and user training sessions. Post-deployment, the application undergoes monitoring and evaluation to track performance, user feedback, and adherence to project objectives. Maintenance activities, including bug fixes, updates, and enhancements, are performed as needed to ensure the continued success and usability of the application.

Keywords and subject code

Digital transformation, MBA administration, Johannesburg Business School (JBS), Shiny enterprise application, data security, key performance indicators (KPIs).

Assignment questions

1. What are the challenges associated with the complex nature of MBA administration, including the handling of diverse information related to students, examiners, and supervisors?
2. What is the importance of effective administration in higher education institutions such as JBS (use the readings provided)?
3. Analyse how the introduction of the Shiny web application reflects the institution's commitment to innovation in overcoming administrative challenges in MBA programmes.

4. How does the Shiny web application enhance the management of contact class students and adjudicated examiners in the MBA administration process?
5. What are the key features of the Shiny web application that make it an indispensable tool for administrators?
6. How does the modular design contribute to the efficiency of the MBA administration process?

Additional materials

Iqbal J and Asghar A (2020) Effect of university administrative and academic processes on student engagement: an empirical evidence from Pakistan. *UMT Education Review* **3(2)**: 85–109, 10.32350/uer.32.05.

Ruth D (2017) What is your MBA for? What's the story? *Management Learning* **48(1)**: 7–22, 10.1177/1350507616646302.

Wiley A (2023) *How MBA Programs are Changing with the Market*. AACSB, Tampa, FL, USA. https://www.aacsb.edu/insights/articles/2023/03/how-mba-programs-are-changing-with-the-market (accessed 15/01/2024).

Suggested core readings

Alvarez-Sández D, Velázquez-Victorica K, Mungaray-Moctezuma A *et al.* (2023) Administrative processes efficiency measurement in higher education institutions: a scoping review. *Education Sciences* **13(9)**, 10.3390/educsci13090855.

Moquillaza Henríquez SD, Quispe Manrique RA, García Quiroz KM *et al.* (2021) Framework to manage academic administrative processes, validating it from the students' perspective: case study – Faculty of Engineering of a public university. In *Proceedings of the International Conference on Industrial Engineering and Operations Management*. IEOM Society International, Southfield, MI, USA, pp. 3940–3952.

Ukpong N (2020) Administration of school business. In *Educational Management: A Guide for Practitioners*. University of Calabar Press, Calabar, Nigeria, pp. 588–604.

Teaching plan and timing

This teaching plan is designed for a comprehensive workshop spanning two sessions, combining theoretical understanding with practical, hands-on training.

Understanding the challenges and introduction to Shiny (50 min)

Challenges in MBA programme administration

- Overview of challenges in MBA administration (as outlined in the literature).
- Discussion on the complexities of handling diverse student, examiner, and supervisor information.
- Interactive session on real-world examples from the participants' experiences.

Introduction to Shiny

- Overview of the Shiny web application and its significance.
- Explanation of the modular design and specialised tabs.
- Discussion on the expected impact on curriculum development, faculty management, and student diversity.

Shiny web application features

- In-depth exploration of each tab: Master, Contacts, Online, Examiners, Supervisors, Appointment Form.
- Live demonstrations of how each feature addresses specific challenges in MBA administration.
- Q&A session for clarifications and concerns.

Hands-on workshop

- Participants will have access to the Shiny web application.
- Guided exercises on navigating through the different tabs.
- Practical scenarios to simulate real-world challenges and how to address them using Shiny.

Practical implementation and optimisation (40 min)

Implementing Shiny in curriculum development

- Detailed walkthrough on using the Master tab for curriculum development.
- Strategies for keeping the curriculum relevant and adaptive to industry changes.
- Collaborative exercises on designing a sample curriculum using Shiny.

Faculty management and collaboration

- Utilising the Examiners and Supervisors tabs for efficient faculty management.
- Encouraging collaboration among faculty members through the application.
- Discussion on addressing the tension between research expectations and teaching commitments.

Assessment

- Participants will be assessed through a hands-on project where they apply Shiny to address a specific challenge in their respective roles.
- A short quiz at the end of each session to ensure understanding of key concepts.

Analysis of assignment questions

Question 1. Challenges associated with MBA administration

The challenges associated with the nature of MBA administration stem from the multifaceted responsibilities involved in managing student, examiner, and supervisor information. The dynamic nature of MBA programmes, evolving industry trends, and the diverse needs of students contribute to the complexity. Administrators encounter challenges adapting the curriculum to stay relevant, recruiting and retaining qualified faculty, addressing cultural and learning diversity among students, and effectively integrating technology into the programme.

Question 2. Effective administration

An effective administration ensures the smooth operation of various administrative processes, including admissions, registration, academic advising, and financial aid. By managing these processes efficiently, the administration facilitates a seamless experience for students, faculty, and staff, ultimately contributing to a positive learning and working environment.

Question 3. Institutional commitment to innovation

The introduction of the Shiny web application reflects JBS's commitment to innovation in overcoming administrative challenges in MBA programmes. Recognising the need for a comprehensive solution, the institution took proactive steps to develop an in-house application tailored to address specific challenges. The commitment to innovation is evident in the Shiny web application's modular design, which targets key aspects of MBA programme administration, offering a dynamic and efficient solution to the complexities posed by managing students, examiners, and supervisors.

Question 4. Enhancement of contact class students and adjudicated examiners management

The Shiny web application enhances the management of contact class students and adjudicated examiners through specialised tabs designed to streamline these processes. The Contacts tab centralises student information, addressing the challenge of managing a diverse student body by providing data on cultural backgrounds, learning preferences, and personal goals. The Examiners tab focuses on efficient faculty management, providing tools for managing teaching commitments, research activities, and collaborative spaces for examiners to discuss academic standards.

Question 5. Key features of the Shiny web application

- Master tab: Serves as a central hub for programme coordination with real-time analytics.
- Contacts tab: Centralises student information, fostering a personalised approach to support services.
- Online tab: Manages virtual classrooms and e-learning resources, streamlining the integration of technology into the curriculum.
- Examiners tab: Enhances faculty management, providing collaborative spaces for examiners.

- Supervisors tab: Facilitates effective communication and collaboration between supervisors and students.
- Appointment form tab: Digitalises appointment scheduling, reducing administrative overhead.

Question 6. Modular design and efficiency

The modular design of the Shiny web application contributes significantly to the efficiency of the MBA administration process. Each specialised tab addresses a specific aspect of administration, allowing administrators to focus on particular tasks without unnecessary complexity. This modular approach facilitates targeted management, providing a more streamlined and efficient solution to the challenges associated with curriculum development, faculty management, student diversity, and technological integration.

Key learnings

The key learnings from this case study lie in the proactive approach taken by JBS to address the intricate challenges in MBA programme administration. The institution demonstrated a commitment to innovation and a keen understanding of the evolving landscape of business education by introducing the Shiny web application. The specialised tabs within the application, such as Master, Contacts, Online, Examiners, Supervisors, and Appointment form, strategically target critical aspects of administration, offering a dynamic and efficient solution to managing the complexities of student, examiner, and supervisor information. The modular design of the Shiny web application emerges as a pivotal factor in enhancing the efficiency of MBA programme administration, allowing administrators to focus on specific tasks without unnecessary complexity. This case study highlights the institution's dedication to providing innovative tools that redefine the administration process, ultimately positioning JBS as a leader in addressing the challenges inherent in MBA programme management.

Prof. Arnesh Telukdarie
ISBN 978-1-83662-645-9
https://doi.org/10.1108/978-1-83662-642-820251019

Chapter 15
Exploring historical data trend analysis in the Chemical Sector Occupational Skills Forecast Tool

Arnesh Telukdarie and Mpho Godfrey Nkadimeng

Workforce planning in the chemical sector: crafting a future-ready skills forecast tool

The research team from the University of Johannesburg, led by Professor Arnesh Telukdarie, convened to analyse the latest labour market data for the chemical industry in January 2024. Their mission was clear: to develop a comprehensive skills forecast tool for the Chemical Industries Education and Training Authority (CHIETA). Utilising data from 2016 to 2023, they aimed to predict future trends and inform strategic planning. The stakes were high, and the challenges ahead were formidable. The multidisciplinary research team, renowned for their expertise in digital transformation and data analytics, embraced this project with a shared vision. Professor Telukdarie, an authority in digital transformation, led the team with a commitment to integrating advanced analytics into workforce planning. Their goal was to create a tool that would not only forecast future skills requirements but also provide actionable insights to guide the sector's educational and training initiatives. Key collaborators in this project included the CHIETA industry research unit and industry stakeholders. These stakeholders were instrumental in providing essential data and insights, ensuring that the tool would address real-world challenges and opportunities within the chemical sector. Their involvement was crucial for aligning the tool's functionality with the practical needs of the industry.

Predicting future skills needs and strategic alignment in the chemical sector

The chemical sector faced several pivotal questions. How many jobs will be created in the coming years? Where would these jobs be located? What levels of education and skills would be required? The research team saw this project as an opportunity to answer these questions comprehensively. By doing so, they aimed to provide strategic guidance that would align educational programmes with industry needs, fostering a well-prepared and resilient workforce. The development process of the Chemical Sector Occupational Skills Forecast Tool was characterised by a multidisciplinary approach, drawing on a diverse array of methodologies and expertise. At its core, the process began with a comprehensive analysis of historical data obtained from the sector skills plan (SSP) framework, mandated by the CHIETA. This framework served as a rich source of information, capturing detailed insights into occupational profiles, skills demands, shortages, and training initiatives within the chemical sector. The challenge was to synthesise this vast amount of data into a functional tool that could accurately forecast future skills requirements. The CHIETA, established to facilitate skills development in the chemical sector, partnered with the University of Johannesburg for this pivotal project. The research team, having a history of successful collaborations with the CHIETA, leveraged their extensive experience and methodological rigour to ensure the project's success. By developing the Chemical Sector Occupational Skills Forecast Tool, they aimed to provide a robust platform for exploring and understanding occupational trends within the sector. As the team delved into the data, they confronted the complexities of labour market dynamics. They needed to account for variables such as gender disparities, employment types, and skills migration. Their objective was to produce a forecast that would empower policymakers and industry leaders to make informed decisions, ensuring the sector's growth and resilience. As they pored over the data, the team realised that their findings could shape the future of an entire industry, transforming challenges into opportunities and ensuring a thriving, skilled workforce for years to come.

The evolution and importance of skills forecasting

Skills forecasting is a critical tool for aligning workforce capabilities with future labour market demands. Its origins can be traced back to the mid-twentieth century when governments and educational institutions began recognising the need to anticipate changes in the labour market to better prepare the workforce. Early efforts were largely qualitative, relying on expert opinion and limited data. Over time, the advent of advanced statistical methods and data analytics transformed skills forecasting into a more precise science. Golden standards in skills forecasting emerged as methodologies evolved. These standards include rigorous data collection, comprehensive labour market analysis, and the use of sophisticated modelling techniques. Key institutions such as the Labor Market Information Council (LMIC) and the Future Skills Centre (FSC) in Canada and the Occupational Information Network (O*NET) in the USA have set benchmarks for best practices in skills forecasting. These organisations employ large-scale surveys, administrative data, and advanced analytics to predict occupational trends and inform policymaking. In South Africa, skills forecasting is critical for addressing the unique challenges faced by the country's labour market. The Sector Education and Training Authorities (SETAs) play a pivotal role in this process. SETAs are responsible for promoting skills development and ensuring that training programmes are aligned with the needs of various economic sectors. The context of skills forecasting in South Africa is shaped by several factors, including high unemployment rates, rapid technological change, and socio-economic disparities. These factors create a complex landscape for forecasting future skills needs.

Challenges and implications of skills forecasting in South Africa

One of the primary challenges in skills forecasting is the availability and quality of data. Accurate forecasting requires comprehensive and up-to-date labour market information, which can be difficult to obtain in some regions. The rapid pace of technological advancements adds another layer of complexity. As industries evolve, so do the skills required by workers. Forecasting must therefore be dynamic and flexible, capable of adapting to new trends and technologies. In South Africa, significant socio-economic disparities impact the labour market. These disparities must be considered to ensure that skills development initiatives are inclusive and equitable. Effective skills forecasting plays a pivotal role in policymaking, with far-reaching implications. It guides decisions regarding education and training programmes, ensuring their alignment with future labour market demands. This alignment is essential for cultivating a proficient and adaptable workforce capable of propelling economic growth and societal advancement. Several policies significantly influence skills forecasting in South Africa, including the Employment Equity Act, Skills Development Act, Skills Development Levies Act, Youth Employment Service (YES), women's empowerment programmes, and quota systems. These policies shape the analysis and utilisation of labour market data, thereby influencing the strategic direction of workforce development initiatives. For instance, the Employment Equity Act necessitates a focus on skills gaps among designated groups, requiring detailed demographic data to identify specific deficits and tailor training programmes accordingly. This ensures equal opportunities and fair treatment in training and employment. The YES programme targets job creation for young people, necessitating skills forecasting to identify high-demand industries for entry-level positions. By aligning the programme with labour market needs, young workers receive training and job placements in sectors with growth potential. Similarly, women's empowerment programmes aim to boost women's participation in the workforce, especially in male-dominated industries. Skills forecasting identifies sectors where women are underrepresented and the skills required to increase their participation, supporting the design of effective empowerment programmes and promoting gender equality. Quota systems mandate job allocations to historically disadvantaged groups, requiring skills forecasting to meet employment targets. Accurate skills forecasts ensure that training programmes address quota requirements, facilitating the integration of disadvantaged groups into the workforce. Overall, these policies underscore the need for accurate and comprehensive skills forecasting to ensure the relevance, targeting, and effectiveness of training programmes. By aligning educational initiatives with policy objectives, skills forecasting fosters economic growth, social development, and the creation of a resilient workforce prepared for future challenges.

Chemical Sector Occupational Skills Forecast Tool

The development process of the Chemical Sector Occupational Skills Forecast Tool was characterised by a multidisciplinary approach, drawing upon a diverse array of methodologies and expertise. At its core, the process began with a comprehensive analysis of historical data obtained from the sector skills plan (SSP) framework, mandated by the CHIETA. This framework served as a rich source of information, capturing detailed insights into occupational profiles, skills demands, shortages, and training initiatives within the chemical sector. The Chemical Sector Occupational Skills Forecast Tool represents a significant advancement in the field of workforce planning and analysis within South Africa's chemical industry. Developed to address the growing need for accurate and comprehensive occupational forecasting, this tool provides users with a robust platform to explore and understand trends within various occupational specialisations in the chemical sector. The tool offers users the flexibility to choose from a range of occupational specialisations within the chemical sector through a user-friendly drop down menu. This feature ensures that users can focus their analysis on specific areas of interest or relevance to their needs. The selection of occupations and specialisations was a critical component of the tool's development. Agencies and industry bodies typically define occupations and specialisations using standardised codes, such as the standard industrial classification (SIC) code. These codes provide a systematic way to classify industries by their primary economic activity, ensuring consistency and comparability across different data sets. The tool integrates data from various sources, using SIC codes and occupational classifications to organise and analyse the information. This ensures that the analysis aligns with industry standards, allowing for a detailed understanding of trends within specific segments of the chemical sector. The use of occupational codes enables the tool to differentiate between various job roles and specialisations within the chemical industry, providing the granularity needed to identify specific skills demands and shortages. The use of standardised codes ensures data accuracy and consistency, which are crucial for making reliable predictions and informed decisions. Once a specialisation is selected, users can access the 'Trend' tab, which presents a comprehensive time series plot of the selected specialisation's data. This feature allows users to visualise historical trends, identify patterns of growth or decline, and gain insights into the trajectory of occupational demand over time. The 'Correlations' tab provides users with a valuable opportunity to delve into the intricate web of time-dependent relationships within the dataset. In the dynamic landscape of the chemical sector, understanding these relationships is essential for informed decision-making and strategic planning. Autocorrelation and partial autocorrelation analyses serve as powerful tools for uncovering hidden patterns and potential correlations between variables. By generating autocorrelation function (ACF) and partial autocorrelation function (PACF) plots, users can visualise the strength and direction of these relationships, gaining insights into the underlying dynamics driving occupational trends. The 'Forecast' tab employs an autoregressive integrated moving average (ARIMA) model to generate forecasts for future values of the selected specialisation's data. Users can view forecast values, confidence intervals, and forecast residuals to assess the reliability and accuracy of the forecasts. This feature equips users with valuable insights into potential future trends, enabling proactive decision-making and planning.

Exploring historical data trend analysis in the Chemical Sector Occupational Skills Forecast Tool

The trend analysis feature of the Chemical Sector Occupational Skills Forecast Tool is designed to provide users with a comprehensive view of historical data and trends within the chemical sector. This feature is accessible through the 'Trend' tab after users select their desired specialisation from the dropdown menu.

Visualisation of historical trends

The 'Trend Analysis' tab presents a detailed time series plot of the selected specialisation's data. This visual representation allows users to observe historical trends, making it easier to identify patterns of growth or decline over time. For example, the COVID-19 pandemic significantly impacted certain occupations within the chemical sector. During the pandemic, demand for chemical sales representatives and chemists remained relatively constant owing to their crucial roles, regardless of the size of the organisation. However, other occupations saw notable shifts (Figure 15.1). The pandemic accelerated the need for new roles, such as mechanical engineering technologists and computer engineering mechanics, which did not exist in 2016 when the data analysis began.

Identifying patterns

Understanding the trajectory of occupational demand is crucial for workforce planning and development. The trend analysis feature helps users identify recurring patterns, seasonal variations, and long-term trends within the dataset. For instance, traditional mechanical engineers have experienced a steady decline, highlighting a shift in the industry towards more specialised roles. In contrast, key occupations such as chemical sales representatives, chemists, and engineers have remained consistent owing to their indispensable nature within the sector. These insights can inform strategic decisions regarding training programmes, hiring practices, and resource allocation. The trend analysis feature enables users to base their decisions on data. This reduces uncertainty and enhances the ability to forecast future needs with greater accuracy. For instance, recognising the rise in demand for mechanical engineering technologists and computer engineering mechanics can help educational institutions and training providers develop relevant programmes. Additionally, it allows users to benchmark current performance against historical data, helping to set realistic goals and expectations. The Sankey plot effectively illustrates the flow of roles from traditional occupations, like mechanical engineers, to emerging specialisations such as mechanical engineering technologists and computer engineering mechanics. This visual tool helps to highlight the transition paths and quantify the movement of workforce segments, providing a clear picture of how the sector's occupational landscape has evolved (Figure 15.2).

Correlation analysis in the Chemical Sector Occupational Skills Forecast Tool

The 'Correlations' tab within the Chemical Sector Occupational Skills Forecast Tool provides users with a valuable opportunity to delve into the intricate web of time-dependent relationships within the dataset. In the dynamic landscape of the chemical sector, where trends evolve rapidly and external factors exert significant influence, understanding these relationships is essential for informed decision-making and strategic planning.

Gestation lags in the labour market

Gestation lags, a fundamental concept in labour market analysis, denote the time delay between changes in labour market conditions and their reflection in employment outcomes. These delays vary in duration, with some changes yielding immediate effects while others may take months or even years to materialise. For instance, shifts in demand for specific skills may not immediately translate into changes in employment levels owing to factors such as recruitment processes and training periods. Recognising gestation lags is essential for interpreting correlation analysis results accurately. By factoring in these lags, users can better understand the long-term implications of workforce trends for the chemical sector. The 'Correlations' tab within the Chemical Sector Occupational Skills Forecast Tool employs autocorrelation and partial autocorrelation analyses to uncover intricate relationships between variables over time. Autocorrelation examines the relationship between a variable and its past values, capturing any systematic patterns or trends. Conversely, partial autocorrelation measures this relationship while controlling for intermediate values, providing insights into the direct influence of past occurrences. These analyses serve as potent tools for identifying the underlying dynamics driving occupational trends within the chemical sector. By scrutinising autocorrelation and partial autocorrelation plots, users can discern patterns of dependency and anticipate future workforce needs with clarity and confidence. Both gestation lags and correlation analyses play pivotal roles in understanding the temporal dynamics of the labour market within the chemical sector. While gestation lags shed light on the time delay between changes in labour market conditions and their manifestation in employment outcomes, correlation analyses unveil the underlying patterns and dependencies driving occupational trends. Together, these insights enable users of the Chemical Sector Occupational Skills Forecast Tool to make informed decisions and strategic plans, navigating the dynamic landscape of the chemical sector with foresight and precision.

Forecasting future trends predictive analysis in the Chemical Sector Occupational Skills Forecast Tool

The 'Forecast' tab of the Chemical Sector Occupational Skills Forecast Tool utilises an ARIMA model to generate forecasts for future values of the selected specialisation's data. This predictive analysis is grounded in careful parameter selection and model optimisation, utilising a variation of the Hyndman-Khandakar algorithm. The selection process is guided by the Akaike information criterion (AIC), a key criterion that balances the trade-off between model fit and complexity. By prioritising parsimony while ensuring effectiveness, the chosen forecasting model optimally captures the underlying dynamics of the chemical sector labour market (Figure 15.3).

Understanding what matters: variable importance in skills forecasting

In this case study, the goal is to understand what influences the number of employees across different categories, such as gender, race, and employment status. To analyse this, a generalised linear model (GLM) with a Poisson distribution and a log link function was used. While this might sound complex, it is actually a strategic choice that helps get accurate insights from the data.

Why this model is appropriate for count data

This case study deals with numbers of people – such as how many employees fall into certain groups. This type of data is known as 'count data,' which simply means counting occurences, such as how many male employees or how many employees are permanent against temporary. The Poisson regression model is perfect for this because it is designed to handle data that involve counts. It assumes that these counts are naturally distributed in a way that fits real-world scenarios, making the results more reliable and accurate.

Making predictions understandable: the log link function

Take the example of predicting how many employees belong to a certain category based on other factors, like their employment status. The log link function helps do this in a straightforward way. It transforms the data so that the predictions make sense – keeping all predicted counts positive, as they should be. Moreover, it shows the relationship between different factors and the employee count in a way that is easier to understand. For instance, it can show how much the expected number of employees might change if one factor, like race or gender, changes while everything else stays the same. The real power of Poisson regression is in how it shows what is important. The results demonstrate which factors have the biggest impact on the number of employees in different categories. This is critical for making decisions in workforce planning. For example, the model might show that permanent employees significantly outnumber temporary ones, or that certain demographic groups are more prevalent in the workforce. The model also shows how well it fits the data using specific metrics, allowing the user to evaluate how much trust can be put into the predictions it makes.

Dealing with real-world data issues: handling overdispersion

In some cases, the data may not perfectly adhere to the assumptions of the model. For example, the actual variation in employee counts might exceed what the model anticipates, a situation known as 'overdispersion.' When this occurs, the model can be adjusted to better reflect the real-world data, ensuring that the analysis remains robust and that the insights are accurate. Table 15.1 details the results of the GLM, including any adjustments made to account for overdispersion.

The forecasting model considers a range of important variables that influence the number of employees in the chemical sector. These variables include gender, race, disability status, and employment status, among others. The coefficient estimates, standard errors, z-values, and p-values for each variable are presented, providing insights into their significance and impact on employment levels. The forecasting model employs a GLM with a Poisson log link function, effectively capturing the relationships between various factors and the number of employees in the chemical sector. The model's fit is assessed using measures such as the Pseudo-R (Cragg-Uhler) and Pseudo-R (McFadden), providing insights into the model's explanatory power. Additionally, the AIC and Bayesian information criterion (BIC) are used to evaluate model performance and guide parameter selection.

Gender

The coefficient for 'Gender Male' suggests a positive impact on employment levels, indicating that being male is associated with higher employment rates in the chemical sector. This finding highlights potential gender disparities in employment opportunities, which may be addressed through policies promoting gender equality and diversity in the workforce. Initiatives under South Africa's Employment Equity Act aim to address such disparities by promoting equal opportunities for employment and advancement, irrespective of gender.

Type of employment

The coefficients for different types of employment status, including 'Permanent', 'Temporary', 'Terminated', and 'Unemployed', provide insights into the varied nature of employment arrangements within the chemical sector. Policies such as the Employment Equity Act and Skills Development Act seek to promote fair labour practices and address issues of precarious employment by advocating for stable, secure employment opportunities for all workers.

Race

The coefficients for different racial categories, including 'Coloured', 'Indian', and 'White', indicate the differential impact of race on employment outcomes within the chemical sector. These findings underscore the importance of policies aimed at promoting racial diversity and inclusion in the workforce. South Africa's Employment Equity Act mandates the implementation of affirmative action measures to address historical imbalances in employment opportunities and promote equitable representation of all racial groups in the workforce.

By considering the influence of gender, type of employment, and race on employment trends within the chemical sector, policymakers and industry stakeholders can develop targeted interventions to promote diversity, equity, and inclusivity in the workforce. These efforts align with broader national initiatives aimed at building a more inclusive and resilient economy that benefits all segments of society.

Data-driven migration trends in the chemical sector

The past eight years have witnessed a significant transformation in the skill landscape of the chemical sector. During this period, a significant migration occurred owing to technological advancements, evolving market demands, and changing job roles. This transition has transformed the skill sets needed for various occupational specialisations, placing a clear emphasis on information-centric capabilities and digital literacy. A key trend observed in this migration is the transition of certain occupational roles to positions prioritising data analysis and management. For instance, individuals previously employed as drivers have seamlessly transitioned into roles centered on data capture, leveraging their logistical expertise in data-driven environments. Similarly, call centre agents have embraced opportunities in data capturing, utilising their communication skills to effectively collect and process information. Additionally, roles traditionally associated with assistance have evolved into positions within information technologies, highlighting the growing significance of digital proficiency. This migration has been propelled by the increasing reliance on technology-driven solutions across administrative support, data management, and process optimisation tasks. Moreover, professionals such as software architects and clerks have shifted towards more information-centric positions, reflecting the heightened demand for individuals adept at leveraging technology to drive innovation and enhance organisational efficiency. Overall, the migration of skills within the chemical sector underscores a broader trend towards digitalisation and information-centricity. In an era defined by complexity and competitiveness, the ability to harness data, technology, and information-driven insights has become indispensable for success. By embracing this migration and investing in the development of relevant skills, stakeholders within the chemical sector can position themselves for sustainable growth and resilience in the evolving landscape ahead (Figure 15.4).

Conclusion

This chapter has detailed the collaborative effort between the University of Johannesburg and the CHIETA to develop a robust skills forecast tool for the chemical sector. Guided by Professor Arnesh Telukdarie, this initiative stands as a notable example of how multidisciplinary research can effectively address complex challenges in labour market forecasting.

Integration of theory and practice

The development of the Chemical Sector Occupational Skills Forecast Tool illustrates the practical application of labour market theories and data analytics in a real-world setting. The project leveraged historical data and current labour market dynamics to create a predictive model that will aid in strategic planning and educational alignment. This aligns with theoretical concepts related to workforce planning and educational responsiveness, showcasing how academic frameworks can be operationalised to meet industry needs.

Academic and sector-wide implications

This tool's development has significant implications for academia and the chemical industry. It serves as a case study in the effective use of data analytics for predictive modelling, providing a template that can be replicated or adapted in other sectors. For educational institutions, this tool underlines the importance of curriculum agility and the need to continuously align educational offerings with emerging industry requirements.

Future perspectives

As we look to the future, the adaptability of the skills forecast tool will be crucial in responding to rapid technological changes and market shifts. Ongoing research and development, enhanced by real-time data integration and advanced analytical methodologies, will be essential to maintain the tool's relevance and efficacy.

Educational continuity and closing thoughts

In conclusion, the Chemical Sector Occupational Skills Forecast Tool not only reflects a significant achievement in bridging the gap between education and industry but also emphasises the critical role of continuous innovation in educational tools. By understanding and applying these principles, educators and policymakers can better prepare the workforce of tomorrow, ensuring that education remains responsive to the evolving demands of the global labour market.

Exhibits

Figure 15.1 Visualisation of historical trends

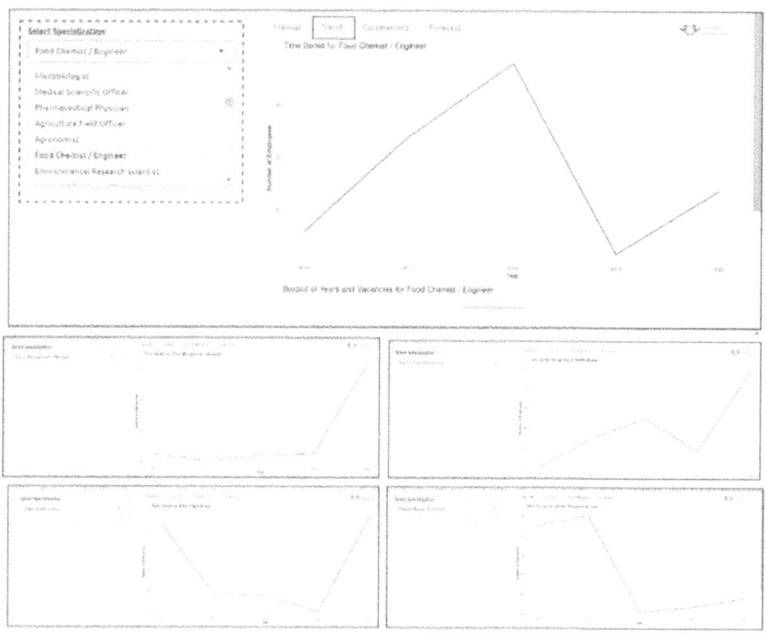

Figure 15.2 Evolution of engineering workforce

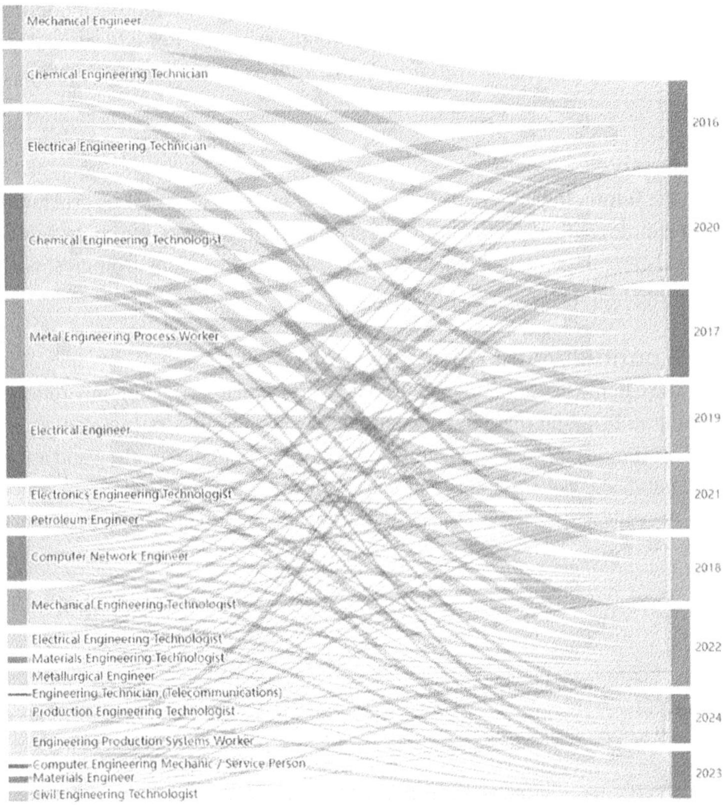

Figure 15.3 Skills forecast

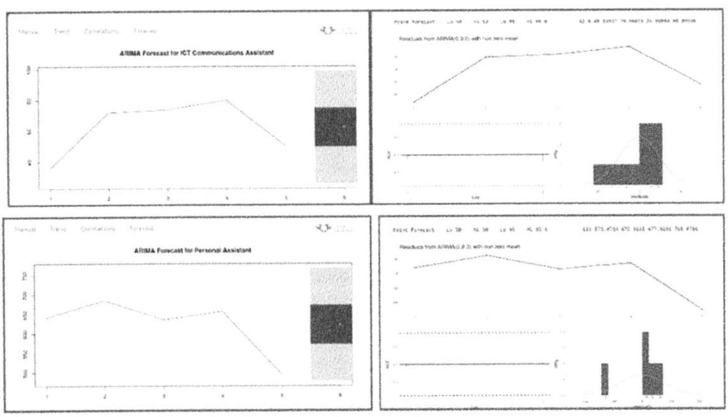

Figure 15.4 Skills migration chart (2016–2023)

Table 15.1 Summary of the GLM used for analysing the number of employees, including model fit statistics, estimated coefficients, and their statistical significance across various demographic and employment-related predictors (continued on next page)

Model information and results

Variable	Estimate	Standard error	*z*-value	*p*-value
Intercept	1.76	0.01	290.55	0.00
Gender: Male	0.41	0.00	212.89	0.00
Race: Coloured	−1.09	0.00	−356.09	0.00
Race: Indian	−1.34	0.00	−371.05	0.00

Table 15.1 Continued

Model information and results

Variable	Estimate	Standard error	*z*-value	*p*-value
Race: White	−0.75	0.00	−314.55	0.00
Disability: Yes	−1.59	0.01	−210.69	0.00
Employment status: Permanent	1.16	0.01	194.83	0.00
Employment status: Temporary	−0.08	0.01	−10.09	0.00
Employment status: Terminated	−0.07	0.01	−9.60	0.00
Employment status: Unemployed	0.91	0.01	125.40	0.00
Contracted for learning programme	0.13	0.01	50.84	0.00
Year L	−0.02	0.00	−6.70	0.00
Year Q	0.08	0.00	32.21	0.00
Year C	−0.07	0.00	−26.42	0.00
Year 4	0.01	0.00	3.45	0.00
Year 5	0.06	0.00	22.84	0.00
Year 6	0.10	0.00	37.23	0.00

Model fit information

Model fit metric	Value
Observations	108 028 (692)
Model type	Generalised linear model (GLM)
Family	Poisson log
Pseudo-R (Cragg-Uhler)	1.00
Pseudo-R (McFadden)	0.13
AIC	3 994 997.18
BIC	3 995 160.21
Standard errors	MLE

*Note: All *p*-values are significant at $p < 0.001$.

Exploring historical data trend analysis in the Chemical Sector Occupational Skills Forecast Tool

Teaching notes

Synopsis

The chapter explores the development and implementation of an advanced skills forecasting tool designed specifically for South Africa's chemical industry. Initiated by a research team from the University of Johannesburg in collaboration with the Chemical Industries Education and Training Authority (CHIETA), this project is a response to the pressing need for accurate and actionable labour market insights in a rapidly evolving industry. The chapter details the multidisciplinary approach employed by the research team, which integrated data analytics, digital transformation, and industry expertise to create a robust platform capable of predicting future occupational trends. It highlights the importance of aligning workforce development with industry needs, ensuring that the sector remains competitive and resilient. Key topics covered include the analysis of historical labour market data, the use of generalised linear models (GLM) for forecasting, and the collaboration with industry stakeholders to ground the tool in real-world applicability. The chapter also discusses the challenges faced during the tool's development, including addressing issues of data variability and overdispersion, and the strategies used to overcome them. Ultimately, this chapter serves as both a case study and a strategic guide for workforce planning, illustrating how targeted, data-driven approaches can empower industries to anticipate and meet future skills demands, thereby supporting sustainable growth and economic development.

Teaching objectives

This case study is designed for master's and MBA students interested in business forecasting, human resource management, and strategic planning within the chemical sector. It assumes a basic understanding of business forecasting techniques and aims to delve deeper into the development of sector-specific skills and forecast tools. It is suitable for courses in Strategic Human Resource Management, Business Forecasting, and Industry Analysis. The case study is also applicable to professionals seeking to understand the intricacies of forecasting workforce needs and aligning them with industry trends. It is a valuable resource for those interested in developing strategic tools for workforce planning and skills development.

- Exploring the challenges of workforce forecasting in the chemical sector: expand students' understanding of the challenges and complexities associated with forecasting workforce needs in a rapidly evolving industry.
- Developing a sector-specific skills forecast tool: initiate discussions on how to create a comprehensive and adaptable skills forecast tool tailored to the unique needs of the chemical sector.

Research methods

The study employs three main research components.

(a) *Literature review*: a thorough review of existing literature on workforce forecasting, skills development, and industry trends specific to the chemical sector.
(b) *Development of the skills forecast tool*: identifying key skills and competencies required in the chemical sector and developing a framework to project future workforce needs.
(c) *Application of forecasting techniques*: Utilising quantitative and qualitative forecasting methods to create a dynamic skills forecast tool that can adapt to industry changes and technological advancements.

Keywords and subject code

Workforce forecasting, chemical sector skills, strategic HR management, industry trends, skills development, forecasting techniques.

Assignment questions

1. Assess the limitations of traditional workforce forecasting methods in the chemical sector. Propose alternative strategies and discuss their potential impact on workforce planning.
2. How can industry-specific trends and technological advancements be integrated into the development of a skills forecast tool for the chemical sector? Discuss the theoretical foundations supporting this integration and evaluate their practical implications.
3. Explore the role of qualitative and quantitative forecasting methods in developing the skills forecast tool. Assess the benefits and potential challenges associated with these methods.
4. Analyse how the skills forecast tool can influence strategic decision-making within chemical companies. Consider the implications for HR managers aiming to align workforce planning with industry trends and discuss potential implementation challenges.
5. Critically evaluate existing skills forecasting frameworks used in other industries. Discuss the strengths and weaknesses of these frameworks in addressing the specific needs of the chemical sector. Consider whether a tailored approach is necessary and its potential impact on workforce development.

Additional materials

Michie J (2018) *Skills and Jobs in the Future: New Economy Strategies*. Palgrave Macmillan, London, UK.

Schwab K (2018) *The Global Competitiveness Report 2018*. World Economic Forum, New York, NY, USA.

Suggested core readings

Mason JL, Johnston E, Berndt S *et al.* (2016) Labor and skills gap analysis of the biomedical research workforce. *The FASEB Journal* **30(8)**: 2673.

McKenney MJ and Handley HA (2020) Using the design science research method (DSRM) to develop a skills gaps analysis model. *IEEE Engineering Management Review* **48(4)**: 102–119.

Pillay R (2010) The skills gap in nursing management in South Africa: a sectoral analysis: a research paper. *Journal of Nursing Management* **18(2)**: 134–144.

Teaching plan and timing

Introduction (15 min)

- Engage students with the question, 'How effectively do current workforce forecasting methods capture future skills needs in the chemical sector?'
- Conduct a quick vote and discuss key responses to highlight the complexity of workforce forecasting.

Contextualising the study (5 min)

- Share a brief YouTube clip illustrating current challenges and trends in workforce forecasting within the chemical sector.
- Facilitate a discussion on how this context aligns with the case study.

Discussion on Assignment Question 1 (25 min)

- Present Assignment Question 1 and introduce an academic model from the teaching notes.
- Facilitate a class discussion on how the case aligns with the academic model.
- Encourage students to propose alternative strategies to address limitations in traditional workforce forecasting methods.

Small group activity: Assignment Question 2 (35 min)

- Introduce Assignment Question 2.
- Form small groups to develop an action plan for integrating industry-specific trends and technological advancements into the skills forecast tool.
- Have two groups present their plans and provide a framework for decision-making.
- Facilitate a class discussion on the action plans and peer feedback.

Conclusion and key learning points (5 min)

- Summarise key insights and emphasise integrating industry trends, technological advancements, and advanced forecasting techniques.
- Highlight the complexity of workforce forecasting and the need for innovative solutions.

Optional postscript or additional materials (5 min)

- Conclude with a brief reading of the postscript or additional materials related to the case.

Analysis of assignment questions

Question 1. Assess the limitations of traditional workforce forecasting methods in the chemical sector and propose alternative strategies.

Model answers

- Traditional workforce forecasting methods often fail to account for rapid technological changes and industry-specific trends.
- *Human Capital* (Becker, 1964) and *Dynamic Capabilities* (Teece, *et al.*, 1997) highlight these limitations.
- Propose using *Scenario Planning* (Schoemaker, 1995) to develop flexible and adaptive workforce forecasting strategies that consider multiple future scenarios.
- Suggest incorporating *Real Options Theory* (Amram and Kulatilaka, 2000) to provide flexibility in workforce planning.
- The instructor should encourage elaboration on practical application and the challenges of balancing flexibility with planning accuracy.

Question 2. How can industry-specific trends and technological advancements be integrated into the development of a skills outlook tool for the chemical sector?

- Integrating industry-specific trends and technological advancements requires a thorough understanding of the sector's evolving needs.
- Use the technology acceptance model (Davis, 1989) to ensure effective adoption of new tools and methods.
- Employ Porter's Five Forces (Porter, 1979) to analyse industry dynamics and identify key skills needed for future competitiveness.
- Propose using the 'Delphi Method' (Linstone and Turoff, 1975) to gather expert opinions on future skills requirements.
- The instructor should explore the implications of the Delphi Method for the chemical sector and discuss potential challenges in implementation.

Key learnings

This chapter has provided an in-depth overview of the collaborative initiative undertaken by the University of Johannesburg in partnership with the CHIETA. The primary objective of this collaboration was to create a comprehensive skills forecast tool tailored specifically for the chemical sector. Under the expert guidance of Professor Arnesh Telukdarie, this project exemplifies the power of multidisciplinary research in tackling intricate challenges present in labour market forecasting. Throughout the course of this initiative, several significant insights and key learnings have emerged.

These findings underscore the importance of integrating diverse academic perspectives and practical industry knowledge to develop effective strategies that respond to the dynamic demands of the chemical industry workforce. By leveraging expertise from various fields, the team has been able to identify current trends, predict future skills requirements, and propose actionable solutions aimed at enhancing workforce readiness in this critical sector.

REFERENCES

Amram M and Kulatilaka N (2000) Strategy and shareholder value creation: The real options frontier. *Journal of Applied Corporate Finance* **13(2)**, pp.15–28.

Becker GS (1964) *Human Capital: A Theoretical and Empirical Analysis, with Special Reference to Education*. Columbia University Press, New York, USA.

Davis FD (1989) Perceived usefulness, perceived ease of use, and user acceptance of information technology. *MIS Quarterly* **13(3)**: 319–340, doi: 10.2307/249008.

Linstone HA and Turoff M (1975) *The Delphi Method: Techniques and Applications*. Addison-Wesley, Reading, MA, USA.

Porter ME (1979) How competitive forces shape strategy. *Harvard Business Review* **57(2)**: 137–145.

Schoemaker PJH (1995) Scenario planning: a tool for strategic thinking. *Sloan Management Review* **36(2)**: 25–40.

Teece DJ, Pisano G and Shuen A (1997) Dynamic capabilities and strategic management. *Strategic Management Journal* **18(7)**: 509–533, doi: 10.1002/(SICI)1097-0266(199708)18:7<509::AID-SMJ882>3.0.CO;2-Z.

Sustainability and Digital Engineering Management

Prof. Arnesh Telukdarie
ISBN 978-1-83662-645-9
https://doi.org/10.1108/978-1-83662-642-820251020

Chapter 16
Transformative innovation: harnessing artificial intelligence for sustainable research practices

Arnesh Telukdarie and Thabile Dube

Introduction

On 23 January 2024, Thabile Dube, Research Manager at Johannesburg Business School (JBS), launched a fascinating project to revolutionise academic research. With a clear goal of integrating artificial intelligence (AI) tools into traditional research processes, Thabile's vision is to streamline and improve research methods. She aims to replace manual tasks such as data collection, analysis, and interpretation with an automated, AI-based workflow. This strategic shift aims to shorten research timelines and increase research quality by merging established methodologies with cutting-edge AI technology. Thabile's motivation comes from recognising the challenges researchers face owing to the slowness of traditional processes and the lengthy manual processing of large datasets. Thabile's AI systems are already making big waves, automating research tasks and making them more efficient and manageable. Working with diverse teams and partners across disciplines, Thabile tackles the complexities of AI and data science with a relentless commitment to driving innovation in academic research.

This initiative is transforming academic research and contributing significantly to the broader goals of industry, innovation, and infrastructure development. Leveraging AI into research, JBS sets a precedent for how advanced technologies can be harnessed to enhance the capabilities of various industries and academic fields.

Industry background: academic research in South Africa

Academic research is critical to promoting innovation, economic development, and social progress (Veugelers, 2014). South Africa has a diverse and vibrant educational landscape, with numerous universities, research institutions, and centres of excellence spread across the region (Uetela, 2017). These institutions are dedicated to imparting knowledge across a wide range of disciplines, from science and technology to the humanities and social sciences. South Africa's historical context is an essential aspect of academic research's industry background. The country has a rich academic heritage, with institutions that have been at the forefront of groundbreaking research for decades. Despite the challenges presented by apartheid-era policies, South African universities have remained resilient and have increasingly become centres of excellence in research in various fields.

The academic research sector in South Africa has undergone significant changes in recent years. The government has significantly emphasised research and development (R&D) investment in its broader national development agenda. Initiatives such as the National Research Foundation (NRF) and the Department of Science and Innovation (DSI) have been instrumental in funding research projects, fostering collaboration between academia and industry, and promoting scientific excellence. Thabile's project aligns with national priorities by integrating industry, innovation, and infrastructure into academic research through the development of AI tools tailored to industry needs. This initiative not only fosters practical research applications but also drives innovation by ensuring that South African institutions remain at the forefront of global research trends (Chubb *et al.*, 2021; Netherton *et al.*, 2020). Researchers can focus on more complex and creative aspects of their work, leading to more innovative solutions and products by automating repetitive tasks.

Leveraging AI, this case study pushes the boundaries of what is possible in academic research. AI technologies such as machine learning, natural language processing (NLP), and predictive analytics revolutionise the methods of data collection, analysis, and interpretation. These innovations enable researchers to uncover patterns and insights that would be impossible to detect manually. Furthermore, the integration of AI encourages a culture of continuous improvement and experimentation, fostering an environment where new ideas and methodologies can flourish. Enhancing research infrastructure with AI technology, the project contributes to advancing new methods, processes, and management in research, which is an area that is still relatively under-explored (Chubb *et al.*, 2021). Moreover, using AI in research can increase productivity by enhancing creative abilities and acting as a general-purpose tool for innovation (Zirar, 2023). The incorporation of AI tools in academic research also raises important considerations regarding academic integrity, student learning, and the quality of peer review processes. Transparency in declaring the use of generative AI in academic writing can enhance the quality of peer review and combat unethical practices like plagiarism (Tang, 2023). Additionally, the limitations of AI tools can drive innovative work behaviour by encouraging workers to leverage their creative abilities alongside AI capabilities (Zirar, 2023). The implementation of AI-driven research methodologies requires a robust digital infrastructure. Thabile's project spearheads the development of this infrastructure at JBS, equipping the institution with essential tools and systems to facilitate advanced research activities. This includes high-performance computing facilities, secure data storage solutions, and reliable communication networks. JBS supports its research initiatives and sets standards for other institutions by building this infrastructure. This infrastructure is critical for sustaining long-term innovation and a competitive edge in the global academic landscape. Another notable aspect is the increasing internationalisation of academic research in South Africa. Collaboration with international partners, participation in global research networks, and exchange programmes for researchers have become increasingly important, enriching the research landscape and increasing the quality and impact of research results. Despite these advancements, the academic research industry in South Africa continues to face significant challenges. Issues such as limited funding, infrastructure constraints, and skills shortages continue to hinder the sector's realisation of its full potential. Furthermore, addressing disparities in research capacity between institutions and regions remains a priority to ensure equal access to opportunities and resources.

University of Johannesburg: a prominent institution of higher learning

JBS is considered a preeminent institution in South Africa's academic landscape and is known for its commitment to excellence in business education and research (Suleman, 2022). Founded to promote innovation, leadership, and sustainable business practices, JBS has become a hub for cutting-edge academic research and industry collaboration. At the core of JBS's academic culture is a robust research framework to address real-world challenges and produce meaningful results. The school's faculty and researchers are actively engaged in various research endeavours in business, management, and economics. These efforts include theoretical investigations and applied research, generating actionable insights influencing business practices, policy formulation, and societal development.

Moreover, JBS recognises the transformative potential of AI in reshaping business landscapes and driving innovation across all sectors. In response to the growing importance of AI in business and society, JBS collaborated with the Tshwane University of Technology (TUT) to establish the AI Institute of South Africa (AIISA) as a dedicated centre of excellence for AI research, training, and collaboration.

AIISA catalysed the advancement of AI-related knowledge and expertise in South Africa and beyond. Through its interdisciplinary approach, the Institute brings together experts from academia, industry, and government to explore the frontiers of AI research and its applications in diverse areas such as finance, healthcare, marketing, and supply chain management. Additionally, AIISA is crucial in nurturing talent and the next generation of AI leaders by conducting groundbreaking research. The Institute offers specialised programmes, workshops, and training initiatives designed to equip students, professionals, and entrepreneurs with the skills and knowledge they need to harness the power of AI for innovation and growth.

Furthermore, AIISA is an industrial partnership cooperation platform that enables meaningful collaborations between science and the private sector. The Institute aims to close the gap between research and practice and advance the adoption of AI-driven solutions to address complex challenges and promote sustainable development by facilitating knowledge exchange, technology transfer, and joint innovation projects.

Integral to JBS's mission is its focus on enhancing industry, innovation, and infrastructure. The school designs its efforts in these areas to stimulate economic growth and enhance societal well-being. Through its research and collaboration with AIISA, JBS contributes to the development of innovative solutions that enhance industrial processes, improve efficiency, and foster sustainable practices. The emphasis on AI research and its applications supports the creation of intelligent infrastructure systems that can adapt to changing needs and optimise resource use. Promoting innovation and integrating cutting-edge technologies into various sectors, JBS and AIISA work towards building resilient infrastructure that supports sustainable industrialisation and economic development. JBS represents excellence in business education and research in South Africa, strongly emphasising innovation, leadership, and sustainable practices. JBS, through its collaboration with the AIISA, is at the forefront of advancing AI research and its applications, fostering industry collaboration, and promoting the development of robust and sustainable infrastructure. These efforts contribute to individuals' academic and professional growth and play a crucial role in driving economic and societal progress in South Africa.

Project management

Thabile Dube brings a diverse background and wealth of experience to her role as research manager at JBS. With a master's degree in tourism from the University of Johannesburg (UJ), she initially established herself as a competent researcher in the field of tourism before venturing into the exciting realm of AI. Recognising the transformative potential of AI in reshaping industries, driving innovation, and enhancing infrastructure, Thabile boldly decided to pursue a PhD in artificial intelligence from UJ. This decision reflects her commitment to remaining at the forefront of technological advances and using them to address complex challenges in research.

AI-based tools are developed to automate and improve academic research processes, providing a best-in-class, journal-worthy solution tailored to the needs of researchers. These tools offer a structured workflow for managing academic research and integrating advanced workflows and automation technologies. Furthermore, AI-based tools provide a tailored workflow designed explicitly for managing academic research projects, seamlessly integrating with existing research methodologies to ensure a smooth and efficient process.

The workflow includes a web scraping tool capable of extracting relevant academic data from platforms such as Google Scholar and automatically populating research projects according to the specific problem, saving researchers a lot of time and effort. Based on the research query or task, these tools create a domain-specific agent responsible for generating research questions that form an objective opinion on the given task. For each research question, the domain-specific agent triggers a crawler agent that searches online resources for relevant information, ensuring comprehensive data collection from various sources. Figure 16.1 illustrates the approach taken by the web scraping tool, showing a code snippet that encodes a query string for a Google Scholar search. The query string, which retrieves academic data from the years 2014 to 2024, includes terms like 'Project Management', 'KPI', 'Key Performance Indicator', and 'Performance'.

Figure 16.1 illustrates how the web scraping tool initiates a search query for academic literature related to project management and performance indicators. The encoded query string ensures that the search covers a comprehensive range of related terms, while the specified timeframe ensures the relevance and recency of the retrieved data. The workflow begins with researchers generating research questions, which set the direction for the study. The 'Generate Research Questions' image at the initial stage represents this process. The system then proceeds to search online resources using a web scraping tool, depicted as 'Search Online

Resources'. This tool automates the collection of relevant academic materials from various online sources, ensuring comprehensive data gathering. Once the resources are collected, the system moves to the summarisation stage, labelled 'Summarise Resources'. Here, each resource is summarised based on its relevant information. After gathering the resources, the system advances to the 'Summarise Resources' stage. The system summarises each resource based on its relevant information.

The next key stage in the workflow is 'Structured Workflow Management', which involves organising the collected data and ensuring a structured workflow. This stage is supported by the integration of domain-specific agents and crawler agents. The domain-specific agent gathers tailored information pertinent to the research field, while the crawler agent continuously searches for updated and relevant resources. These agents enhance the efficiency and effectiveness of the research process by automating data collection and management. Following structured workflow management, the system extracts academic data, as shown in the stage labelled 'Extract Academic Data'. This process focuses on isolating the most pertinent information from the collected resources, ensuring that only the most relevant data are considered for further analysis.

The workflow then advances to 'Advanced Workflows and Automation', where advanced techniques are implemented to automate repetitive tasks. This includes filtering, categorising, and aggregating data to streamline the research process further. The automation of these tasks significantly reduces the time and effort required for manual data handling, allowing researchers to focus on more critical aspects of their work.

Finally, the system generates a 'Final Research Report'. The system filters and aggregates all summarised sources to produce a comprehensive overview of the research findings. Data from various credible sources support this final report, ensuring a cohesive and reliable presentation of the research outcomes. By integrating these advanced features, the AI-based tools modernise research workflows, ensuring that researchers have access to the most current and comprehensive information available. This approach significantly reduces the time and effort required for manual research tasks, allowing researchers to focus on generating insights and advancing knowledge in their fields.

Research: University of Johannesburg

Traditional methods have long been the cornerstone of academic endeavours, including at JBS. However, research needs to evolve and, as data volumes increase exponentially, these traditional approaches face significant challenges. Traditional research processes are often characterised by a slow pace, limited capacity to process large datasets, and an inability to adapt to the dynamic environment of AI and data science (Henke et al., 2016). In response, JBS launched a groundbreaking initiative to integrate AI-based tools into research workflow.

JBS has a reputation for innovative teaching rooted in African perspectives while embracing the complexities of the digital age. Thabile Dube's extensive experience in research management at JBS allowed her to identify the significant inefficiencies of traditional methods, particularly their inability to handle massive datasets and adapt to technological advances. Recognising the need for a paradigm shift, Thabile embarked on integrating AI tools into the research framework at JBS to drive innovation, increase efficiency, and maintain JBS's competitive edge.

The slow pace of traditional research at JBS impeded the timely delivery of results and restricted the institution's ability to remain competitive in the evolving academic landscape. Moreover, the limited capacity to process vast volumes of data constrained researchers, narrowing the scope of their studies. As industries and academic fields increasingly adopt AI, JBS risked falling behind if it did not evolve its research processes. Thabile recognised that AI was not just an enhancement but a necessity for advancing research capabilities.

A thorough review of JBS's existing research practices was conducted. The potential for AI to revolutionise workflows, reduce manual efforts, and allow for the rapid processing of academic data was uncovered. Through a strategic approach, the school's research operations were streamlined while maintaining its commitment to academic rigour and excellence.

Overview of the AI-driven workflow

This project's vision materialised in a five-stage system designed to integrate AI into traditional research processes. As depicted in Figure 16.2, the first stage, Input Academic Data Sources, involves identifying and integrating various academic sources like journals, conference proceedings, and online repositories. This stage automates the collection of data – a significant improvement over traditional manual entry.

In the Extracted Academic Data stage, web scraping tools automatically gather data from the selected academic sources. The extracted data are then formatted and structured to ensure they are ready for processing. The heart of the system lies in the Automated Data Processing phase, where the data are cleaned, analysed, and organised using AI algorithms. This step may also incorporate NLP, allowing the system to interpret the textual content and categorise the data.

Input academic data sources

As depicted in Figure 16.3, the user interface for adding new data sources includes fields for Data Source Name, Data Source Type, and Data Source URL. For instance, a user might input 'Project Management' as the source name, select 'Conference' from the dropdown menu, and provide 'Google Scholar' as the URL. This automation reduces manual effort and ensures that diverse academic resources are integrated into the research process.

Extracted academic data

The Extracted Academic Data stage (Figure 16.4) uses web scraping to automatically collect and structure data, displayed in a table with column headings such as Publication Title, Author, Abstract, and Publication Date. This structured format enables efficient analysis and helps eliminate the manual sorting of large datasets.

Automated data processing

At the system's core, Automated Data Processing (Figure 16.5) involves several key steps. Initially, the Analyze step evaluates the extracted data for patterns, trends, and anomalies. Next, in the Clean phase, the system removes inconsistencies and irrelevant data, ensuring that the dataset is reliable and ready for the Process stage. Here, data are organised into usable formats using advanced algorithms. Lastly, Apply NLP incorporates NLP to interpret the content and derive deeper insights from academic texts, including text classification and sentiment analysis.

Populated research projects

Figure 16.6 demonstrates how the system can automatically populate research projects. Through web scraping, relevant academic data are extracted and used to generate project ideas. The interface includes fields for Project Name, Project Description, and Project Lead. By reducing the manual burden of data collection, researchers can focus on critical aspects of their projects.

Generated insights

Generated Insights (Figure 16.7) provides a user interface that allows for synthesising and aggregating research findings. The 'Generated Insights' button triggers an automated process that filters, analyses, and organises the data into a comprehensive report. This feature simplifies complex data, providing users with actionable insights and improving collaboration within the academic community.

Success and broader impact

The integration of AI into JBS's research processes proved to be a resounding success. The AI-driven workflows enabled researchers to process large datasets quickly and efficiently, accelerating the timelines for key findings. In addition, AI's flexibility allowed JBS researchers to adapt to the constantly changing academic landscape, ensuring continued relevance in their fields. Beyond enhancing JBS's research operations, this case study also contributes to broader societal goals related to industry, innovation, and infrastructure. Integrating AI, JBS strengthens its research infrastructure, fostering an innovative ecosystem that supports not only academic progress but also industrial applications. For businesses, AI-driven insights from academic research inform strategies and drive innovation, while for industries, these advancements provide a competitive edge with enhanced operational efficiency. Improving JBS's research infrastructure, this initiative has created a benchmark for other institutions in South Africa, demonstrating how modern technology can elevate traditional research processes. The success of this integration supports the creation of a resilient research environment capable of responding to evolving demands across sectors.

Conclusion

This case study which integrates AI-based tools into academic research at JBS has led to significant progress and promising results. AI-driven solutions have been used to improve research processes and outcomes. JBS successfully implemented an AI-based workflow automation tool, revolutionising traditional research methodologies and accelerating research timelines. The integration of AI has streamlined data collection, analysis, and interpretation processes, enabling researchers to leverage vast amounts of data more efficiently and effectively. Furthermore, Thabile's collaborative efforts with interdisciplinary teams and industry partners have fostered a culture of innovation and knowledge exchange, driving impactful research outcomes and real-world applications of AI technology.

This case study aligns with the broader goals of industry, innovation, and infrastructure, which are crucial components of the sustainable development goals (SDGs). By bridging the gap between academia and industry, JBS has positioned itself as a leader in AI-driven research and education, contributing to advancing knowledge and developing AI talent in South Africa. Integrating AI tools in research processes supports SDG 9, which focuses on building resilient infrastructure, promoting inclusive and sustainable industrialisation, and fostering innovation, exemplifying the intersection of industry, innovation, and infrastructure in several ways.

- Industry: The AI-driven research processes at JBS provide valuable insights that can inform business strategies, drive innovation, and improve operational efficiencies. This fosters closer collaboration between academia and industry, ensuring that research outcomes have practical applications and can contribute to economic growth and competitiveness.
- Innovation: Incorporating AI in research fosters a culture of continuous improvement and creativity. By exploring new methodologies and solutions, JBS researchers are encouraged to push the boundaries of traditional research, leading to groundbreaking discoveries and technological advancements. This case study highlights the importance of innovation in achieving sustainable development and addressing global challenges.
- Infrastructure: The advancements in AI integration at JBS set a benchmark for other institutions, showcasing how modern technology can enhance traditional systems to create a more efficient and forward-thinking research environment. By building a robust infrastructure for AI-driven research, JBS contributes to developing a resilient and adaptive academic ecosystem that can respond to the rapidly changing demands of various industries.

This case study led to a transformative change towards a more AI-driven future for academic research at JBS. While challenges persist and the journey continues, the success of this initiative is a testament to the power of innovation and collaboration to drive positive change and shape the future of academic research, demonstrating how integrating AI into research processes can contribute to sustainable development, enhance industry collaboration, and build resilient infrastructure that supports continuous learning and adaptation.

Under Thabile's leadership, JBS is advancing academic research and playing a pivotal role in achieving the SDGs by promoting sustainable industrialisation, fostering innovation, and building resilient infrastructure. This initiative serves as a model for other institutions and highlights the critical role that academia can play in driving global progress toward a more sustainable and inclusive future.

Exhibits

Figure 16.1 Code snippet for Google Scholar search

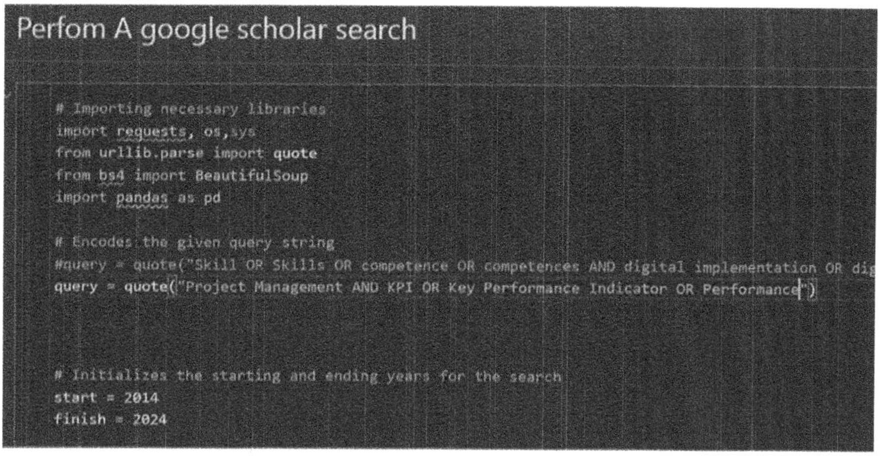

Figure 16.2 AI workflow diagram

Figure 16.3 Input academic data sources

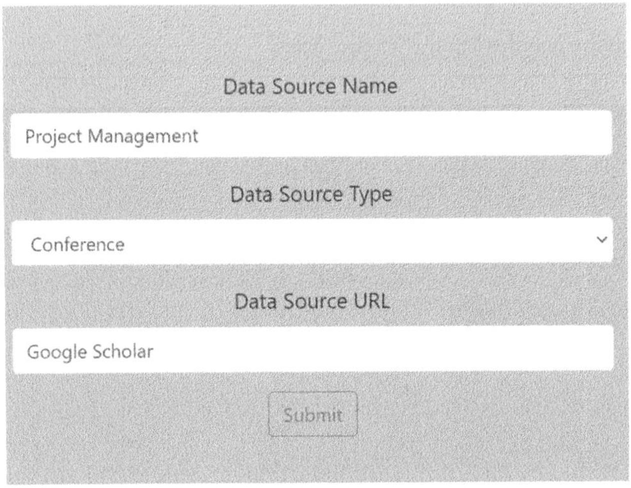

Figure 16.4 Extracted academic data

Extracted Academic Data

Publication Title	Author	Abstract	Publication Date	Source URL
AI in Healthcare	Bajwa et al.	Artificial intelligence in healthcare: transforming the practice of medicine.	2021-07-02	link
Quantum Computing Advances	Wang et al.	Recent Advances in Quantum Computing for Drug Discovery and Development.	2023-03-27	link

Figure 16.5 Automated data processing

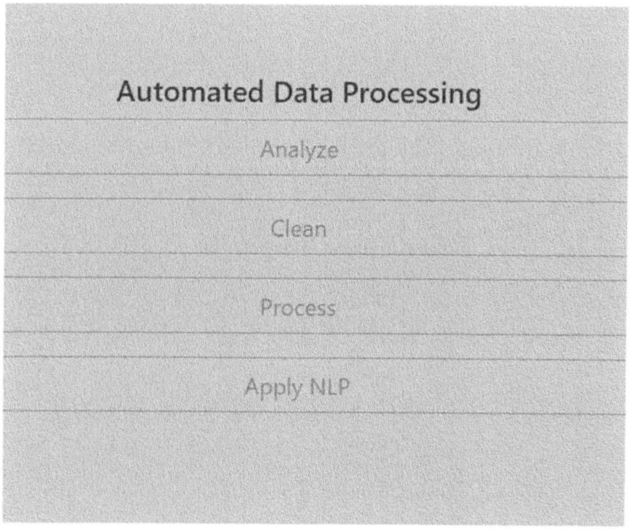

Figure 16.6 Populated research projects

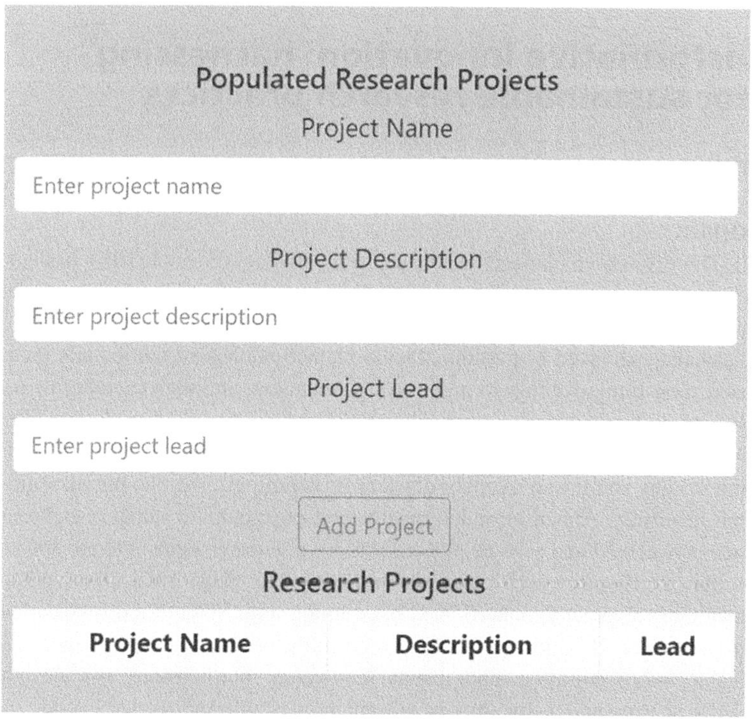

Figure 16.7 Generated insights

Transformative innovation: harnessing AI for sustainable research practices

Teaching notes

Synopsis

Thabile Dube, research manager at Johannesburg Business School (JBS), has launched a groundbreaking initiative to integrate artificial intelligence (AI) tools into traditional research processes. Set in South Africa and spanning the present, this case study examines the challenges faced by traditional research methodologies, particularly their inactive pace, their limited ability to process large data sets, and their inability to adapt to dynamic AI and data science environments. By using AI-based tools, Thabile aims to revolutionise academic research. This strategic deployment not only accelerates research studies and ensures timely delivery of meaningful results but also increases research flexibility. Aimed at undergraduate and postgraduate students exploring the interface between AI and science, the case study provides insights into advanced techniques that improve research processes and automate tasks. This initiative supports sustainable development goals (SDGs) by aligning with SDG 9: industry, innovation and infrastructure. It builds resilient research infrastructures, promotes inclusive and sustainable industrialisation, and fosters innovation. Furthermore, it contributes to knowledge sharing and technology advancement in South Africa and beyond.

Teaching objectives

The case study provides an ideal opportunity for undergraduate and postgraduate courses in project management and data science. This allows students to assess the extent to which AI-based tools can be integrated into traditional academic research methods. Additionally, emphasis is placed on evaluating factors such as research efficiency, timeliness, adaptability, particularly when dealing with dynamic datasets, and the broader impact on industry, innovation, and infrastructure.

- Identify traditional research challenges: Participants will identify and comprehend the limitations and inefficiencies of conventional research methods. They will analyse how the challenges hinder progress in various research projects and understand the necessity for innovative solutions.
- Assess the impact of AI on research proficiency: Participants will evaluate the practical benefits of AI integration in research endeavours, such as accelerated study timelines, timely dissemination of significant findings, and increased flexibility in adjusting to dynamic datasets. They will also consider how these improvements can enhance industry standards and drive innovation.
- Cultivate critical thinking and decision-making skills in research innovation: Participants will engage in discussions that will encourage them to consider the potential for AI to bring about significant changes in academic research, and to reflect critically on the consequences of these changes. They will explore how AI-driven innovations can support the development of robust research infrastructures and promote sustainable industrial practices.
- Integrate AI tools into practical academic research scenarios: Participants will have the opportunity to gain practical experience by applying various AI tools, including flow optimisation, workflow automation, natural language processing (NLP), web scraping, and AI-driven contextualisation, to address specific challenges that arise in academic research. They will learn how these tools can enhance research efficiency and contribute to the development of advanced, sustainable research methodologies that align with SDG 9. This section teaches students how AI can revolutionise academic life while giving them the tools to make a difference in shaping future research paradigms. Through this comprehensive understanding students will be better equipped to contribute to sustainable development and technological advancement in their respective fields.

Research methods

The methodology draws on a combination of primary and secondary data sources to provide a comprehensive examination of this initiative. Primary data are obtained through structured interviews, providing an in-depth understanding of the vision, challenges, and strategic decisions made in integrating AI into academic research. Additionally, organisational documents, including internal reports, strategic plans, and relevant communications, serve as valuable primary sources. Secondary data include industry reports and academic publications that provide context to the challenges of conventional research methods and support the rationale for AI integration. In-depth field interviews within the case organisation provide qualitative insights into the implementation of AI tools.

Keywords and subject code

Artificial intelligence (AI), academic research, research automation, natural language processing, web scraping, sustainability.

Assignment questions

1. How are the disadvantages of traditional research methods addressed by integrating AI tools and what specific inefficiencies need to be overcome?
2. What factors should be considered when strategically adopting AI-based tools for academic research and how do flow optimisation, workflow automation, NLP, web scraping, and AI-driven contextualisation help accelerate research studies and ensure timely impactful results?
3. Assess the potential impact of AI on research agility. How can the introduction of AI-based tools improve the adaptability of academic research methods in a dynamic landscape? Provide examples and discuss potential challenges associated with this transformative approach.
4. Provide practical examples of how AI tools can be used to address challenges in academic research. Discuss the impact of these applications on the depth, efficiency, and relevance of research studies.

Additional materials

Ng A (2023) Opportunities in AI. *Stanford Online*, 26 July. https://www.youtube.com/watch?v=5p248yoa3oE (accessed 27/12/2023).

UNESCO (2023) Ethics of AI: Challenges and Governance. 4 March. https://www.youtube.com/watch?v=VqFqWIqOB1g (accessed 27/12/2023).

Suggested core readings

Archibald A, Hudson C, Heap T *et al.* (2023) A validation of AI-enabled discussion platform metrics and relationships to student efforts. *TechTrends* **67(2)**: 285–293, 10.1007/s11528-022-00825-7.

Balel Y (2023) The role of artificial intelligence in academic paper writing and its potential as a co-author. *European Journal of Therapeutics* **29(4)**: 984–985.

Irfan M and Murray L (2023) Integration of artificial intelligence in academia: a case study of critical teaching and learning in higher education. *Global Social Sciences Review* **8(1)**: 352–364, 10.31703/gssr.2023(VIII-I).32.

Lauriola I, Lavelli A and Aiolli F (2022) An introduction to deep learning in natural language processing: Models, techniques, and tools. *Neurocomputing* **470**: 443–456.

Viganò E, Loi M and Yaghmaei E (2020) Cybersecurity of critical infrastructure. *International Library of Ethics, Law and Technology* **21(December)**: 157–177, 10.1007/978-3-030-29053-5_8.

Teaching plan and timing

Introduction (15 min)

- Begin with a thought-provoking question: 'In your opinion, what are the key drawbacks of traditional research methods in academia?'
- Conduct a quick class vote on the stability of integrating AI in academic research, followed with a discussion on the reasons behind each choice.
- Provide a brief overview of Thabile's initiative and the challenges posed by conventional research methods.

Setting the context (5 min)

- The instructor should show a YouTube clip illustrating the current landscape of AI in academic research, highlighting its potential and challenges. It stimulates interest and sets the stage for deeper exploration.

Assignment Question 1: Discussion (25 min)

- The instructor should jump into Assignment Question 1: 'How are the drawbacks of traditional research methods addressed through the integration of AI tools?'
- Introduce the relevant academic model from the exhibits, encouraging students to discuss how the case aligns with the model.
- Facilitate a class discussion on the implications for academic research efficiency and the potential benefits of overcoming identified challenges.

Assignment Question 2: Group activity (35 min)

- Transition to Assignment Question 2: 'What factors should be considered when strategically adopting AI-based tools for academic research?'
- Have students work in pairs to develop action plans. Randomly select two pairs to present their plans to the class.
- Introduce a framework from the exhibits to assist students in making decisions related to AI adoption in academic research.
- Encourage discussion on the risks and benefits associated with flow optimisation, workflow automation, NLP, web scraping, and AI-driven contextualisation.

Conclusion and key learning points (10 min)

- Recap the key learning points from the discussion, emphasising the transformative potential of AI in reshaping academic research.
- Consider reading the postscript or sharing final thoughts on the significance of this case study.

Board plan

- Use a board plan to visually organise key concepts, frameworks, and insights discussed during the class.
- Ensure the board plan highlights the journey from traditional research challenges to the strategic integration of AI tools, fostering a clear understanding for students.

Analysis of assignment questions

Question 1. How are the drawbacks of traditional research methods addressed through the integration of AI tools?

In this case study, AI tools are used to overcome the inefficiencies of traditional research methods. Traditional methods often suffer from limitations, such as time-consuming data collection, manual analysis prone to bias, and the inability to process large data sets efficiently. By integrating AI tools, this initiative aims to streamline these processes, improve data analysis capabilities, and ultimately improve the quality and reliability of research results. It intends to address specific inefficiencies such as the following.

- Time required: Traditional methods often require manual data collection and analysis, resulting in long research times. AI tools can automate data collection, analysis, and interpretation, significantly reducing the time it takes to conduct research.
- Bias in analysis: Human analysts can introduce biases into the research process that affect the objectivity and validity of the results. AI-based algorithms can minimise bias by objectively analysing data and identifying patterns or trends that may be missed by human researchers.
- Limited data processing: Traditional methods may struggle to effectively process large amounts of data, resulting in incomplete or superficial analysis. AI tools like machine learning algorithms can process large amounts of data quickly, allowing researchers to gain deeper insights and make more informed decisions.
- Scalability: Traditional research methods may lack scalability, making it difficult to adapt to changing research needs or scale projects. AI integration allows researchers to scale their studies more efficiently by automating repetitive tasks and accommodating larger data sets.

The ethical implications of AI integration must be considered, such as privacy concerns, algorithmic bias, and the potential for unintended consequences. In addition, the reliability and accuracy of AI tools should be assessed to ensure that they produce trustworthy results consistent with research objectives.

Question 2. Strategic adoption of AI-based tools for academic research

Several factors should be considered when strategically using AI-based tools for academic research. Flow optimisation, workflow automation, NLP, web scraping, and AI-driven contextualisation can all help accelerate research studies and ensure timely impactful results.

- Flow optimisation: By optimising research workflows through AI automation, processes can be streamlined, bottlenecks reduced, and overall efficiency improved. This includes identifying key stages of the research process where AI tools can be integrated to increase productivity and effectiveness.
- Workflow automation: AI-based tools can automate repetitive tasks such as data collection, cleaning, and analysis, allowing researchers to focus on higher-level activities such as hypothesis generation and result interpretation. This automation accelerates the research timeline and frees up valuable time for researchers.
- NLP and web scraping: Researchers can analyse and extract insights from unstructured data sources such as academic papers, social media posts, and online forums. Web scraping techniques can collect data from the web and provide researchers with access to a variety of information for analysis and contextualisation.
- AI-driven contextualisation: AI algorithms can contextualise research findings within the broader scientific literature and identify connections, contradictions, or gaps in existing knowledge. This contextualisation increases the relevance and impact of research studies by embedding them in the broader scientific discourse.

Potential challenges include data quality issues, algorithmic biases, and the need for interdisciplinary collaboration when integrating AI-based tools into research workflows. It is important to carefully evaluate the capabilities and limitations of AI tools and ensure that they match the specific needs of the research project.

Question 3. Impact of AI on research agility

The introduction of AI-based tools has the potential to significantly improve the agility of academic research methods in a dynamic landscape. Through using AI algorithms for data analysis, pattern recognition and predictive modelling, researchers can adapt more quickly to changing research needs and new trends. For example, AI-powered sentiment analysis can track public opinion on social media in real time, influencing research directions and hypothesis formulation. However, challenges such as algorithmic bias, privacy concerns, and the need for expertise in AI implementation can hinder the widespread adoption of AI tools in academic research. Additionally, researchers must critically evaluate the validity and reliability of insights generated by AI to ensure that they contribute meaningfully to the research process.

Question 4. Practical examples of AI tool applications in academic research

AI tools can address various challenges in academic research across disciplines, improving the depth, efficiency, and relevance of research studies as outlined below.

- Medical research: AI algorithms can analyse medical image data to detect patterns indicative of disease, leading to earlier diagnosis and better patient outcomes.
- Environmental science: AI-based predictive models can analyse large data sets to identify trends in climate change, thereby informing policy decisions and conservation efforts.
- Social sciences: NLP techniques can analyse social media data to study public opinion on political issues, social trends, or consumer behaviour.

These applications demonstrate how AI tools complement traditional research methods, enabling researchers to tackle complex problems, work more effectively, and generate actionable insights with greater efficiency. However, researchers must be vigilant and consider ethical issues such as privacy, algorithmic bias, and transparency in AI decision-making to ensure the integrity and validity of their research results. Collaboration between researchers, data scientists, and subject matter experts is critical to the effective and responsible use of AI tools in academic research.

Key learnings

The instructor should encourage students to reflect on both the major challenges that they successfully identified during the case discussion and those that may have escaped their attention. In addition, the lecturer should conclude by providing insight into possible solutions to the challenges discussed in the assignment questions. By providing guidance and perspective on how to address these challenges, the instructor provides students with actionable insights. This approach ensures that the case discussion not only highlights the intricacies of the scenario but also provides students with practical strategies and solutions relevant to the real-world application of AI in academic research.

REFERENCES

Chubb J, Cowling P and Reed D (2021) Speeding up to keep up: exploring the use of AI in the research process. *AI & Society* **37(4)**: 1439–1457, 10.1007/s00146-021-01259-0.

Henke N, Bughin L, Chui M *et al.* (2016) *The Age of Analytics: Competing in a Data-Driven World*. McKinsey Global Institute, McKinsey & Company, New York NY, USA.

Netherton T, Cardenas C, Rhee D, Court L and Beadle B (2020) The emergence of artificial intelligence within radiation oncology treatment planning. *Oncology* **99(2)**: 124–134, 10.1159/000512172.

Tang A (2023) The importance of transparency: declaring the use of generative artificial intelligence (AI) in academic writing. *Journal of Nursing Scholarship* **56(2)**: 314–318, 10.1111/jnu.12938.

Uetela P (2017) *Higher Education and Development in Africa, Higher Education and Development in Africa*. Springer, Cham, Switzerland, 10.1007/978-3-319-31141-8.

Veugelers R (2014) *The Contribution of Academic Research to Innovation and Growth*. WWWforEurope Working Paper. https://www.econstor.eu/bitstream/10419/125727/1/WWWforEurope_WPS_no071_MS65.pdf (accessed 15/01/2025).

Zirar A (2023) Can artificial intelligence's limitations drive innovative work behaviour? *Review of Managerial Science* **17(6)**: 2005–2034, 10.1007/s11846-023-00621-4.

Prof. Arnesh Telukdarie
ISBN 978-1-83662-645-9
https://doi.org/10.1108/978-1-83662-642-820251021

Chapter 17
A conclusion to *Sustainability and Digital Engineering Management: African case studies in sustainable digital transformation*

Arnesh Telukdarie and Mahure Hlobisile Natasia

Conclusion

Adopting sustainability principles in digital engineering management is crucial in today's fast-evolving technological environment. Sustainability entails meeting current needs without compromising the ability of future generations to meet their own needs. It encompasses environmental conservation, economic viability, and social accountability. On the other hand, digital engineering management focuses on employing digital tools and methods to streamline engineering processes. The convergence of these two fields fosters innovation, competitive edge, efficiency, and long-term value generation in engineering, while reducing environmental impact and promoting the sustainable development goals (SDGs). Despite the consensus on the importance of adopting a sustainable approach, there remains a significant disparity between awareness and implementation. Although digitalisation presents significant advantages, there is a lack of practical strategies for aligning it with SDGs. Therefore, the book featured comprehensive case studies from diverse industries, illustrating how digitalisation can be leveraged to advance sustainable development. It provided firsthand insights into applying digitalisation to establish sustainable businesses, urban and rural communities, the education sector, and a range of other industries, offering a practical roadmap for implementing sustainable digitalisation in these sectors.

Table 17.1 provides a comprehensive summary of the chapters covered throughout this book, highlighting their significant contributions to the achievement of the SDGs. Theme one focused on industrial digitalisation for sustainable operations. This section offered in-depth insights and practical guidance on the implementation of a variety of technologies across different industries, including healthcare, mining, water and energy. By presenting detailed case studies, this theme illustrated how these innovations can lead to more sustainable practices, ultimately contributing to the attainment of SDG 6, SDG 7, and SDG 3. Theme two explored sustainable business best practices, showcasing a range of technologies and strategies aimed at promoting economic sustainability. This segment emphasised the importance of responsible consumption and production patterns, aligning closely with SDG 8 and SDG 12. The third theme addressed sustainable urban and rural development, examining the vital role of

Table 17.1 The contribution of this book toward the attainment of the SDGs (continued on next page)

Chapter	Area of focus	Relevant SDGs
Chapter 2	Smart water management for sustainability	SDG 6: Clean Water and Sanitation
Chapter 3	Green hydrogen industry and NLP techniques	SDG 7: Affordable and Clean Energy SDG 13: Climate Action
Chapter 4	Sustainability and uranium digitalisation	SDG 7: Affordable and Clean Energy SDG 12: Responsible Consumption and Production
Chapter 5	Satellite imagery for water leak detection	SDG 6: Clean Water and Sanitation
Chapter 6	Digital twins in healthcare	SDG 3: Good Health and Well-being SDG 9: Industry, Innovation, and Infrastructure
Chapter 7	Business process for digital sustainability	SDG 8: Decent Work and Economic Growth SDG 12: Responsible Consumption and Production
Chapter 8:	Advancing ESG metrics	SDG 12: Responsible Consumption and Production SDG 13: Climate Action
Chapter 9	RPA and AI for digital systems	SDG 9: Industry, Innovation, and Infrastructure SDG 12: Responsible Consumption and Production
Chapter 10	Sustainable rural development through collaborative innovation	SDG 1: No Poverty SDG 11: Sustainable Cities and Communities
Chapter 11	Sustainable waste management for urban settlements	SDG 11: Sustainable Cities and Communities SDG 12: Responsible Consumption and Production
Chapter 12	Green learning management systems	SDG 4: Quality Education SDG 9: Industry, Innovation, and Infrastructure
Chapter 13	Vocational welding education and VR integration	SDG 4: Quality Education
Chapter 14	Sustainable business intelligence in Master of Business Administration programmes	SDG 4: Quality Education SDG 8: Decent Work and Economic Growth

Table 17.1 Continued

Chapter	Area of focus	Relevant SDGs
Chapter 15	Chemical Sector Occupational Skills Forecast Tool	SDG 4: Quality Education SDG 8: Decent Work and Economic Growth
Chapter 16	AI for sustainable research practices	SDG 9: Industry, Innovation, and Infrastructure

sustainable practices in enhancing living conditions in communities. It focused on case studies and methodologies that contribute to achieving SDG 1, SDG 11, and SDG 12. Finally, the last theme centred on digital transformation in education and skills development. This section highlighted innovative tools and approaches designed to enhance the quality of education and foster opportunities for decent work. The insights presented in this section contribute to the attainment of SDG 4 and SDG 8, demonstrating how digital advancements can be leveraged to improve educational outcomes and economic prospects for communities. Overall, the book encapsulates a multifaceted approach to sustainable development, illustrating how diverse themes and strategies can interconnect to drive meaningful progress towards the SDGs.

To sum up, the book thoroughly examined sustainability principles and their seamless integration into digital engineering processes across diverse industries. It highlighted the importance of reducing environmental impact, promoting social equity, and ensuring economic viability. The book presented real-world case studies that illustrated successful implementations of sustainability-centric digital engineering practices, emphasising how digitalisation effectively addressed industry-specific challenges while adhering to sustainability principles. Moreover, the book discussed the barriers encountered and the opportunities that arose during the implementation process, providing practical insights into implementing sustainability-centric digital engineering management in various industries. This makes it an invaluable resource for those seeking to achieve sustainability goals while leveraging digital engineering practices for growth and innovation in their industry. The book provided a thorough and in-depth examination of how sustainability and digitalisation intersect within the field of engineering management. Unlike many other works that focus solely on digitalisation in engineering, this book went further by offering practical guidance on integrating sustainability strategies in the pursuit of digitalisation in engineering management. Drawing on real-world industry experience and success stories, it presented a practical roadmap for professionals to follow. The book caters to practitioners such as engineers and project managers who wish to effectively integrate sustainability principles into their engineering projects.

Prof. Arnesh Telukdarie
ISBN 978-1-83662-645-9
https://doi.org/10.1108/978-1-83662-642-820251022

Index

Printed and bound by CPI Group (UK) Ltd, Croydon, CR0 4YY

15/09/2026

14962488-0008